CSR, Sustainability, Ethics & Governance

More information about this series at http://www.springer.com/series/11565

Michael D’heur
Editor

Sustainable Value Chain Management

Delivering Sustainability Through the Core Business

Springer

Editor
Michael D'heur
Shared.value.chain
Munich
Germany

ISSN 2196-7075 ISSN 2196-7083 (electronic)
ISBN 978-3-319-12141-3 ISBN 978-3-319-12142-0 (eBook)
DOI 10.1007/978-3-319-12142-0

Library of Congress Control Number: 2014958010

Springer Cham Heidelberg New York Dordrecht London

Printed on acid-free paper

Springer is part of Springer Science+Business Media (www.springer.com)

Foreword

The Role of CSR in the Value Chain's DNA

Sustainability and Corporate Social Responsibility are valuable economic components emerging as hot topics in public discourse. Today's immense social and environmental challenges provide new opportunities and the need for integrative management techniques, economic specialization, and innovative technologies. Academics and management practitioners discuss this trend with both scientific and practical applications in this series on CSR, Sustainability, Ethics, and Governance published by Springer. The goal is to explain the complex issue of Corporate Social Responsibility (CSR) by providing a breakdown of its application on specific areas of business administration such as Supply Chain Management.

For too long, the subject of sustainability and the assumption of corporate responsibility were depicted in management literature as a predominantly defensive strategy and, therefore, applied practically as a mere cost factor. Until recently, social and environmental issues and the related business opportunities remained generally underexposed. However, CSR and sustainability are now being integrated into a company's DNA, i.e., in the core processes of value creation, innovation management, and organizational development. Therefore, the release of the publication "Sustainable Value Chain Management" has come at just the right time for worldwide discussion.

Throughout economic history the concept of value has been central to business operations. Companies were focused on the efficient organization of value chains.

Value-added processes, however, have become increasingly complex in the wake of globalization and the associated intensified division of labor within businesses. Furthermore, organizational transparency and pressures from various stakeholder groups (customers, employees, NGOs, etc.) are constantly increasing. Value chains need to be newly designed and organized to fit within this context. In addition to economic factors, environmental and social issues play an increasingly large role.

Companies can prepare for the future by developing a sustainable business model in which the creation of business as well as social value is essential.

In this publication, necessary aspects of sustainable value creation are described along with practical examples. Readers are, therefore, cordially invited to use the techniques discussed to aid in their own professional challenges and to partake in intense CSR discussions with the editors, authors, and supporters of this series. Last but not least, I would like to extend my sincerest thanks to the editor Michael D'heur for his extensive commitment, Christian Rauscher from Springer for his assistance, as well as to all the supports of this truly global CSR series. Finally, I wish all of our cherished readers an interesting journey through the various aspects of Sustainable Value Chain Management.

Cologne, Germany René Schmidpeter

Foreword

Sustainable Economics: More of Doing "Good"

What is an inherently good corporation? A "good" corporation has to be successful in economic terms but simultaneously contribute to the environment and strengthen social coherence. Within its various spheres of influence, it has to work according to universal ethical principles. Many corporations already act as a "good" one and without a lot of discussion about it. Yet, there are others, which do not work according to these values at all.

However, how do you distinguish a "better" corporation from a "normal" one? To tell them apart is difficult because there is currently no frame of reference for sustainability which can be applied to all corporations. Such a frame of reference, which enables competition between sustainable business models and value chains, is still to be established. Having a system of equal requirements for all business in place would enable the discussion about sustainable management to be more tangible, comprehensible, and invigorated. Otherwise sustainable economics remains too vague and a matter of publicity rather than actual facts, which in the long run would cause more discontent than trust.

We are a long way from understanding sustainable management as a matter, which requires attention also in economic and business-related settings. In fact, syllabi of economics or business management courses at university currently do not give an account of the current challenges faced by businesses or economic policies, including sustainability. Even better that the book at hand is a practical one, devoted to sustainability in products and value chains and showcasing the problems of sustainability, CSR, and market penetration: diverging juridical spheres, growing global population, demographic change in Europe, and different societal, ecological, and economic conditions. These are the facts that make it problematic to establish one universal frame of reference for sustainability.

The question of whether corporations can be made subject to a more sustainable ecological, societal, and economic development at all poses a further challenge.

20 years ago sustainability was defined mainly in environmental terms; nowadays it considers also societal issues and economic aspects. The globalized economy and worldwide communications make the world a global village. This amplifies the request for peaceful coexistence and social justice from a national to a global context. Therefore, we need a global approach and a framework for comparing economic, societal, and ecological development – in short: for comparing sustainability. As soon as its target, the stakeholder, and tools are more clearly defined, a new dynamic will arise to change the existing markets. Subsequently, investments will be made in corporations that align their strategy, their core businesses, corporate processes, and their governance with sustainable development.

However, this movement has started already. We need stakeholders and investors who decide themselves whether a corporation is active and engaged in sustainability. The public sector, too, needs to align its procurement strategy according to ecological and social criteria and purchase goods and services only from corporations, which act according to sustainable principles in the long term and in a reliable manner.

I am certain that many corporations are well prepared for competition under the premises of sustainability. I am convinced that corporations can benefit from competition which also takes ecological and social criteria into account. But for this, enhanced visibility and critical analysis with one's own sustainability accomplishments are prerequisite. Recognizing and discussing conflicting aims allows one's perspective to widen, which is essential for future market success.

The solutions to the challenges of corporates' responsibility for sustainable development are extremely diverse. There is no "one-size-fits-all" solution. This book discloses many different approaches and offers a practical orientation for other corporations and organizations on their way toward sustainability. I hope this book will address many readers and strengthen the business-related debate and research about sustainability and "doing good."

Berlin, Germany Marlehn Thieme

Foreword

Moving from a Postindustrial to a Sustainable Age

Already in 1969 Alain Tourrain pointed toward the fact that a new form of society, which he labeled "postindustrial," is coming into being. With his statement he wanted to indicate that it was time to see off the industrial age. Yet, at that point it was not clear where exactly the journey would lead. Today, more than 40 years later, the new era seems to materialize increasingly. The process is accelerated by intensifying crises – may they be economic, related to climate, scarcity of resources, or demographic change – showing that our focus on economics and the development model based on Gross Domestic Product (GDP) is no longer sustainable.

At least in Europe there is a growing consensus that our future economic and social model needs to reach a balance between economics, ecology, and social values. In order to attain such balance in the long term, a new understanding of growth is required. The discussion increasingly focuses on qualitative growth, which the European Union understands to be innovative, inclusive – meaning to take in all social groups, resource efficient, as well as environmentally friendly. Such kind of growth can no longer be expressed in straightforward GDP terms.

This does not merely call for constraints, but mainly offers chances that can be used in an innovative way. A new dynamic comes into being, which is not primarily focused on GDP growth, but is connected to an augmentation of quality of life – the new notion of prosperity.

In European societies, this basic concept has already been accepted widely and its basic ideas have even made its way into the Treaty of Lisbon. Nevertheless, as long as corporations, the most important economic players, do not adopt this concept it is not of much practical use. The concept of CSR, which basically describes and measures to what extent a corporation has committed itself to the new understanding of economic action, has existed for a long time now. It is the responsibility of our society and the state to make CSR standards comparable with each other, and furthermore to ensure that existing incentives and tax legislation support those corporations that implement CSR and that do not hold on to old principles.

For corporations it has to be profitable to act according to CSR. It should be in our interest to remove any obstacles on the journey toward a more sustainable society.

Corporations which have committed themselves to CSR or which have embedded CSR in their management strategies have to face many limits of implementation. On the one hand, this is because they do not find themselves in a fair condition for competition; on the other hand, because the increasingly interwoven structure of our economy makes it difficult for a single corporation to practice CSR in its complexity. Because of this, it is a logical step to merge production and value chains in order to use the numerous synergies within these chains.

This book offers a solid analysis of what is possible to achieve in terms of CSR and effective supply chain management. Numerous practical examples intend to encourage entrepreneurs and businesses to move toward the new economy.

Alpbach, Austria Franz Fischler

Contents

shared.value.chain: Profitable Growth Through Sustainable Value Creation

Michael D'heur

1 Definitions and Context

The question how Corporate Social Responsibility (CSR) (Schneider 2012: 19 et seq.) can be incorporated with integrity in a company's core business (i.e., products and supply chains) is subject to public debate and different levels of interpretation. Embedding CSR (or sustainability to use a more current term) requires a range of decisions and subsequent implementation steps across all management levels and departments of an organization. In this article the term "**sustainable value creation**" is used to describe the desired target/ideal situation, where sustainability aspects are considered in all dimensions of conducting business. The fundamental orientation of sustainable value creation is based on a combination of three individual concepts, namely "**sustainability**," "**sharing**," and "**value creation**." Given that these individual concepts are interpreted in many different ways in public debate and when used by companies, it is important to describe each one of them, before we eventually combine them to set the context of shared value creation.

1.1 The Concept of "Sustainability"

The public perception of the term "**sustainability**" covers a wide range of definitions, some of which are extremely vague. The scope is broad, ranging from environmental protection to conservation of resources, habitat preservation, biodiversity, recyclable/pollution-free products, sustainable (in the sense of stable) operations, and fair working conditions. Even the concept of sustainable profit is

M. D'heur (✉)
shared.value.chain, Adams-Lehmann-Str. 56, 80797 Munich, Germany
e-mail: mdheur@sharedvaluechain.com

M. D'heur (ed.), *Sustainable Value Chain Management*, CSR, Sustainability, Ethics & Governance, DOI 10.1007/978-3-319-12142-0_1

used. In 2012, Siemens AG defined sustainability primarily as the achievement of the goals of the "One Siemens" initiative. This initiative defined sustainability as the achievement of revenue growth, capital efficiency/profitability, and capital structure. Sustainability relates here to sustainable profit and increasing the value of the company (Siemens 2010: 12)[1]

Despite many different views on what sustainability is and how it can be achieved, the definition that currently describes "**sustainability**" in the most concise way and the one most often quoted was formulated by the Brundlandt Commission in 1987 when the term "sustainable development" was introduced:

"Sustainable development is development that meets the needs of the present generation without compromising the ability of future generations to meet their own needs. An economy where natural resources are used only to the extent that they can regenerate" (United Nations 1987).

As result of the Brundlandt report the world's first Earth Summit in Rio de Janeiro in 1992 was held, to put the recommendations into action. Since then numerous definitions and interpretations were created, clarifying partial aspects of sustainability and expanding on the definition provided by the Brundlandt Commission. To name one example, the Federal Republic of Germany issued its National Strategy for Sustainable Development, in which it addresses fiscal sustainability, sustainable growth, climate and energy and sustainable water policies as our current challenges to sustainability (Bundesregierung 2012). The strategy also offers guidance, indicators, and goals in order to make sustainability a driver of growth and development.

Critics argue that the term sustainability has been twisted and used by governments and business in variations to conduct business as usual. In an undertaking to renew its commitment to promoting sustainability, the European Union Commission revised its "Europe 2020" strategy to provide a new, more concrete definition (where the concepts of Corporate Societal Responsibility and sustainability are interchangeable) in October 2011:

"CSR is the responsibility of enterprises for their impacts on society"

With this definition, the EU Commission has for the very first time moved away from the purely voluntary definition of CSR/sustainability, placing corporate responsibility at the forefront. For companies to be able to adopt a responsible approach across the board, it is necessary to take economic, ecological, and societal

---

[1] Please find more detailed information on how Siemens AG is embedding sustainability in their value chain in the chapter "Siemens: Managing Sustainability Along the Entire Value Chain to Benefit Our Customers."

goals into consideration. Human rights and consumer concerns also need to be incorporated as part of management procedures and corporate strategy through close collaboration with stakeholders. Companies are encouraged by the EU Commission "to adopt a long-term, strategic approach to CSR, and to explore the opportunities for developing innovative products, services, and business models that contribute to societal wellbeing and lead to higher quality and more productive jobs" (European Commission 2011: 8). The Commission recognizes the promotion of societal and environmental responsibility within the value chain and the consideration of non-financial indicators as an important cross-functional requirement (Schneider 2012: 21). Sustainability therefore needs to be addressed strategically. The aim is to achieve competitive advantage on the market via new products and services and innovative business models. Economic efficiency and sustainability are therefore no longer opposites, but rather two sides of the same coin (Schmidpeter 2013: 16). Countless innovations are required to enable companies to "take responsibility for their impact on society" – this does not mean societal commitment outside the core business, but responsible management of the core business and a departure from the voluntary approach advocated thus far.

One common factor shared by all definitions of sustainability is that they describe the requirements of societal responsibility for organizations in general and companies in particular, in a logical and intuitive way. However, in terms of recommendations for the practical implementation of sustainability, the definitions and concepts remain often very vague. Current approaches to sustainability have been mostly voluntary and have led companies to constantly emphasize that they are committed above and beyond the legal requirements. However, sustainability activities often remain superficial, not necessarily addressing the products, value chains, and services of a company.

In this article the terms "**corporate societal responsibility**" and "**sustainability**" are used interchangeably, while sustainability stands for the more recent term that is used. Many practitioners in companies believe that CSR is already an outdated concept and that the understanding of the issues at hand has moved on to use sustainability as a more comprehensive approach. From a business perspective, both terms should be inseparably linked; indeed, over time the meanings of these terms have coalesced (see Schneider 2012: 11f and Crane et al. 2008, who do not see sustainability as a separate topic, but as a concept that can be subsumed under CSR).

1.2 The Concept of "Shared"

The term "**shared**" likewise is perceived by the general public as well as businesses to have a variety of meanings and expectations. The concept of "shared" in a value chain context means involving all direct and indirect stakeholders consciously and deliberately in the product creation process and operational value creation. From a company's viewpoint, stakeholders are not restricted to business customers and end consumers in their role as primary customers for goods and services. It is much more a case of maintaining an active dialog with investors, suppliers, employees, business partners, and above all the communities where companies operate their value chain. Dialogue is a driver for product innovation and improved value creation. This ultimately generates value-add for all parties involved.

Due to mutual dependencies, this type of collaboration requires a systemic approach and an understanding that sustainability in the core business cannot occur solely within a company's "own 4 walls." Whereas in the past companies had extensive control over their own value chains due to a high degree of vertical integration, today's globalized economy is characterized by mutual dependencies and interrelated effects. Even medium-sized companies now often have global value chains. Opportunities and risks depend on the intensity of collaboration. This requires a change of perspective to adopt a network approach on the basis of transparency, collaboration, and flexibility. This network approach forms the basis of a company's flexibility and adaptability to new circumstances. The conscious removal of previous barriers to collaboration presents a challenge, as a great deal of trust must be built up between the partners. However, it is this very collaboration based on trust that makes it possible to explore new avenues, create value, and build a stable base for future growth.

1.3 The Concept of "Value Creation"

Due to its many different applications in a range of different sectors of the economy such as business management, finance, and economics (particularly macroeconomics), the term "value creation" is hard to define. The basic principle consists of generating the highest possible level of operational value-add, i.e., generating profit on a regular basis and increasing the value of the company. This definition of value creation is currently implemented in most profit-oriented companies, often driven by the demands of the capital markets to achieve continuous growth and the necessity of showing a profit every quarter.

Based on the increased recognition of value creation being more than a linear process, the circular economy model is becoming increasingly popular. A company's product responsibility does not end with its responsibility for the waste generated by the production process; companies also need to take into account the safe disposal of their products after use. In many countries this is not a voluntary decision but rather a statutory act, as for example in the German Closed Substance Cycle and Waste Management Act. An ecological corporate strategy therefore requires the flow of materials and information to be circular.

The realization that a linear economy, where "disposable products" – many with harmful constituents – are produced on a large scale, is not compatible with natural cycles, caused architect Bill McDonough and chemist Michael Braungart to develop the Cradle-to-Cradle approach (Braungart and McDonough 2002). The Cradle-to-Cradle approach is aligned with nature: its aim is for product design and manufacturing methods to be structured in such a way as to ensure that the highest possible percentage of a product can be returned to a biological or technical cycle at the end of the product's lifecycle. As there is no concept of waste in nature, McDonough and Braungart call for the symbolic elimination of the concept of "waste" in order to pave the way for adopting a corresponding change of perspective. With this approach, economic activity and environmental protection are not opposing concepts, but closely intertwined.

1.4 *Sustainable Value Creation*

The concepts "sustainable," "shared," and "value creation" jointly form the basis of "sustainable value creation," which we define as follows:

Sustainable value creation stands for a company's commitment to structure all aspects of its core business (i.e., products and supply chains) in ways that deliver economic, ecological, and societal value-add at the same time.

Sustainable value creation builds upon the basic understanding that economic, ecological, and societal value-add can only arise where the approach is purposefully embedded within the company's core business by the Senior Management Team and is adopted at all management levels. In this context, the term "core business" means "the combination of customers, sales channels, products, internal capabilities, and markets enabling companies to grow through sustained profits. From the customer's perspective, this is synonymous with differentiation from the competition and therefore signifies a company's unique market positioning. This is where a company's specific capabilities play a role, such as special production systems and technology, first class marketing concepts, customer-aligned innovation systems, and sophisticated supply chain management" (Bain and Company 2010).

The value chain forms the company's backbone. All of the important decisions and parameters laid down in the corporate strategy are ultimately implemented in the value chain. As a result, the value chain accounts for a significant proportion of a company's success in economic, ecological, and societal terms. The interaction between customers, business planners, buyers, suppliers, internal/external production facilities, logistics, and operational control has a significant role in determining a company's success. A radical restructuring of production processes to make procedures "greener" or "less harmful" is not enough. Companies need to adopt sustainability as a core business requirement, necessitating collaboration along the entire extended supply chain (Lee 2010). Products and supply chains are no longer merely a means to achieving an economic goal. They are the manifestation of the implementation of a sustainable corporate strategy – one aimed at creating value for all concerned stakeholders. Sustainability that is driven "inside-out" from a company's core business entails a continuous assessment of the type of economic, ecological, and societal value being created. Decisions are made on the basis that there will be positive outcomes for profitability, the environment, and the people involved. Sustainable value creation means that intentions and words are followed up with tangible actions, so that it is transparent what is "beyond the label" of sustainability: it is the tangible implementation of a sustainable corporate strategy that is supported by all management levels and linked by means of an effective internal and external communications strategy.

In this context, special attention must be given to the internal and external "interface" with regard to value creation. New opportunities (extending to new business models) arise when in-house collaboration takes on board the potential for innovation offered by customers and suppliers. A systematic approach is essential for understanding dependencies and identifying opportunities. This leads to economic, ecological, and societal value creation and provides a platform for profitable and sustainable growth. Growth is not measured in uniquely quantitative terms, ranging from increased turnover, market share gain, and GDP. Ecological factors (environmental protection, biodiversity, etc.) and societal factors (societal activity, cultural activity, and long-term effects) also have a role to play alongside economic factors. All of these factors combined determine our quality of life (Braungart and McDonough 2002: 37).

Above and beyond this interpretation, the term "sustainable value creation" is used in this article with a focus on sustainable product design on the one hand and the application of sustainable practices across the entire supply chain on the other hand.

The value chain consists of the product development and supply chain processes of an organization. It covers all stages of the lifecycle from idea/ concept, raw material sourcing, production, distribution, end customer use to the point where the product goes back to a biological or technical cycle, thus closing the loop.

It should be the aim of product design to ensure that products can be reused in subsequent cycles at the end of their initial lifecycles. This means that during the design phase, the aim should be to achieve positive societal and ecological value alongside pure economic utility. Similarly, the effects on customers and the environment need to be thought through as part of a lifecycle analysis *before* the product itself is created. Therefore, product components need to be selected for their minimal impact on people and the environment, and ideally for their capacity to be returned into a closed cycle. Should this not be possible, the input of resources should at least be continuously minimized, with the intention of achieving the greatest possible level of efficiency. Additionally, the use of harmful materials and substances should be totally eliminated. Sustainability also means that there are no negative effects when a product is being used. Alongside the aspect of product design, the way in which the product is sourced, manufactured, and distributed through the company's value chain is of major importance. Products need to be selected and business processes designed with cost-efficiency in mind and with the least possible impact on employees and the environment. This operational aspect should include the entire value chain through all production stages, from raw materials to customers and back.

Alongside theoretical concepts and stakeholder viewpoints, valuable information and inspiration can in essence only be provided by practical examples of corporate implementation. Sustainable value creation is not a concept reserved solely for the corporate world. Public institutions also need to apply it, playing a pioneering role in spreading its use. For example, the procurement practices of public institutions can be aligned more consistently with sustainability and shared value creation. Even NGOs, many of which operate or influence value chains, should have a greater emphasis on shared value creation. With this interpretation, sustainable value creation can form the basis of current and future growth for companies, the environment, and society. Despite the many efforts made by companies and society to establish value creation on a more sustainable and shared basis, we are still in the early stages of a major but necessary change in terms of core business sustainability.

2 The Gap Between Sustainability Ambition and the Core Business

2.1 The Consequences of a "Linear" Economy

A multitude of rapid changes in the economic and societal environment has made leaders in charge of companies and governments realize that new flexible concepts are required to keep pace with increasing market volatility. Many studies conducted by internationally recognized academics have confirmed that the way global business is conducted today is not sustainable in the long run (Randers 2013). Although sustainability is a permanent topic of public and corporate discussion, the majority of global economic activity is still oriented toward chasing the paradigm of perpetual growth and accelerating a linear economy: "Bigger, better, faster, more" rules!

The call for continuous growth and regular (mostly short-term) success is driven specifically by the global finance sector. Because investors have the opportunity to transfer massive flows of funds in a short time, companies with a national and international presence are continuously exposed to the demand for short-term growth and profit. Driven by business, investment banks, and hedge funds, the hunt for short-term profits, where credit risk is seen as just another form of merchandise, has led to a financial crisis of an unprecedented degree. Over the last decade, we have seen a change which is part of a comprehensive process labelled "financialization." This refers to the increased importance of the financial sector over the "real" production of goods and services (Nölke 2012). Emanating from the USA and the UK, this phenomenon has now reached the German financial system, although savings banks and cooperative banks have been less affected in comparison. The process began after the collapse of the Bretton Woods Monetary System at the beginning of the 1970s and the consequent liberalization of the financial sector, encompassing the deregulation of financial transaction controls, the concomitant intensification of transactions between banks and an increase in profits due to financial activity. However, financialization does not merely mean that profits in the financial sector have risen more steeply than in the "real" economy, but also that the power of the financial sector over the "real" economy has increased, to an extent that companies aligning themselves with the expectation for short-term yields commonly expected by financial markets.

Banks and the real economy are inseparable, as the real economy would be unimaginable without banks. At the same time, there has been a perceptible change, with an ever-widening gap between banks and the real economy. This is a problematic process, especially considering the underlying vulnerability of banks to crisis, which has increased even further due to financialization (Nölke 2012). There are not only effects and risks for companies; anyone who is reliant on crude oil or food and agricultural products will ultimately be affected by speculation in these markets.

As a consequence of such a dominant commercial orientation, prosperity and growth have widely become the norm in industrial countries. Even emerging and developing countries have benefited from this development. Life expectancy increases with a higher standard of living. Medical supplies and education become widely available. Agricultural productivity is increased through new methods and food storage is improved. Electricity and telecommunications raise the standard of living (Braungart and McDonough 2002: 26). In contrast, however, the negative consequences of today's "linear" economy are becoming increasingly apparent. According to McDonough/Braungart, these negative consequences are the result of a design fault in the globally deployed "production system" (Braungart and McDonough 2002: 18), that creates the following results (among others):

- Millions of tonnes of poisonous substances are deposited into the atmosphere, water sources, and the soil.
- Materials are produced that are so dangerous that they have to be monitored for generations to come.
- Huge mountains of waste are produced.
- Valuable materials are buried in landfills and nothing can ever be recovered from them again.
- Thousands of complex regulations are needed to restrict the negative impact of the economy.
- Productivity is measured by how few people are employed.
- Prosperity is achieved through the depletion of natural resources, only for them to be buried or burned at the end.
- Biodiversity is diminished and cultural practices are threatened with extinction.

The design fault manifests itself in an economy that is oriented toward perpetual growth, optimization, and profit maximization, resulting in products that are made according to the principle of "Take–Make–Use–Throw Away," causing major environmental and societal problems.

The consequences of the current linear economy and its globally distributed value creation are complex and diverse (see Fig. 1). The unprecedented growth phase in the world economy that started in 2004 was followed by a global financial

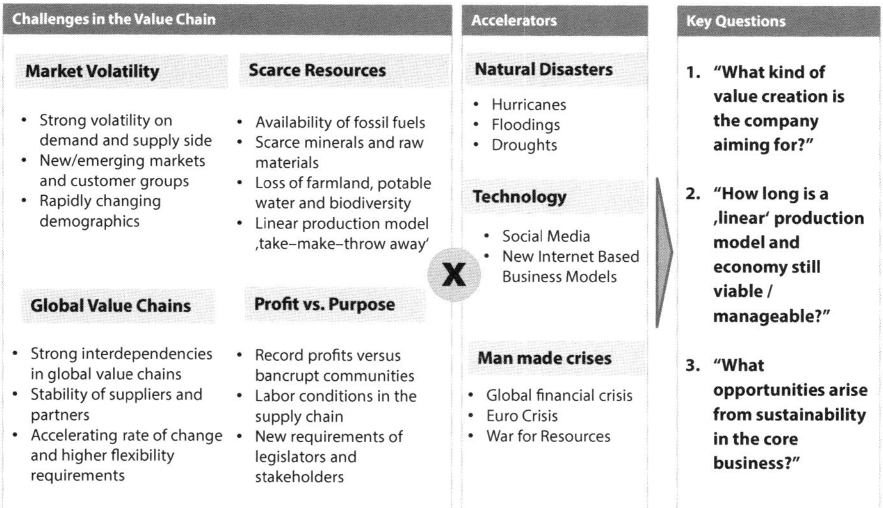

Fig. 1 The "new normal" – the challenges for value chain ecosystems are accelerating (shared.value.chain 2012)

crisis in 2008 and a series of natural disasters. The continuing euro crises, is placing heavy demands on companies, society, and consumers. The consequences of today's economy become increasingly visible in the form of gradually scarce raw materials, global warming, more frequent natural disasters, overburdening of eco-systems, environmental pollution, and harmful product constituents. In addition there are social consequences, such as the outsourcing of employment to low-wage countries, food speculation, and the under-funding of communities—often alongside record profits for companies that minimize their contributions to the communities where they operate through tax dodges.[2] Shocks and crises are occurring with greater frequency and their impacts are becoming more severe and longer-lasting.

In this particular context, it is interesting to observe that although there has been heated discussion about the need for behavioral change among groups of companies, the financial sector, stakeholders, customers, and academia, historically these groups have generally been unable to reach a consensus. The recent past has seen a great deal of reaction, but very little real action. Even the Rio+20 Climate Summit led by the United Nations will be remembered more for its failure than any success in solving the problems caused by uncontrolled growth.

Studies of corporate attitudes to CSR and sustainability reveal a number of interesting differences: firstly between countries. The view held by Milton Friedman: "the societal responsibility of business is to increase its profits" [see Wirl (2012) and The Economist (2011)] is dominant in developing and emerging

[2] Thanks to a sophisticated but legal tax avoidance model, Apple pays only 2 % tax in the USA and Ireland.

countries. In Germany and a number of northern and western EU countries however, there is greater emphasis on CSR. Secondly, attitudes to CSR vary between industries. Whereas oil companies in particular are strongly oriented toward CSR (which is not always a successful strategy, as the case with BP), other extraction industries refer less frequently to CSR (Wirl 2012: 2). Looking at the sustainability efforts of companies the following observation now generally applies: the closer a product is to the everyday needs of the consumer, the greater the effort made by companies to position them with a sustainability message.

Until now, the link between sustainability and the core business has not been sufficiently recognized by companies as an opportunity: this is an area of great potential for companies, the environment, and society alike. The process of embedding sustainability within the core business means embedding it in products and the supply chain. This is both a challenge and an opportunity for companies. This particularly involves working with stakeholders, suppliers, and society itself. Even nowadays, sustainability and core business are still not considered as an automatic coupling based on mutual dependency. One of the reasons for this lack of understanding is the complexity posed by sustainability in companies and society as a whole. In fact, the context in which sustainable value creation takes place could not be more complex. It runs through the whole of society: from sole traders to multinational groups, governments, interest groups, and NGOs. The multiplicity of sectors, sizes, legal forms, national/international relations, etc. has meant that up until now, there has been a lack of uniform and practical regulations/systems on the necessary scale (Brix et al. 2006). Commercial practices over the last 100 years have been strongly affected by the impact of the Industrial Revolution, which in particular views the environment as an unlimited source of resources. With its focus on raising operational efficiency, the Industrial Revolution placed value on increasing yields, improving product quality, lowering operations costs, and improving service and supply. Over the years, the continuous pressure for improvement has brought considerable economic success. As a result of this, a number of business optimization methods and models have appeared, such as Theory of Constraints, Lean, the Toyota Production System and Six Sigma, to support the continuous improvement process. The complexity of the necessary business optimization has also been marked and accelerated by increased globalization. The importance of value chains as well as a basic understanding of how to manage them, have changed significantly across all industry sectors over the last 30 years. Technological progress, the amount of available capital, and the need to generate further growth have been the main drivers for corporate globalization. In a series of studies conducted by business consultants PRTM, supply chain managers said that they assumed that over 50 % of a company's value creation would be distributed globally in future. Study participants also said that they employed sustainability practices merely to meet legal requirements or in response to explicit instructions from their customers [see PRTM (2008) und PRTM (2010)].

In the field of product development, the World Climate Conference in Rio in 1992 marked the changing point when ecological criteria were taken seriously for the very first time. The "Changing Course" report delivered by the Business Council for Sustainable Development (today: WBCSD) did indeed set the course toward reduced resource consumption and a sharper focus on environmental aspects with its concept of "eco-efficiency."

However, the recommendations issued by the Council, an association of 48 industrial sponsors (including Dow, Conagra, and Chevron), referred to those aspects that would deliver value-add for companies if they focused on "eco-efficiency." No suggestion was made for a radically different approach to product development in this context and the process of exploiting the environment was merely slowed down and solving the problem transferred to future generations (Braungart and McDonough 2002: 53). Even the strictest eco-efficiency paradigms fail to challenge basic practices: a shoe, building, factory, car, or shampoo can still be designed badly, even if the materials and processes used in production become increasingly "efficient" (Braungart and McDonough 2002: 76).

With the current recycling systems in place, some products are indeed returned to a cycle. However, as these products are not designed for recycling from the outset, the result of the recycling process is often a material that is of lower quality and has lost some of its properties compared to the starting material. This means that primary materials still need to be sourced. The philosophy of "eco-efficiency" certainly addresses the process, but only leads to a deceleration of it. A further problem lies in the "disposal" of waste that contains problematic substances. "Disposal" often consists of exporting waste to far-away areas, often developing countries. The problem is "out of sight, out of mind."

Right from the product design phase, most of the products available today are conceived to be thrown away at the end of their lifecycle. McDonough and Braungart call them "Cradle-to-Grave" products (Braungart and McDonough 2002: 27). In many instances it is easier for consumers to buy a new product or the latest technology instead of repairing or overhauling the existing product. Companies face continuous criticism for purposefully designing products in ways that lead to malfunctions/defects after a certain amount of time and hence requiring to buy a new product. Specialist manufacturers of electronic consumer goods are often suspected of this practice. In his 2006 book "Made to Break" Giles Slade reviewed the practices of planned obsolescence in the US. For Slade "planned obsolescence is the catch-all phrase used to describe the assortment of techniques used to artificially limit the durability of a manufactured good in order to stimulate repetitive consumption" (Slade 2007: 5). Why is it not possible to replace a smartphone battery? Why is the circuit board in a television designed so that a heat-sensitive capacitor is located right next to a heat conductor – even when other design options are possible? Planned obsolescence, is regrettable in terms of sustainability, as the production of another device requires considerably more resources than the replacement of a single component. The situation is aggravated by the fact that the majority of defective devices are not recycled, ending up in landfills. Valuable raw materials are lost or transformed into toxic substances via waste incineration. Besides the design aspects that lead to technical obsolescence, the particular way electronic consumer products are marketed has conditioned customers for "psychological or fashion-based obsolescence" (Slade 2007: 27) – with ever new features/functions. While the approach to stimulate repetitive buying has been invented in the US Automotive industry in the 1950s, it is now the standard among almost all consumer products, pushed by marketing and media campaigns as well as subsidies from service providers. This leads to situations where properly functioning products like mobile phones or MP3 Players end up in drawers at home. Well-working products and their auxiliaries like cables, chargers, and headphone end up as electronic waste. A survey commissioned in 2014 by the German Federal Association for Information Technology, Telecommunications and New Media (BITKOM) found that in Germany approximately 106 million operable working mobile phones end up as electronic waste in drawers at home, just because a new device was bought shortly after the initial one. This is an increase of 24 % compared to 2013 (86 million operable mobile phones ending up as electronic waste) and represents a number far greater than there are residents in Germany (BITKOM 2014).

Due to the current orientation of the economic system toward regular and (ever-increasing) commercial profits, companies produce predominantly according to "Cradle-to-Grave" designs. However, such an orientation does not just have consequences for product development alone. As a result of growth and margin pressure, products are manufactured with the cheapest raw materials, components, and ancillary materials available on the global market – which means that prohibited and regulated product constituents find their way into the production process and end up in the hands of consumers. While this practice can lead to problems during the product processing phase, critical substances are a particular problem during the utilization phase and at the end of the product lifecycle. The increase in the incidence of cancer, allergies, asthma, and other "unspecified" diseases is only the tip of the iceberg.

Toy manufacturer Mattel is a well-known example of this, as paint containing lead was used by a Chinese subcontractor in its production process. This practice was not only poisonous to production workers, but led to a wide-ranging product recall of the Mattel toy in the USA because the product posed a risk to small children. Not only was Mattel's reputation damaged – the Chinese factory owner also committed suicide.

Ayres and Neese assume that 90 % of consumer goods produced in the USA immediately become waste (Ayres and Neese 1989: 93). It is therefore difficult to understand why valuable raw materials are "disposed" at landfills and incineration facilities, when they have been obtained under difficult conditions and costs have been incurred in sourcing and processing. A further problem is that many products and their constituents are not appropriate for landfill or incineration. McDonough and Braungart call products, that are not designed from the outset to be useful to people or the environment "crude products" or "products plus." What that means is that consumers obtain not only the product they wish to acquire and use, but also obtain a range of possible side effects into the bargain, which are often undeclared (or do not have to be declared) (Braungart and McDonough 2002: 37, 40).

Developing countries (particularly countries such as China, India, Brazil, and Russia) have also adopted the growth mantra that has been prevalent in industrial countries for decades. Because of their high population levels, these countries are attractive growth markets for companies. This makes it necessary to achieve economic growth and revenue so that citizens are able to access consumer goods.

The booming economic growth in China is a good example of this. The Chinese government has succeeded in achieving an annual economic growth average of 8–10 % in recent years. At the same time, several hundred million people are now able to live above the poverty line. In order to build the necessary economic growth, China has not only captured an extensive share of global value creation (especially in the manufacturing sector), but has also stepped up in the role of a global investor to secure access to raw material reserves in foreign countries for itself. In addition to a strong demand for products and services from the West, this necessity for further growth in China has also led to a rapidly increasing consumption of natural resources, leaving a heavy environmental footprint. Many of the country's fresh water reserves are now contaminated and in numerous cases, working conditions do not meet the standards of western countries. Water and air pollution have resulted in a rapid increase in the instance of diseases. Air quality in Beijing was the "worst on record" in February 2013, as the city's pollution monitoring center warned residents to stay indoors with pollution 30–45 times above recommended safety levels (Reuters 2013). Aside from the increased risk of respiratory and cardio diseases, the increase in air pollution has also been linked to other health issues including disorders in neurological developments. In addition to water and air, food is at the center of frequent issues. One food scandal follows another.

Due to the globalization of procurement markets, natural resources throughout the world are being drawn into this system and mutual dependencies are strengthening. The restriction on the export of rare earths imposed by the Chinese government in 2010 led to a drastic rise in global raw material prices and a crisis in the supply of electronic components. In industries such as the electronics sector that are dependent on rare earths, this led not only to price increases but also begged the question of how this dependency could be mitigated. Replacing or substituting rare earths used in products and new concepts for recycling electronic scrap are being discussed as possible solutions. The result of this is that the discussion on sustainability in the value chain is receiving new impetus out of economic necessity. However, these concerns are still operating within the old paradigm. An approach that is "less harmful" or "consumes fewer limited resources" is definitely to be welcomed as a first step, but does not go far enough toward avoiding the use of such resources right from the start in terms of design.

2.2 *Educated Consumers and Their Awareness of Greenwashing*

Despite the globalization of business activity and the effects associated with constant economic growth, there is also a very interesting aspect on the demand side: the globalization of value creation, the availability of Internet technology, and social networks has led to the creation of new and well-informed customers groups. These global groups (Edelman 2010) are linked throughout the world, equipped with a wealth of information at hand and the ability to closely scrutinize corporate messages. Exposing product scandals or similar situations in company value chains is not the sole right of activists. Bad news travels more frequently, faster, and further than before. The demand for transparency and credibility is increasing.

According to studies conducted by the Edelman Market Research Institute in the US, companies can no longer differentiate themselves in the eyes of their customers solely on the basis of their products (Edelman 2012). Edelman also says that confidence in companies and their management has been on the wane since 2008. High quality products, good working conditions, efficient operating processes ("Operational Excellence") and a leading position in a market are no longer the basis for consumer confidence in a company, but have become an essential basic prerequisite. Companies are expected not only to function well, but also to act with integrity, "do good" in the world and, above all, not cause any harm. Faced with this level of expectation, the interpretation of a company's corporate purpose coupled with the way in which a company communicates and implements this purpose have come to constitute the competitive advantage of the future.

The public expression and communication of a company's purpose extends far beyond corporate social responsibility. It is part of a company's DNA, the company's reason for existence. Jeremy Galbraith, CEO of the public relations company Burson–Marsteller in Europe, regards corporate purpose as an important element of differentiation: "Companies that embed corporate purpose strongly within their corporate strategy and communicate it well both internally and externally, enjoy significant competitive advantages. Communication of the corporate purpose is becoming an important tool for managers wishing to build their reputation and a relationship of trust with stakeholders." These changed expectations pose a challenge to companies, especially as company information that has been publicly disclosed is surely of great interest to competitors. However, the challenge is not so much one of transparency, but more one of credibility. "If you cover up problems, close yourself off and fail to work systematically and transparently on the solution to a problem, you will only struggle from one crisis of confidence to the next," says Georg Lahme, transparency expert for strategic communication consultants Klenk & Hoursch.[3] With access to around-the-clock supply of information that causes them to be globally connected, consumers increasingly make their purchase decisions on the basis of corporate purpose and the visible support of good works. For the same product features and prices, consumers opt for a brand supporting a good cause. They will either recommend or penalize (Edelman 2012). Issues such as corporate purpose, sustainability in product creation and usage, recycling, and societal justice have become critical factors in sourcing decisions for these buyers.

As more and more companies start reviewing their approach to value creation, the concept of shared value is also moving from products and value chains toward brand building itself.

This extension of the widely accepted triple bottom line approach (focus on profit, people, and planet) toward inclusion of brand messaging, shows the importance of putting sustainable value creation at the heart of the brand (see Box 1 for an example).

[3] See also chapter "Telling the Backstory: Transparency in Global Value Chains."

Case in Point: Shared Value as Brand Building Constituent (BBMG)
Raphael Bemporad, Principal of the New York based Brand Building agency BBMG explains the growing importance of Shared Value for Brands as follows:

"The practice of creating shared value is fundamentally about capitalizing on the connections and mutual interdependencies between business and society. A business needs the community to provide demand for its product, natural resources, a supportive regulatory environment and the employees to bring their product and services to market efficiently and effectively. A community needs successful businesses to provide helpful products, jobs and wealth creation opportunities for its citizens.

However, without considering the power and influence of brands and the full participation and co-creativity of consumers, community members and other stakeholders, we're leaving tremendous opportunities for engagement, collaboration, cultural influence and value generation off of the table.

At BBMG, we focus on branding for shared value because we believe it's a transformational way of doing business – combining the foundational purpose and core values of brand building with the environmental imperative of sustainability and the creative potential of innovation.

Branding for shared value must consider the full set of relationships in every part of the value chain – consumer, product, brand, community and planet – and allow for the integration of mutually beneficial roles that we can play as individuals, organizations and as a society (see Fig. 2).

Bringing the full meaning and influence that brands have in society to the forefront of business design and innovation strategy, helps us generate disruptive business solutions and delightful brand experiences that enable shared value creation.

By harnessing the promise of branding, sustainability and innovation, we can meet the needs, hopes and aspirations of new consumers; build more respectful, collaborative and enduring relationships with all stakeholders; and unleash our collective co-creativity to bring better, smarter and more impactful ideas to life in ways that create shared value for all."

Fortunately, against the background of current economic, environmental, and societal issues, the efforts made by companies to position themselves around the subject of sustainability have now increased. However, much of what is currently communicated under the umbrella of CSR and sustainability is actually a sham, as marketing and sales interests are far too often the driving factors. "Green" products can be found everywhere and there are even more copycat products that are sold to unwary customers. False claims of commitment to sustainability are merely "Greenwashing." The term "Greenwashing" mainly refers to companies priding themselves on ecological or societal efforts that are either nonexistent or minimal compared to the negative socioecological effects of their core business. Anyone using advertising or PR activity to "green up" individual products, companies, or

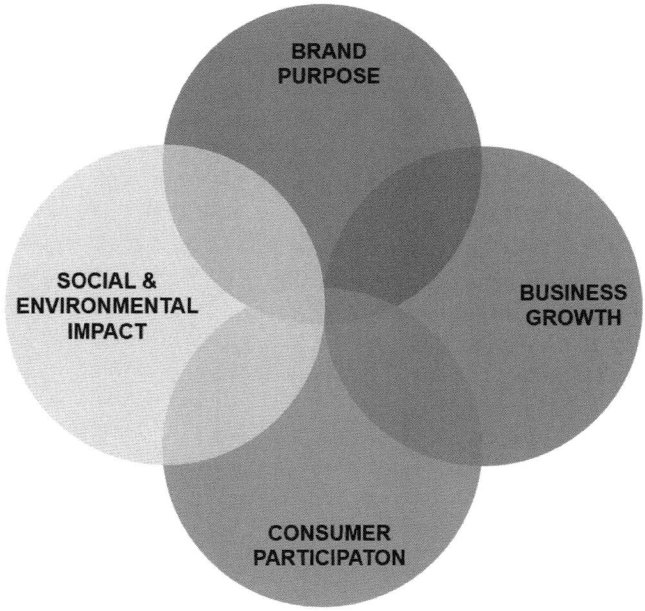

Fig. 2 Brand purpose and shared value as extension of the 'triple bottom line' concept (BBMG 2013)

political strategies is primarily aiming at creating the impression of being particularly environmentally friendly, ethical, or fair. Such an approach is absolutely right if it can be backed up with integrity and transparency. However, in many cases, those who talk about their "green conscience" actually only fulfil the basic requirements, if at all. Consumer confidence can only be won when products and corporate commitment go actively and convincingly beyond the interests of profit (Edelman 2013). "Greenwashing" is the wrong approach in this context and constitutes a bigger risk to companies than previously acknowledged. Neglecting the issue of sustainability has become a serious risk factor for companies: scandals such as environmental pollution, contaminated products, and poor working conditions lead to loss of reputation. Where listed companies are concerned, this can lead to plunging share prices and damage to the brand. If a company is tainted by "Greenwashing," both its reputation and economic basis are under threat.

It is for this reason that large companies in particular adopt practices that have a positive influence on public perception of the company in terms of having a "green conscience." Wal-Mart, the world's biggest retailer, has a deliberate policy of appearing regularly in interregional daily papers, as well as niche and specialist media. Its aim is to send regular targeted "green" messages to a somewhat sceptical specialist audience. Whereas every step toward sustainability is to be warmly welcomed, it still remains to be seen what role sustainability actually plays in Wal-Mart's core business. The business model includes building malls, which are usually located far from city centers and therefore only accessible by car. The construction of malls and car parks, totalling an area of 60,000 ha in the USA alone by 2012, has resulted in habitat fragmentation and isolation (Gang 2012). The product portfolio also focuses on the cheapest products, which are manufactured according to the "Cradle-to-Grave" philosophy and will certainly add to the growing volume of landfill waste. "Wal-Mart claims that the company is committed to ensuring that the pollution associated with product manufacturing is reduced. This sounds good at first but, at the same time, all Wal-Mart activity is aimed at reducing the shelf life of consumer goods, speeding up the flow of products from factory to landfill and encouraging consumers to make purchases," according to Sandy Mitchell of the US Institute of Self-Reliance (Mitchell 2012). PR-driven stances on sustainability that lack any real substance within a company's core business will always be perceived by consumers as Greenwashing.

For many companies, sustainability and Corporate Social Responsibility raise a considerable number of new and complex challenges ranging from a responsible approach to resources, the ecological consequences of a product, CO_2 emissions, fair working conditions, the promotion of women's rights, anticorruption practices, transparency, and societal commitment – all these elements must be taken into consideration. "Relevance, transparency, and clarity" are the defining factors. Generally accepted standards for products are now abound (organic guidelines, eco-label, and Fair Trade). However, these labels do not currently provide customers with any transparency in terms of the product design itself or the way in which the company's value chain is operated (this applies to mobile phones, food, and any other type of product). Although companies like Nestlé[4] have intensified their efforts to achieve greater transparency, nevertheless there has still been no major breakthrough. It is either impossible or too costly to obtain detailed information. Most of us are familiar with clicking through "greened-up" websites that hide more than they reveal or reading a sustainability report that runs for more than 100 pages.

The case of the British oil company BP is one of the biggest examples of Greenwashing in the world. It devised the "Beyond Petroleum" slogan to have an effect on its target audience and adopted a sun logo; nevertheless, despite record profits, it left oil extraction facilities to fall into disrepair. This made BP jointly responsible for the destruction of the Deepwater Horizon oil platform in 2010 and the biggest environmental disaster in US history. Costs amounted to at least USD 41 billion, the group's share price collapsed and the damage to its image is permanent. Ecological and societal labelling fraud is now being exposed and publicly denounced at an increasing pace. The current reporting practices of companies likewise fail to meet the requirements of all stakeholders (customers, investors, lenders, employees, consumer organizations, NGOs, etc.). Claims still fail to match reality, as internal regulations and organizational structures have not been established in many cases. It is rare that people responsible for sustainability and those responsible for operational activities work together in the same area or even communicate regularly. This shortcoming constitutes a danger to credibility, with

[4] For more details about the Nestlé approach to sustainability, please refer to the chapter "Nestlé: Sustainable Value Chain Management from the Farm to the Fork."

concomitant economic risks. We still see a large number of reports on sustainability and Corporate Societal Responsibility that appear to discuss only "good deeds."

Greenwashers are running the risk of boycotts, delisting, and warning notices. Some greenwashers even face legal action because of unfair competition or consumer fraud, as witnessed in 2010, when Opel and VW issued misleading green statements to attract customers, when the German Atomic Forum used wind turbines in a promotional image and when Lidl, the retail chain, demonstrably failed to comply with societal and employment standards. The Clean Clothes Campaign even confirmed inhumane working conditions at Lidl suppliers. Lidl issued a cease-and-desist order and the situation did not result in legal proceedings. If companies get involved in Greenwashing, they can expect the same fate as Vattenfall, the electricity group. In 2008, Vattenfall used the print media, the Internet, cinemas, and public places to call on the general public to sign up for climate protection. Given that Vattenfall actually operates climate-damaging coal-fired power stations, NGOs gave the company the "Climate Greenwash Award 2009" (climategreenwash.org 2009).

Customers expect companies to take environmental protection and societal standards seriously – any company that fails to do this, will quickly attract dissent: disappointed customers spread their knowledge of doubtful business practices like wildfire over the Internet and NGOs expose scandals. Even icons like Apple cannot escape the consequences of bad publicity. When the working conditions at the Apple contract manufacturing company, Foxconn, were suspected to be violating basic principles, young members of a Chinese group of activists secured jobs at Foxconn so that they could report on the working conditions from inside a company (after all earlier external requests to review the working conditions were refused by Foxconn). Apple was forced to rethink their value chain practices after activists revealed inhumane working conditions. Apple became a member of the Fair Labour Association as result of the scandal and started to make their supplier base and working conditions transparent. When we consider that Apple, the best-known brand of consumer electronics in the world, is currently in the spotlight, we need to ask ourselves how other electronics industry manufacturers run their value chains.

Even signatories of the Global Compact UN initiative have fallen short to such an extent that caused over 3,100 companies (as of mid-2012) to being excluded from this UN initiative. The approximately 8,700 members in 135 countries, 6,000 of them companies (around 200 in Germany), are now required to demonstrate what they are doing to implement the ten principles of sustainability for ecological and societal responsibility. Collaboration with experienced environmental organizations may be of benefit here. However, environmental and development organization logos are not always certain proof that companies are running their core businesses responsibly. In many cases, these logos are merely used for Greenwashing. In the face of a confusing range of different NGOs, expert Frauke Fischer, biologist at the University of Würzburg, Germany, warns: "It is not enough just to work with any long established environmental organisation, as there is no independent auditing of the performance of nature conservation organisations." As long as this is the case, companies need to analyze carefully which organization delivers "the best product for their money." Analysts in research agencies add that they rate collaborative research projects positively if they contribute toward increasing business sustainability performance in an important area. This is also the case if there are measurable objectives and deadlines for environmental protection, for example, and if collaboration runs right through the company. On the other side of the coin, collaboration is a sensitive issue for NGOs too. "Collaborating with multinational companies in particular involves the risk of being suspected or even blamed, and betraying your own ideals," says Bernward Geier, Director of Colabora, which supports the process of dialogue that the Rainforest Alliance (RA) enters into with other organizations. Tchibo, Kraft Foods, and Chiquita procure goods from agricultural companies that are certified by the Rainforest Alliance: for example, Chiquita buys RA-certified bananas. However, the Alliance faces the continuous criticism that it is not stringent enough and that employment laws have been breached on certified plantations.

Many NGOs pull the plug when the risk to their reputation becomes too great. WWF Netherlands ended its collaboration with the energy provider Essent in 2009 when it was taken over by RWE, the biggest emitter of CO_2 in Europe (Aachener Stiftung Kathy Beys 2013a).

Indeed, there are different ways of measuring the manner and speed with which changing customer requirements and global value creation impact on companies, depending on the industry sector, product portfolio and company size. One thing is certain, however, namely that a purely reactive approach does not go far enough. The approaches historically adopted by companies regarding the issue of Corporate Social Responsibility are inadequate in terms of meeting the expectations of well-informed customers. As described in the standard works "Corporate Social Responsibility" by Schneider and Schmidpeter (see Schneider and Schmidpeter 2012), the practical implementation of sustainability in companies and society is still either from a strongly philanthropic viewpoint or alternatively, from pure cost considerations. It is noticeable that CSR has been adopted in company departments, but that these departments do not work in close connection with actual core business, i.e., operations or the value chain.

2.3 *Going "Beyond the Label": What's in a Value Chain*

But how do we get "beyond the label," to really fill the concept of sustainable value creation with life?

In order to understand the constituents of a company's value creation process, one needs to understand which processes are interacting to deliver value. Every company that wants to generate the necessary/expected profit, does so in combining product development and all other value creating processes (the supply chain) in an integrated fashion, because "all companies are a collection of activities whereby a product is designed, manufactured, distributed, delivered and supported... these activities can be described as a value chain" (Porter 1989: 37). The concept of the value chain (also referred to as supply chain or service chain) coined by Professor Michael E. Porter from Harvard Business School describes corporate core and support processes. Figure 3 shows a schematic representation of the linked business activities making up the commercial goods production process.

According to Porter, there are five primary activities that describe the actual value creation process: internal logistics, production, external logistics, marketing & sales, and services. There are also four support activities supplementing the value creation process: corporate infrastructure, human resources, technology development, and sourcing. Each individual business activity constitutes an opportunity for differentiation, contributing toward the company's cost position in relation to the competition (Gabler 2013). Harting describes the "value chain" as "the stages of a transformative process that a product or service goes through, starting from the basic raw material to end use" (Harting 1994). The current interpretation of the **value chain is as a series of closely-linked linear steps leading to a seamless flow of products – from raw material to customer**. The ultimate goal is maximum operational efficiency for planning and business processes, ranging from sourcing to production and distribution of the end product to the customer. This goal is achieved through improved coordination of resource implementation in the value chain, based on the interaction between people, financial data, systems, and operational facilities (So et al. 2012).

There have been many different versions of operational implementation since Porter introduced the concept of the value chain and Harting provided the interpretation. The Supply Chain Council (SCC) has carried out pioneering work in this field. It is an independent, nonprofit organization that has taken on responsibility for developing the Supply Chain Operation Reference Model (SCOR), a process reference model to describe supply chains. The Supply Chain Council was founded in 1996 by Pittiglio Rabin Todd & McGrath (PRTM) and AMR Research, two business consultancies in Boston. The SCC started out with 69 members; since then the number has rise to over 1,000. Most of the members are companies from various sectors, industries, and stages of the supply chain. Although the emphasis is on exchanging experiences about the practical application of the model, the work of the SCC also includes input from scientists and advisors. The creation of the SCOR

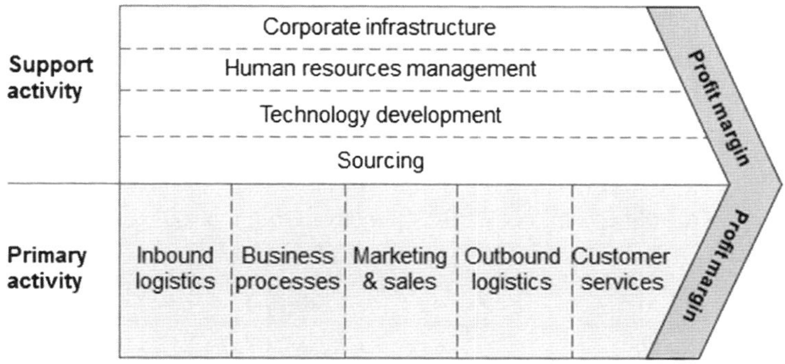

Fig. 3 Schematic value chain model according to Porter (1989)

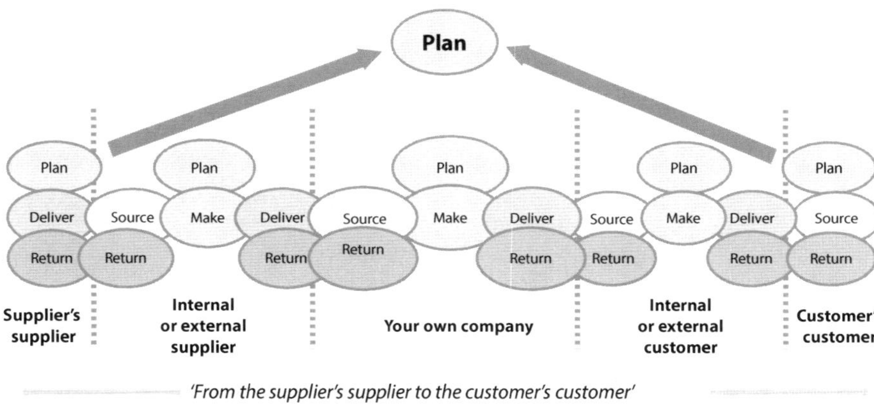

Fig. 4 Integrated end-to-end supply chain according to the SCOR® model (SCC 2006)

model has provided a reference framework that is now used in many companies to model their supply chains and supports further their development.

The main aim of the Supply Chain Council is to provide a reference model for efficient supply chain management. Developing the model further through the exchange of information between practitioners is acknowledged as an essential part of building the "body of knowledge" for state of the art supply chain management. Knowledge obtained in this way is fed into the model in the form of new or extended process steps and best practices. The standardized SCOR model helps companies to increase efficiency, reduce the input of resources, and accelerate supply chain processes. An important aspect of the SCOR model is to ensure that companies remain consciously aware of their supply chains from an end-to-end perspective, i.e., ranging from the "supplier's supplier" through the company's own organization right up to the "customer's customer."

Figure 4 is a conceptual representation of the interaction between the various processes of PLAN, SOURCE, MAKE, DELIVER, and RETURN. Describing an integrated supply chain end-to-end and then putting it into daily practice entails setting clear targets and bridging the internal divisions between roles and departments. The SCOR model provides a uniform language and clear definitions for achieving this.

The SCOR model provides a definition of the core processes PLAN, SOURCE, MAKE, DELIVER, and RETURN. The process hierarchy underpinning the SCOR model starts with a description of the most important parameters of the Value Creation Strategy (competitive basis: innovation leadership, cost leadership, service, etc.) and proceeds to break down these parameters into four detailed hierarchical levels containing standardized process elements, detailed information, the relevant key performance indicators, and best practices. The best practices of the model also take into account the characteristics of various industries and are developed on a continuous basis.

These process elements can then be used to define and describe any supply chain "configuration" (see Fig. 5). Depending on the industry, market conditions, product portfolio, and competitive situation, it is possible to take various levels of integration and manufacturing strategies into consideration. Make-to-Stock, Make-to-Order, Assemble-to-Order, and Engineer-to-Order are the most modeled configurations. Each different strategy impacts the design of the supply chain and describes interaction with customers, suppliers, and other partners. Various processes in the supply chain can be mapped in a model, resulting in supply chain "configurations." Companies with a global presence and a diversified portfolio generally operate several configurations at the same time.

The advantage of the SCOR model is that it is sufficiently generic to be applied in a range of different industries. It is also flexible enough to be adapted to specific requirements. Using key performance indicators (KPIs), the SCOR model can be

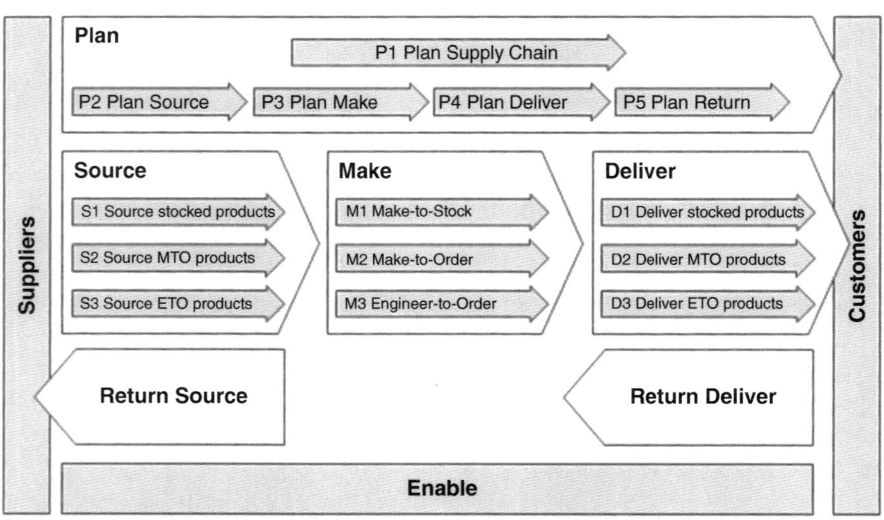

Fig. 5 SCOR® model process elements (SCC 2006)

applied as a basis for benchmarking supply chains. By comparing the performance of a supply chain in a specific industry (or even across industry sectors), company managers can obtain important indicators showing where potential for improvement lies or establish specific performance levels for certain processes. Using the experience gained from application of the SCOR model, companies have been able to refine and develop their core processes, best practices, and KPIs over time.

From the outset, the SCOR model included a description of the product RETURN process, although for a long time this only played a role in terms of handling requests for repairs and the associated logistics management process. The topic of sustainability was not addressed until version 9 of the SCOR model (released in 2008). Thanks to the inclusion of "Green" SCOR in Version 9, it was now possible to map "green" aspects, but also to describe these aspects via the best practices processes and KPIs. The work on Green SCOR itself had started much earlier back in 2002, when it had been developed by a research group in the USA, which then went on to win an academic excellence prize for its work just 1 year later. As a result of the work carried out by this group of Supply Chain Council practitioners, Green SCOR was incorporated as one of the standards in the 2008 model (Wilkerson 2008). This principally involved adding the environmental aspect to the existing process categories. For the very first time, the environmental impacts of a supply chain could be identified via a process model. Green SCOR extends the scope for considering the aspects of customer use and end-of-life recycling. On the basis of the SCOR model, the PLAN, SOURCE, and DELIVER processes are correspondingly applied to these areas. The area of waste management was specifically added to the model. Best practices were also incorporated, such as working with partners on the issue of environmental problems, reducing energy costs and packaging materials. The relevant indicators relate to CO_2 emissions and air pollutants, liquid/solid waste, recyclable waste proportions, energy costs, and units per cargo load (Wilkerson 2008). The Total Environmental Footprint is measured by adding up the total of CO_2 emissions, air pollutants, fluid/solid waste, and then deducting the proportion of recyclable waste across all production stages (MAKE). By standardizing the modeling options available for the SCOR model, companies can implement sustainability within the supply chain systematically (So et al. 2012).

With the introduction of Green SCOR, the link between product design and supply chain structuring became the focal point (see Fig. 6). **The trigger came when the working group realized that product costs as well as environmental impact could largely be determined by decisions that were made at the beginning of the product development cycle.** Close and consistent cooperation between product development and the supply chain department is required in order to address these aspects within product development and the supply chain. However, actual implementation is often difficult, because decisions on product design are typically the responsibility of development departments, whereas all other decisions (from raw material extraction to product end-of-life) are generally influenced by the supply chain department.

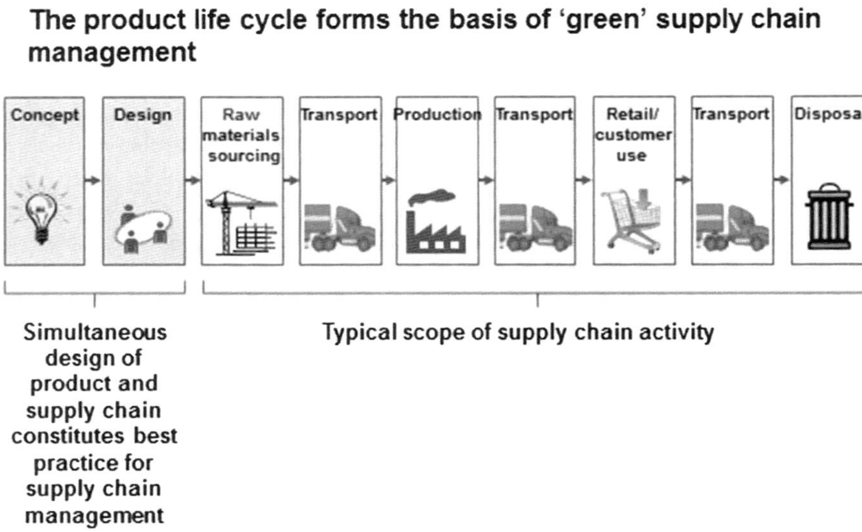

Fig. 6 Green SCOR as a further development of the SCOR model (Wilkerson 2008)

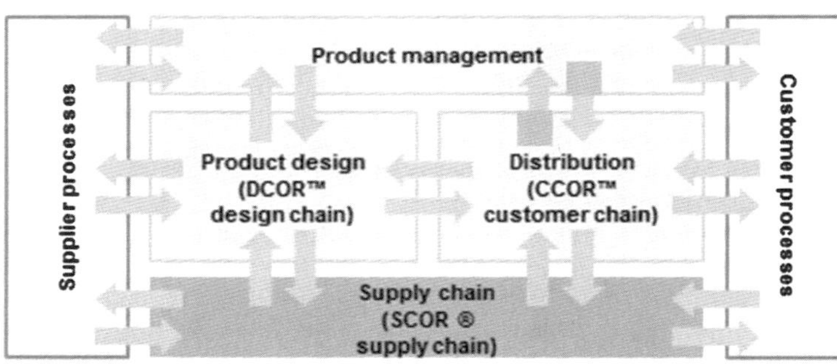

Fig. 7 From "supply chain" to "value chain" reference model (SCC 2012)

Encouraged by the success of implementing the SCOR model within companies, the Supply Chain Council decided to apply the basic idea of a reference model to other business areas (see Fig. 7). Whereas the SCOR model focuses on the supply chain and end-to-end aspects of the flow of materials and information, further reference models were made available for product design (Design Chain Operations Reference – DCOR), sales and customer service (Customer Chain Operations Reference – CCOR), and product management. The models take interactions between customers and suppliers into consideration. An integrated model of a complete value chain is mapped by modeling a supply chain covering all aspects of product creation across operational value-creation processes and customer-aligned processes.

The "value chain" is the combination of products and supply chains of a company.

VALUE CHAIN = PRODUCTS + SUPPLY CHAINS

The value chain is the company backbone – it is the manifestation of a company's value systems and strategy. Implementing sustainability primarily means implementing it in products and supply chains.

SCOR and its extensions provide a means of describing the interaction of processes, people, and systems in the value chain. However, in the past other models were also developed alongside the SCOR model, providing an alternative means of describing value chains.

A good example of this is the Value Reference Model (VRM), developed by the US trade consortium Value Chain Group. The VRM provides an open semantic dictionary for value chain management, where the reference model covers the areas of product development, customer relations, and supplier networks. As a process framework, the VRM is aimed at modeling, designing, and measuring those processes that involve the planning, relationship management, and the customer-related aspects of a business. The Value Chain Group claims that the VRM model is *the* tool for describing the next generation of business process management, resulting in product, operational, and customer excellence. Equally popular is the EFQM model, a Total Quality Management system developed by the European Foundation for Quality Management (EFQM) in 1988.

What all models have in common is that they aim at eliminating waste from operational processes. In recent years, optimization methods such as Six Sigma, Lean, Inventory Management and Sales & Operations Planning have also appeared within the sphere of the value chain. A large number of IT systems have emerged, that provide the supporting technology for the management of integrated value chains. The desired result is an overall reduction in operating costs, thus contributing to company profitability. Now as in the past, these optimization models are also a major aid in helping companies increase their market share and meet customer requirements.

The value chain presented by Porter and interpreted by Harting has developed further in line with changes in the business environment. Günther's "Value-Creation-Cycle" describes a further development of the value chain, where every production system sees the environment as having the role of both supplier (resource supplier) and customer (receiving environment). This creates a direct connection with sourcing, sales, and disposal (Günther 2008). Given that economic activity is not possible without interaction with the environment, there is a need to break away from a linear economy and move toward a closed circle economy. Whereas the value chain model proposed by Porter focuses on in-house process stages, the Supply Chain Council purposely incorporates multi-stage relationships with customers and suppliers, where the role played by the RETURN process is anticipated at an early stage. Günther's Added-Value-Circle also adopts the concept of the end-to-end supply chain, emphasising aspects of a product's lifecycle and the very process of value creation itself.

Public, political, and corporate awareness of the necessity for sustainability has increased in recent years. Companies have intensified their efforts, so far with the emphasis on public image. This is a positive trend, but one with huge potential for improvement. Whereas the public perception of companies also depends on efforts made in the area of sustainability communications, a crisis of confidence occurs when there is a scandal linked to a product or value chain. However, there does not have to be a scandal for consumers to become distrustful of a company's products.

It is therefore increasingly important to base product development and value chains on sustainability criteria.

Public debate on how to achieve greater economic sustainability is characterized by complex issues with a range of different viewpoints. However, what is the current state-of-play in terms of implementing sustainability within companies and value chains?

Aside from the requirements described in terms of sustainability communications, a change has occurred in the value chain in recent years that has often gone unnoticed (see Fig. 8): the value chain has been transformed from a necessary component within a company into a company's strategic capability for flexibility and future competitiveness. In the past, the purpose of the value chain was to bring about cost reductions and continuous improvement of all processes in order to ensure high delivery capacity with simultaneous low inventory levels. The 2008 financial crisis was the pivotal point in transforming value chains and increasing their significance.

2008–2010: In an initial transformation phase marked by great uncertainty, companies introduced a process resulting in the evaluation of value chains. A phase of strong economic growth and market share acquisition was followed by a focus on liquidity, cash flow, and profitable growth. This turnaround was not driven by any

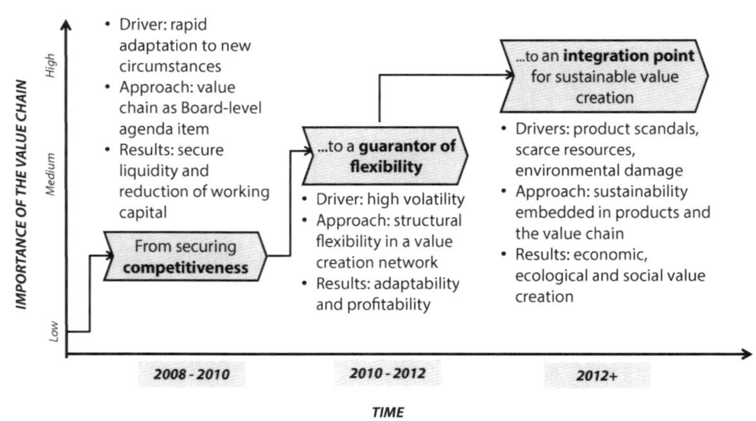

Fig. 8 Value chains are becoming the integration point for sustainable value creation (shared.value.chain 2013)

motive relating to sustainability. In many cases, it was about company survival, with the corresponding focus on securing liquidity and running down inventory levels. Many companies used this phase to make structural changes. In this tough economic environment and period of crisis, sustainability commitment and activity was often driven by short-term goals in the background. At that time, CSR was applied for reasons of absolute necessity (compliance). On a positive note, awareness grew about the enormous importance of the value chain. In many companies, value chain and operations managers were appointed to the Senior Management Team to keep a tighter control on liquidity and inventory risk. This made it possible to react quickly to market requirements. The focus in this case was on current assets, liquidity, and inventory levels.

2010–2012: In the second transformation phase, flexibility within the value chain was raised to the status of a principle factor for competitiveness. The importance of an integrated value chain increased in terms of being able to react more quickly to higher rates of change (in terms of frequency and scope). At that point, Professor Hau Lee of Stanford University coined the concept of "Triple A" supply chains, which have to be Agile, Adaptable, and Aligned in order to meet requirements. New forms of collaboration between participants emerged as a result of this. The necessary consideration from a systemic viewpoint has meant the increased removal of organizational barriers and the transformation of collaboration into the tool for future success. This has led to the appearance of value chains that can be described as "agile, adaptable, aligned, robust, sustainable, and integrated." According to Hau Lee, in future there will be no competition between companies, but between value chains and value creation networks.

Establishing and managing flexibility within a global value chain might almost be described as the "Holy Grail" of optimization: many strive toward it, but only a few companies come close to the ideal. One of the reasons for this is, that very few companies have a clear definition of flexibility. In order to understand flexibility properly, we must firstly establish a clear definition and secondly examine relationships with customers, suppliers, and partners closely. A company's adaptability is determined by the relationships and structures within the overall system connecting all participants. On a critical note, it must be mentioned that in this period, flexibility was mainly risk-driven and not driven so much by market opportunities. Risk management in the value chain – a serious topic of debate and a widely-implemented corporate policy during this period after the 2008 financial crisis – clearly indicates that in this period, a reactive approach was the norm. Many companies cut back their sustainability activity to an absolute minimum requirement in the aftermath of the financial crisis. Apart from legal and internal compliance requirements, no special priority was given to developing sustainability in the value chain. This meant that most sustainability initiatives were prioritised for their contribution to cost avoidance, e.g., the prevention of CO_2 emissions and saving energy in the production process in the sense of environmental management.

2012+: In the current transformation phase, the realization is slowly emerging within companies and society that the economic and political situation remains volatile and that raw materials are becoming increasingly scarce. The political balance has shifted from a few former leading industrial nations (G7) toward new decision-making constellations (G8, G20 or even G0, i.e., there is no real leading country anymore). These new constellations made joint and forward-looking decision making and implementation much more difficult and cumbersome. Flexible and sustainable global value chains are a decisive factor for competitiveness in an environment that is fraught with unresolved monetary crises and an ever-increasing number of serious environmental disasters. The capability to manage a global value chain and the synchronization of product launches together pose a massive challenge for many companies. The complexity of the task requires employees to have the best training, work in several languages, and be able to overcome cultural differences in an international environment with ease. Even the ability to understand and manage a systems approach extends far beyond previous requirements. This involves creating value for customers, avoiding or reducing waste throughout the corporate "system" and accelerating the overall "system" throughput. Corporate value chain optimization – from product development to the procurement, production, distribution, and return of products – has been supported by the application of reference models and information exchange between experts for many years now.

Despite clear changes, 90 % of companies in industrialized countries continue to focus on quantitative growth and maximization of profit. At the moment, securing a foundation that enables sustainable as well as profitable growth appears to be high on the management agenda of only a small number of companies. In fast-growing emerging economies such as Brazil, Russia, India, and China (BRIC), growth and prosperity are part of the local government promise to the people. Despite this, there is still an awareness of the fact that acting according to the past principles of industrial countries leads to consequences that can no longer be controlled. Sustainability needs to be a core component of a viable economic model in industrial and emerging economies alike.

Nevertheless, the focus in the majority of companies when optimizing the value chain is predominantly on efficiency and speed. Sustainability is only of secondary importance and then only of interest if costs can ultimately be reduced through energy, water and waste savings. The financial priorities dictated by shareholder and "lack of interest" in the customer base often referred to by companies are usually nothing more than an excuse.

Speed, flexibility, and sustainability should be the key attributes of any value chain.

However, for reasons of frequent abuse of the term sustainability as "a label," often with no real substance and insufficient transparency on the implementation of sustainability, there is a perceived gap between economic, ecological, and societal value creation

The short-term focus on profits and the necessity of operating a global value chain even in toughest conditions mean that flexibility is a priority within companies, whereas customers are showing an increasing interest in greater sustainability.

The gap is also evident at industry events, value chain conferences, and practitioners meetings, where the agenda often list sustainability as a side or niche topic. At present, the discussion hardly touches on the relationship of sustainability and flexibility as interdependent issues within the core business. Insufficient or imprecise communication in support of "greater flexibility," "lower costs," and "greater sustainability" will not lead to consistent implementation within the relevant company functions. Clear and ambitious targets need to be agreed, internal barriers broken down, and roles and responsibilities clearly defined. It is not enough for the Senior Management Team to recognize the importance of sustainability: all relevant parties also need to be involved in the implementation of sustainability. Marketing and communications departments also need to be encouraged to think outside the box. "Marketing and communications people devise campaigns, messages, and product labels to touch a nerve with customers. Additionally, employees in product development and the value chain work with different goals and time frames. In most cases however, it appears that communication happens too fast, that day-to-day business activity cannot keep pace with the sustainability messages and hoped-for market positioning" (Fig. 9).[5] The rethinking process has begun – initiated by pioneering companies that have started to take responsibility for their entire value chain and that also begin to include external factors ("externalities"), i.e., operations costs that are not borne by the company and that are often neglected in decision making processes.

At this point, we can conclude that an economy with increasingly strong global interdependencies is faced with rapidly changing conditions and customer requirements – companies and governments can no longer make decisions in isolation or hope that there will be no interference. It is true that Senior Management Teams now acknowledge more frequently that sustainability can contribute toward medium- and long-term growth, but in most cases there is still a lack of consistent implementation and embedding within the management system. Continuous optimization of the global value chain is an important prerequisite for commercial success, but it does not go far enough, as it neglects the opportunities provided by sustainability and a shared value creation.

There is a gap between the "need and ambition" for sustainability and its tangible realization within a company's core business (products and value chain).

[5] See also chapter "Telling the Backstory: Transparency in Global Value Chains."

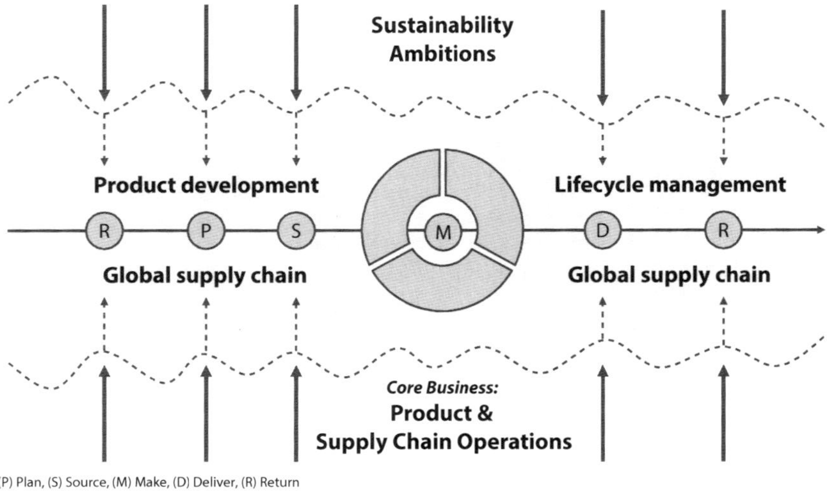

Fig. 9 Gap between sustainability, product development, and supply chain operations (shared.value.chain 2013)

This gap exists because of an economic approach that is primarily focused on commercial goals and a lack of consideration of ecological and societal opportunities/costs. Most consumers cannot see these connections right away. In the meantime, there has been an increasing awareness that we are living beyond our means. Consumers are becoming increasingly critical, asking what contribution companies are making across all value creation areas (economy, ecology, and society). This is where the concept of "sustainable value creation" can make a significant contribution, establishing sustainability within core business, and establishing it "inside-out" via products and supply chains.

3 Sustainability "Inside-Out": Building Blocks for Sustainable Value Creation

3.1 Toward Sustainable Value Chain Management

The recognition that economic activities in their current form are not viable in the long run has been a topic of societal, scientific, and economic debate for some time now. Michael Porter, the academic "father" of the value chain concept, has been advocating the "Shared Value" approach with Robert Kramer since 2006, promoting a radical rethinking of economic activity (Porter and Kramer 2011).

The "Shared Value" concept proposes three principles that form the basis of future growth and value-add in economic, ecological, and societal terms (see Fig. 10).

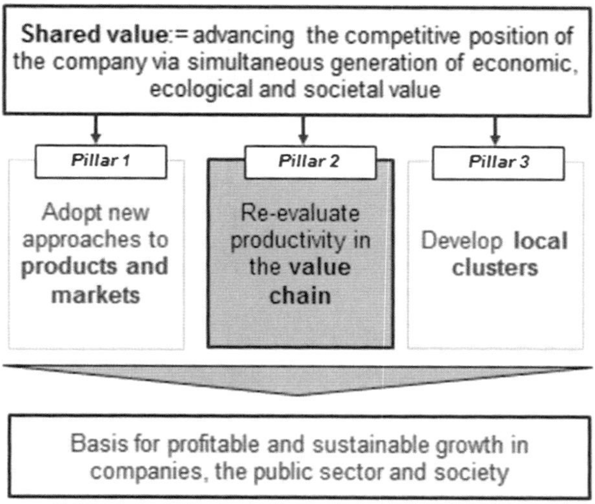

Fig. 10 Shared value is based on three pillars (shared.value.chain, author's own representation with reference to Porter and Kramer 2012)

1. The first principle relates to reconceiving of products and markets. It entails the design of sustainable products that help to solve societal problems or are oriented toward shared value.
2. The second principle relates to a new definition of value chain productivity, leading to a restructuring of global value chains. The objective is to raise quality, quantity and reliability within value chain processes, while at the same time making a commitment to preserving natural resources.
3. The third principle encompasses the integration of the company within local economic clusters which need to operate in a stable manner and interact with each other (Porter and Kramer 2012: 137).

Supplementing the concept of the social market economy in Germany and the interpretation of sustainability, there is a further important aspect to the concept of Shared Value: Shared Value is deliberately aimed at ensuring that companies conduct their business in this area out of self-interest. It is fine for a company to make a profit, as long as there is value creation for the environment and society alike. This eliminates the previous separation between business interests, society, and the environment (as factors that are external to the company). The aim is to create parity between business and society on the basis of mutual dependencies, risks, and opportunities. This approach looks not just at the company itself, but also the very foundation of the company's economic activity (Porter and Kramer 2012: 138, 139).

The deliberate inclusion of external factors in strategy and profitability calculations (e.g., environmental service procurement costs, effects on suppliers, cost of pollution) represents a significant turning point for the management philosophies that have previously been expounded. Porter and Kramer call for a dialogue leading to a reconsideration of value creation, whereby (negative) external factors are brought into central focus and the requirement for active management is made key. External factors are the economic costs incurred by corporate value creation, but that are not borne by the company itself. This takes specific account of the utilization of "natural capital." "Services" provided by nature such as clean drinking water, food, wood, climate regulation, shelter, and much more. Natural capital thus relates to the sum of all mineral, hydrological, fossil, and biological resources. From the business owner's viewpoint, costs and services provided by nature should be evaluated according to the same principles as other raw materials and services that are required for the manufacture of products. Liquidity, profit and loss calculation, and balance sheets can be applied to natural capital in the same way, thereby providing a financial valuation.

With the concept of Shared Value, Porter and Kramer do not argue the case for greater social commitment in the sense of philanthrophy with the concept of Shared Value, but instead the case for profit orientation combined at the same time with the solution of ecological and societal problems. The intention behind combining these two aspects is to protect the natural resource base and to generate shared value for all involved. In the opinion of Porter and Kramer, an economy based on Shared Value principles will form a more stable basis for medium- and long-term growth than any short-term approaches focused exclusively on shareholder value.

The world's biggest food producer, Nestlé, was the first multinational company to adopt the ideas of the Shared Value concept within the company's management system. Its management system, Creating Shared Value (CSV), focuses not only on profits, but also on rural development, water supplies, and the fortification of its products and derisking the food supply chains, that are behind the products. Implementation includes a reconsideration of "externalities" in value creation. Nestlé Global Value Chain Manager José Lopez, explains the reconsideration of external factors by taking the example of forests. For a company a natural forest has no value until it is cut down and becomes part of the Gross National Product. Logging is not seen as a "liability" toward nature. Now being part of a value chain process it is rather seen as "capital." The current pricing structure does not take both aspects into account. However, someone will have to pay for taking out this natural capital sooner or later. The challenge and goal for Nestlé is to evaluate the costs of extraction from the environment and include them into the business accounting process (Lopez 2012).

Taking reference to the Shared Value concept, "Sustainable Value Creation" is linking corporate social responsibility, sustainable product design and creating shared value at all intersection points along the supply chain. "Sustainable Value Creation" provides a healthy basis for prosperity in companies, environment and society alike.

Companies have the biggest leverage to be the driving force behind sustainable value creation. They provide the optimum conditions for creating economic, ecological, and societal value. Economy, ecology, and society need to be connected in order to have a positive effect in precisely those locations where all major decisions are currently being made: in corporate product development departments and the local, regional, and global value chains of companies and public sector institutions.

Achieving economic success means offering marketable products and innovations that create value throughout the entire product lifecycle and that can be returned to biological or technical cycles at the end of their useful life. The ability to meet customer requirements rapidly and efficiently in volatile markets requires adaptable value chains and involving all participating stakeholders. Companies and public sector institutions, their products, and value chains are just as much part of an "eco-system" where the interaction between all participants determines the chances of future long-term success. Sustainability, flexibility, and adaptability within a company are inextricably linked.

The contribution made by the World Business Council for Sustainable Development (WBCSD) through its work on the subject of "eco-efficiency" can be seen as an important step in the right direction. As an organization, assembling a large number of global companies from a range of different industries, the WBCSD represents the corporate perspective of sustainability. Through its work and its specific focus on eco-efficiency, the WBCSD helped to expand purely economic considerations by adding the aspect of the environmental impact of value creation.

In its "Vision 2050" report, the WBCSD makes the explicit proposal of a development path interweaving the economy, ecology, and society on the basis of a shared action plan. The role of economic development (value concept, products, and supply chain) is equally important to the role of ecological and societal development (WBCSD 2010). The "Vision 2050" report was created to form a basis of discussion for companies, governments, and society to work together to find new ways of securing the livelihood of around nine billion people using the planet's available resources.

Moving away from a pure focus on efficiency (improve, avoid, accelerate) toward effectiveness ("do the right thing" – from the start) opens up new avenues of possibility. Risks become opportunities. An orientation on short-term profits transforms into medium- and long-term market opportunities. Ecology, economy, and society reap the benefits during each phase of product creation and at each point in the value chain. The incorporation of this approach within an overall concept from the outset and at every stage of the decision-making and implementation process forms the core element of a sustainable value chain. An open, transparent, and collaborative approach is an important factor for success. Sustainability and global value creation are multilayered issues and current debate still sees many different ways of addressing them. As a management concept, sustainable value creation contributes toward unifying sustainability and optimization within a company's core business; the unified concepts of sustainability and optimization can then be implemented by means of targets, processes, and measuring systems. This makes sustainable value creation suitable not only for companies but also for value chains in public sector organizations, government organizations, and NGOs with a high degree of physical logistics activity.

Sustainable value creation is concerned with strategy, product development, operational processes, and multi-stakeholder collaboration. There are, without doubt, many further important aspects of sustainability, which are not directly attributable to the core business. These aspects include leadership, personnel management, employee training and retention, attractiveness for new employees, and societal commitment (philanthropy). These important aspects come within the overall remit of corporate responsibility. However, in the further explanation of the principles underlying sustainable value creation, the focus is on the changes that need to be made to products and value chains. Successful implementation via an integrated management system based on an "inclusive" and "system perspective" is the only way to achieve value creation with a future.

The desire for transparency in every aspect of value creation needs to become a defining principle, but without overstepping the boundaries of corporate confidentiality. Allowing stakeholders to "look behind the scenes" is a sign of openness and a expresses the desire to work together toward new solutions that generate positive impact for the environment and communities alike.

Given the fact that the use of the terms like "sustainability," "value chain optimization," and "profitable growth" are often misunderstood or used as empty clichés, the concept of sustainable value creation needs to be backed up by real substance. The Triple Bottom Line approach with its focus on "People, Profit, and Planet" so keenly adopted outwardly by companies, often merely represents the existing economic thinking: companies prefer to act according to the usual principles, winning a few bonus points on the way for environmental awareness and societal commitment.

"Go beyond the label" is the call to review what is really behind each term in the practical application in businesses. Companies using the above terms, should be ready to open up and explain, what are the concrete actions and benefits provided within and external to the organization. Organizations that deliver sustainable value creation with competence, integrity, and do employ sustainable practices at all stages of the value chain are the leaders in this field. **Every product decision and every value chain decision is a decision for or against sustainability.** Product lifecycles across all stages of the global value chain should be connected in such a way that the requirements of companies, the environment, and society can all be met at the same time. The value generated in this way for all parties involved constitutes the basis for sustainable and profitable growth.

Unfortunately to date, there has been only a limited basic understanding of sustainable value creation as a basis for future growth, and a number of misconceptions still need to be clarified. Business representatives often react with surprise when asked to implement greater levels of sustainability within their company, referring to existing efforts being made to guide their company toward good practice or the fact that they are already adhering to legal requirements and codes of conduct. Having spent the last 16 years involved in cross-functional project work on value chain optimization for multinational groups and medium-sized companies, my experience is that many opportunities for sustainable value creation have hitherto not been taken up to the core business. Short-term profit orientation continues to dominate and sustainability only appears on the management agenda if a scandal is uncovered or new legislation makes it an absolute necessity. However, the example of companies that are taking a leading role in shared value creation show that people, profit, and planet can be successfully harmonized to deliver profitable and sustainable results.

3.2 *Building Blocks for Sustainable Value Creation*

To enable sustainable value creation through the core business, one can utilize a number of building blocks, which can be deployed depending on the company context, maturity, and starting position. The journey itself is the objective, as customers, employees, suppliers, partners, and investors need to be persuaded. These building blocks cover both the vertical dimension (ranging from a company's strategy, its products, and its business processes to actual integration within the organization) and the horizontal dimension (ranging from the supplier to the customer and back).

The basic building blocks required to embed sustainable value creation are described below. The description primarily refers to manufacturing companies, but the principles can be applied to other organizations as well.

The incorporation of sustainable value creation within the core business requires a systemic approach or "blueprint," mapping out the appropriate building blocks for identifying objectives, allocating resources, and monitoring the progress of implementation. Figure 11 shows the building blocks that form part of the concept of sustainable value creation. In addition to these building blocks, there are additional key areas that contribute toward the performance capability of the "overall system": Community development, philanthropic engagement, financial controlling, as well as human resources and talent development. Given that they do not affect the core

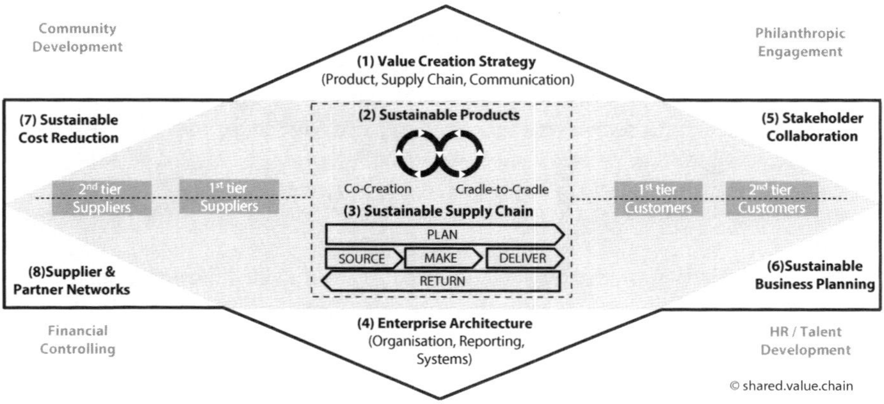

Fig. 11 Key building blocks for sustainable value creation (shared.value.chain 2012)

business directly (i.e., products and value chain), they will not be discussed in this article. The most important aspects will be explained to help illustrate the building blocks.

3.2.1 Building Block 1: Value Creation Strategy

The corporate strategy serves as an important input for defining a company's value creation strategy. The corporate strategy determines the basic objectives and parameters that build the foundation for the company's current and future success. The strategy generally covers a period of 3–5 years and defines the corporate purpose and mission statement, sets the parameters for the company's core business, and differentiates between short-, medium-, and long-term goals.

According to management guru Peter Drucker, the Senior Management Team's specific task is to do the "right" things (in the sense of effectiveness), whereas operations managers have the job of implementing these things in the "right" way (in the sense of efficiency) (Drucker 1986). Clarity is required regarding the type of growth being sought by the company and how a balance can be achieved between economic, ecological, and societal growth.

The primary question is not about how to make the company's existing operations model even more productive, faster, or smaller. Taking Drucker's "focus on the right things" as a starting point, the aim should be to identify the opportunities that enlarge the market for all players involved (i.e., making the "cake" bigger). Clayton Christensen of Harvard University speaks in this regards about the identification of "empowering" innovation, rather than pure sustaining/efficiency innovations, where the focus is on making the product "better." Empowering innovation is required to create a prospering economy where corporate objectives and societies needs are balanced. According to Christensen, we currently live in a Capitalists dilemma, where the rules of corporate finance "lock up" the funds for empowering innovation in their perpetual quest to generate more capital, but leaving the required jobs and opportunities for society behind (Christensen 2012).

The Value Creation Strategy stipulates how a sustainable corporate strategy is translated into sustainable products and sustainable practices along the supply chain. The resulting sub-strategies are Product-, Supply Chain- and Communications strategies.

The **Value Creation Strategy** is significantly influenced by the competitive basis and differentiation specified in the corporate strategy. A company can select cost, innovation, or service leadership as its competitive basis. The result of this decision will determine the way the value chain will be structured and the orientation of internal/external collaboration in order to meet market demand and achieve competitive advantage. The economic, ecological, and societal effects will already be largely determined by the company's selected product philosophy, its value chain, and its interaction with stakeholders. Additional factors to be considered include changing regulatory requirements, local legislation, and the tax regimes. Existing laws and regulations should be interpreted as minimum standards, which there is often a need to exceed.

While defining the value creation strategy, it is essential that the major underlying decisions are addressed. The first decision relates to the interpretation of sustainability. If sustainability is seen as the cornerstone of business activity and actively translated into a commitment by the Senior Management Team, there will be a good chance of embedding sustainability within the company's core business. If a purely reactive or compliance-driven approach is adopted, there is a danger that sustainability will only be implemented where legally prescribed or to meet the company's assumptions of what customers are expecting.

A number of parameters need to be included in the decision-making process in order to clarify the role sustainability plays in the core business: corporate purpose, growth targets, profitability targets, and the "licence to operate" are critical issues for consideration. The term "licence to operate" designates the societal acceptance of companies. This is based on the subjective interpretations of different members of society, making it impossible to ascertain it in any formal sense. In the context of increasing criticism levelled at companies and their value chains, the question of safety is becoming increasingly relevant. The absence of a "licence to operate" is evidenced by the gradual loss of the corporate capacity for cooperation, thus hindering corporate value creation; it can therefore also be described as the basis of corporate value creation (Lin-Hi 2013).

The basic principles of sustainable value creation need to be applied and translated into specific objectives, work packages, and implementation plans. Product Strategy-, Supply Chain-, and Communications-Strategy need to be developed as substrategies of the Value Creation Strategy (Fig. 12).

An analysis of the starting position and applicable areas for action build the basis for describing the Value Creation Strategy. Again, the **Value Creation Strategy** consists of individual substrategies, identifying objectives, scope, tasks, and desired timelines. Innovation, product portfolio, supply chain, and communication parameters are clearly defined and provided with objectives in the relevant substrategies (product, value creation, and communication strategies).

The **Product Strategy** defines the range of products and solutions intended to satisfy customer requirements and win new customers both now and in the future. The product strategy serves as the basis for achieving objectives laid down in the corporate strategy. Target customers, markets, technology, and customer value are not the only aspects to be defined at this point. The question needs to be answered as to whether there is an understanding of the consequences throughout the entire product lifecycle, and how negative consequences can be eliminated or at least minimized from the outset. The issue of returning the product and its components to closed cycles needs to be considered as part of the product design phase. New approaches such as Design-for-Sustainability (D4S), are investigating the use of

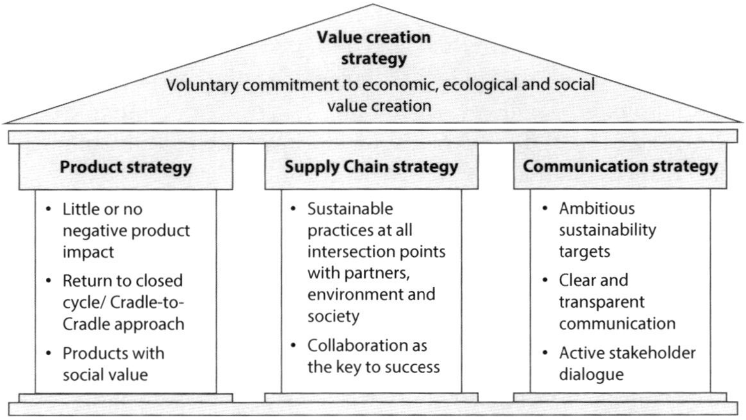

Fig. 12 Value creation strategy with sub-strategies (shared.value.chain 2013)

recyclable and renewable components (e.g., bamboo in the packaging sector). Sustainability and recyclability aspects can be added to existing product development parameters, which are generally oriented toward product cost savings, alternative materials, and simplifying manufacturing operations. Design approaches such as Design-for-Manufacturing and Design-for-Supply Chain have a specific focus on operational requirements.

A regular review of product design in terms of components, costs, and functions to improve competitiveness is a prerequisite for any marketable portfolio. The product philosophy needs to include a consideration of the entire product lifecycle, answering the question of whether the product will have a purely economic value or an ecological and societal value as well. Product designs that consider product recycling at the end of the lifecycle from the outset will also entail a reduced consumption of natural resources. It is generally cheaper and more environmentally friendly to process waste than to extract new raw materials.

In addition to a product strategy, it is also necessary to formulate a sustainable supply chain strategy. The **Supply Chain Strategy** translates the requirements of the Value Creation Strategy into tangible objectives, capabilities, and subsequent work packages. The organization should use the supply chain strategy almost as a "blueprint" for the implementation process. All objectives need to be broken down into individual management levels and brought into line with executive management and departmental incentive programmes. This process serves to describe core processes, identifies resources, allocates budgets, and confirms implementation periods. When describing the supply chain strategy, all the relevant elements of the value chain should be taken into consideration from an "end-to-end" perspective. A detailed assessment should be made to establish how planning, implementation, and control processes are interlinked. The flexibility factor is significantly determined by the entire chain extending from the supplier's supplier, via the company itself and finishing at the customer's customer.

Depending on the industry sector and market position of a company, a number of questions can guide the development of the supply chain strategy:

- How are operations planned and managed?
- What principles are applied for sourcing materials?
- Who are the most suitable suppliers?
- Where should production take place (and should it be done by the company or by an external service provider)?
- How can supply service be kept high and inventory low at the same time?
- What risks and opportunities are associated with the supplier base structure?
- How can risks and opportunities be managed actively?
- Who is responsible for warehousing and dispatch to customers?
- How is the return process organized?
- What are the internal and external service level expectations?

When describing the supply chain strategy, all elements posing a risk to efficient operations management or influencing overall system performance need to be identified. A chain is only as strong as its weakest link. Production system throughput is determined by the weakest components (bottleneck). Scientific approaches based on Goldratt's "Theory of Constraints" (Goldratt 1999) have made a significant contribution to improving supply chains. The supply chain strategy should take into consideration not only throughput but also complexity management and waste avoidance. In the context of the lean philosophy developed by Taiichi Ohno, "Muda" – i.e., "any activity that consumes resources but creates no value" (Hopp and Spearman 2000: 287) – should be eliminated not only from the entire supply chain, but especially from the whole network of all those involved in value creation. Spreading the concept of optimization beyond your own business reality into the intercompany value chain, incorporating the consequences at a number of supplier and customer levels and ascertaining all of the relevant consequences for the environment and society constitutes a complex but necessary task:

- What is the interaction with partners up- and downstream?
- What input and output flows are available for individual processes?
- How can a company's production facility be integrated within the natural environment?

As part of translating strategy intents into concrete implementation steps, particular emphasis needs to be given to applying new KPIs and timelines in the sourcing department with regard both to sourcing and to resource consumption. Sourcing has a key role, as it is necessary not only to manage sourcing costs, but also to ensure a high level of supply and production capability. This is often only possible with the right partners from the supplier base. The major challenge consists of balancing the sourcing role as "the fast route to profit" against a longer-term view and including sustainability criteria in sourcing decision-making. In a period of crisis, people quickly look toward sourcing for short-term contributions, as any amount saved on purchases has a direct effect on profitability. The effect of short-term savings is often not felt for long, as quality, supplier reliability, and on-boarding costs usually become an issue.

The supply chain strategy also determines which parts of the supply chain should be controlled by the company. This decision can bring competitive advantage in the prevailing context of global value chains, provided that it is not seen as purely cost-driven. Examples of companies with a tight hold on their value creation activities are the zip manufacturer YKK and the smartphone manufacturer Apple. Both companies have all aspects of their supply chains under total control. Whereas YKK differentiates itself through product quality and produces a high proportion of its products inhouse, Apple outsources all value creation activity to partners and service providers except for research, product design, and its own stores. Over many years, these subcontractors have become very closely associated with Apple and also bound by strict confidentiality rules. Through strategic investment in suppliers and critical components, Apple has succeeded in acquiring an advantage in the form of preferred supplier status. Tim Cook, current Apple CEO and successor to the legendary Steve Jobs, was responsible for the Apple value chain for 13 years and built his reputation on the successful implementation of this supply chain strategy. Both versions of supply chain control are successful models, the Apple model being more vulnerable to shocks (such as natural disasters).

The third sub-strategy that is derived from the Value Creation Strategy is the Communication Strategy, which represents an essential ingredient for sustainable value creation. The communication strategy should not only focus on persuading all internal participants of the importance of sustainable value creation, but should also focus on all external communication and dialogue requirements. The Senior Management Team and sustainability managers must not only discuss sustainability but also proficiently combine specialist expertise with business sense.

Knowledge of the interrelationship between finance, accounting, marketing, and the value chain within an organization must be coupled with the quick-thinking ability to make wise and sensible decisions that will lead to a positive outcome for the company, the environment, and society. The ability to communicate the business case for sustainability in a language that will be received positively at all management levels throughout the organization is essential for achieving set objectives. In many cases, sustainability is expressed using a special language. Concepts such as carbon footprint and sustainable consumption used by the sustainability department are not necessarily easy for colleagues in operations departments to understand. Production departments talk of "lean production" and not "zero waste," for example. Construction management talks of "reducing energy costs" and not "greenhouse gases." To be successful, sustainability managers need to communicate their messages in a language that can be understood by other company divisions. Desired objectives and consequences should also be formulated so that everyone understands what the relevant KPIs are for the corresponding departments.

Formulating a credible and transparent communication strategy is particularly important in terms of external relationships. Customers and other stakeholders have an increasingly important role in this respect. When all is said and done, it is not just what is good for the company that counts, as customers often question what the benefit is to them. Customer requirements are therefore becoming a determining factor.

If we look at the current situation regarding the implementation of sustainability within the core business, it is noticeable that sustainability often remains limited to philanthropy and personnel development in most large and medium-sized companies. Promoting societal development, philanthropic commitment, educational opportunities, better health, and on-site training do come under the remit of sustainability, but are not directly attributable to the core business. There are still major gaps between CSR-driven corporate image campaigns and actual incorporation within core business.

When product, supply chain, and communications strategy are linked together, they cover all relevant interfaces between a company, the environment, and society. The ideal situation regarding how sustainable value creation is implemented and practised may look like this: companies are successful in economic, ecological, and societal terms at the same time; products and services generate value for the company and society; all aspects are integrated and optimized via the value chain; and results are communicated transparently to encourage others to do the same. Sustainable value creation is the foundation for flexible and adaptable closed cycles, that lead to positive results.

Due to increased public and stakeholder interest in sustainability, companies are also addressing the issue with greater assiduity. It is noticeable that the environmental and societal value aspects are gaining importance as a result of a structured strategy development process. Investment and objectives in these areas are now being seen rather as strategic, and not merely as charitable activity. A study by the Bertelsmann Foundation, examining the presence of sustainability elements in various industries, came to the unequivocal conclusion that the focus of sustainability activity is clearly shifting toward societal and ecological sustainability in the value chain (Riess 2012). A look at how corporate commitment to sustainability has manifested itself in core business reveals focus areas of sustainable product development, corporate environmental protection, sustainable production and combating climate change through the reduction of CO_2 emissions. Levels of maturity vary widely in terms of tangible development.

The targets and directions described in the value creation strategy should result in improved income, education, health, diversity, and biodiversity for all involved. At the same time, it is important to fight actively against any increase in environmental pollution, disease, water shortages, or loss of biodiversity. Sustainability needs to be an integral part of corporate strategy. Leading companies do not therefore need a sustainability strategy – they need to work on the basis of a sustainable corporate strategy.[6] It is great to see that a number of companies have

---

[6] See also chapter "SAP AG & StarShea Limited (Ghana): Sustainable Value Creation Through Collaboration with Companies, NGOs and Intermediaries."

already understood this connection and adapted their focus accordingly. Medium-sized companies as well as an increasing number of multinational companies now have a holistic view of sustainability and value creation. Sustainable value creation is not seen as "nice to have" or "must have due to regulations" – it is increasingly viewed as a positive stance that prioritizes corporate purpose and uses sustainability as a platform for growth.

Because the corporate context and objectives often vary widely, an in-depth analysis of the starting position needs to be carried out as a basis for developing substrategies. A "Shared Value Opportunity Analysis" or "Materiality Analysis" looks at the risks and opportunities of sustainability for the company and the relevance of potential areas for action. Critical questions need to be asked regarding the contribution a company can actually make in terms of sustainability on the basis of its market positioning and core business. A company that cannot earn money by implementing sustainability within its core business will soon put a stop to its efforts. Companies should use the chance to conduct a shared value opportunity analysis, as it offers the potential to prevent negative effects from the outset (Braungart and McDonough 2002: 153f).

As part of a collaborative dialogue, the "shared value opportunity analysis" delivers a sound appraisal and overview of the 'hot spots', in which a company should become involved in terms of sustainable value creation in order to make a contribution and differentiate itself from the competition. For an understanding of the effects of corporate value creation, it is useful to obtain an overview of the risks and opportunities that can be found at the interfaces formed between the core business, the environment, and society.

This analysis requires to consider the economic, ecological, and societal factors in the existing product portfolio and value chain to be given equal weighting. As part of this process, the following questions need to be asked:

- Are economic objectives alone applied to the product portfolio and supply chain?
- What requirements do customers currently have with regards to products and the supply chain?
- Is ecology already a part of the business model?
- How important is adherence to legislation and standards of conduct (compliance)?
- What consequences do the company's products and activities have for society?
- How does the company currently address societal issues?
- Is sustainability embedded in core business?

In addition to having a clear understanding of stakeholder expectations and corporate policy, an analysis of the value chain is particularly important. Embedding sustainability within the value chain involves this analysis at all points of the chain where interaction with the environment and society occurs. Each core process of the value chain – INNOVATION, PLAN, SOURCE, MAKE, DELIVER, RETURN – can be seen as an interface or "value chain point," which will have a simultaneous consequence for the economy, ecology, and society (see Fig. 13). An

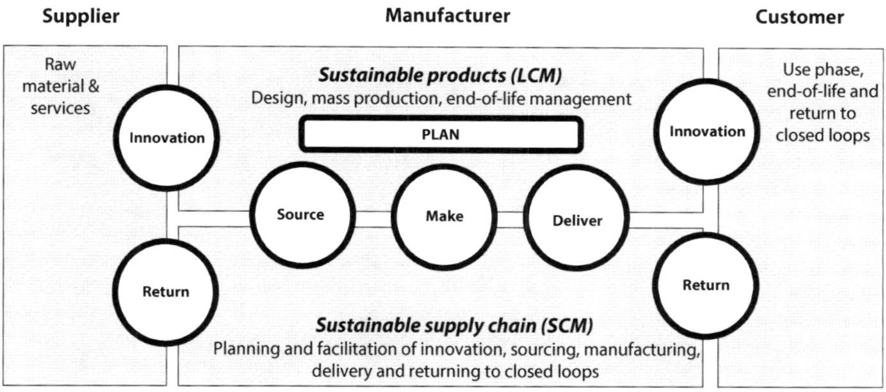

Fig. 13 Value chain points as pivotal interfaces for sustainable value creation (shared.value.chain 2013)

analysis of the value chain points helps to identify opportunities, risks and challenges for sustainable value creation.

Leading companies have a deliberate policy of applying this analysis to identify opportunities for innovation and sustainable value creation at each interface point (Fig. 14). Consequences, opportunities, and risks need to be identified and evaluated for each individual point along the value chain. Wherever possible, these consequences should be made measurable via standardized KPIs. The results of the analysis need to include clear and challenging objectives, encouraging the company to adopt an innovative stance when addressing issues and enabling it to make a positive contribution for future generations.

This does not mean that objectives must always be measurable in hard terms. A mix of measurable "hard" and currently unmeasurable "soft" objectives is useful to push through the necessary innovation. New ways of thinking can also be found outside what we already know. Courage is required to strike out in new directions. The process itself of defining specific, ambitious objectives has proved to be a very promising approach.

As a result from defining the value creation strategy, the Senior Management Team should decide how to communicate the specific objectives and targets that need to be achieved to ensure sustainable value creation and the schedule for achieving the desired objective. It is necessary to set brave and ambitious targets to serve as a guideline for the organization, even if it is still not clear how they can actually be achieved at the time when the strategy is defined. This type of courage is exemplified by International Paper, one of the world's biggest paper manufacturers. A process was conducted to set 12 specific objectives with defined values (energy efficiency, emissions, safety, waste, water, fibre, value chain, and philanthropy) that were intended to be gradually implemented by 2020. It is interesting to note that only a third of the objectives are measurable or have been assessed as feasible. A further third of the objectives have target values applied to them, but it is still not clear how they will be measured. The current approach for the remaining objectives

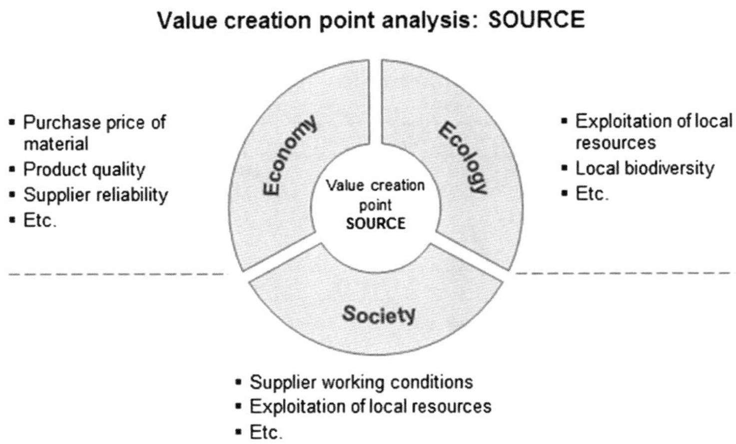

Fig. 14 Example of a SOURCE value creation point (shared.value.chain 2013)

is primarily oriented toward continuous improvement until a concrete roadmap can be established. Nevertheless, Teri Shanahan, Vice President Sustainability at International Paper, states that the Senior Management Team and workforce are confident of achieving the objectives (Shanahan 2013)[7]. Ambitious targets provide the impetus for making the necessary progress toward sustainable value creation, as they often result in close networking and collaboration during the implementation process, which can then lead to new approaches and even new business models. Collaboration produces learning effects for a company's own organization and opportunities for promoting employee motivation.

3.2.2 Building Block 2: Sustainable Products

The innovation of products and services that, in addition to their purely functional aspects, can also be returned to closed cycles and/or deliver societal value is one of the core elements of sustainable value creation. This philosophy of innovation and design is not only a reflection of an evolved understanding of customer behavior but also deliberately aligns itself with the design principles found in nature. In a climate of increasingly rapid technological development, global competition, pressure on profit margins, and relentlessly high investor expectations, the consideration of nonfinancial criteria and sustainability is certainly a challenge. Successful efforts to substitute harmful product components are to be welcomed, but are not enough on their own.

---

[7] For further details, please refer to the chapter "International Paper: Creating Value Through Sustainably Managed Natural Resources."

According to Frank-Martin Belz, Professor of Sustainability Innovation at the Technical University of Munich, sustainable products have the following characteristics (Belz and Peattie 2009):

1. They improve customer satisfaction: products and services that fail to satisfy customer requirements will disappear from the market in the long term.
2. They have a twin focus: in contrast to nonenvironmentally friendly products, sustainable products have both an ecological as well as a societal significance.
3. They focus on the entire lifecycle: sustainable products are environmentally friendly end-to-end, i.e., there is no negative impact on the environment from the moment when raw materials are extracted to the moment when the final product is disposed of.
4. They drive significant improvements: sustainable products help to address societal and ecological problems on a global level or lead to a measurable improvement in the societal and ecological performance of the product.
5. They are improved continuously: as knowledge, technology, and societal expectations are constantly evolving, sustainable products should also constantly adapt to changing conditions.
6. They challenge traditional products: sustainable products still lag behind traditional products – but these products can also provide a benchmark for societal and ecological performance.

There is a need (and ultimate opportunity) to think beyond known contexts and alter one's perspective with regard to features, functions, and efficiency:

- What effects does a product have throughout the entire lifecycle?
- Where do these effects arise and where can they be eliminated or at least minimized?
- How can products be designed so that they can be fully or partially returned to biological of technical closed cycles?
- How can a product also deliver ecological and/or societal value?
- How can we make the switch from limited resources to renewable raw materials?
- How can products deliver value throughout the entire lifecycle?

Product lifecycle assessments during the design phase have become a valuable and necessary tool for answering these questions. In the absence of any specific results from a lifecycle assessment, developers will merely attempt to develop "better" or "less harmful" products – not always with the best possible results. A lifecycle assessment makes it sufficiently clear where and when product effects arise. Armed with this understanding, specific design initiatives can be implemented to reduce these effects.

The results of a lifecycle assessment can also be used to formulate specific requirements of the value chain or sourcing activity. The deliberate inclusion of customers and suppliers in the product creation process can be a source of valuable information and innovation. "Outside-in" product innovation has been the philosophy of consumer goods manufacturers like Procter & Gamble for a number of years now. Closed cycle integration is becoming an important issue for the future, especially for companies whose products have so far been part of a "Take, Make, Throw-away" philosophy. Fashion producers have started to place a consistent emphasis on sustainable recyclable raw materials in the product development process.[8] At the same time, product management at the end of the lifecycle is undergoing a renaissance, with an increasing number of return systems appearing, as the process enables companies not only to reclaim valuable raw materials but also to demonstrate their active environmental and societal responsibility. Managing streams of return products and components is often forgotten when managing the value chain, but are now coming back into focus as a strategic priority. Whereas recycling is not a new concept and has been practised for many years in Germany, anglo saxon, emerging and developing countries have a considerable amount of catching up to do in this respect.

The interface between product development, product launch management, and the supply chain requires further specific examination, extending beyond a mere description of the activities to be carried out at each node of the value chain. Due to technological progress, product lifecycles have become increasingly short in many industries. The combination of highly volatile customer demand and a global customer base raises the challenge of launching the right products in the right markets at the right time. If a mobile phone product is launched only a few weeks late of the planned launch date, the average sales price can end up to be 20 % lower. Good launch management needs to rely not only on smooth internal processes but also on external partners, who should also be seen as part of the overall system and involved accordingly.

The Cradle-to-Cradle design approach developed by Prof. Dr. Michael Braungart and Bill McDonough promotes a view that aligns itself with natural principles. According to McDonough, there are no design problems in nature as there are in human design philosophy (Braungart and McDonough 2002). According to the Cradle-to-Cradle philosophy, products should be designed so that they do more than merely deliver benefits throughout the entire lifecycle. One goal of the Cradle-to-Cradle approach is to eliminate today's interpretation of "waste," replacing it with natural principles. It should be possible to break a product down into its basic materials at the end of the lifecycle and return it to the biological or technical closed cycles as input material (in the sense of "food"). Materials should not just be 'down cycled' (whereby they nevertheless still end up as landfill) but instead, should be endlessly recyclable. Thanks to this design philosophy and the associated return process, resources are protected and a contribution made toward the stability and diversity of the entire system.

According to McDonough, efforts by companies to assess their product portfolio in terms of 'eco-efficiency' criteria are to be welcomed as a first step but do not go far enough, as this does not prevent the further depletion of natural resources,

[8] See also chapter "VAUDE: Sustainable Value Creation as a Corporate Mission Statement for Small and Medium-Sized Companies."

merely delaying it at best. Any new design approach needs to address the issue of how products can have positive effects on people and the environment end-to-end. The challenge lies in combining the short-term usefulness, convenience, and esthetic aspects of a product with the long-term recyclability of the material. This is where a role can be played by innovation, which is not subject to the criterion of efficiency alone but takes into account a whole range of different aspects (Braungart and McDonough 2002: 72).

All those involved in product development and the supply chain need to rethink their relationships if they want to implement new approaches to sustainable product development (Cradle-to-Cradle or Eco-Efficiency) in an effort to move toward a circular economy. A considerable number of different parameters and strategic options need to be taken into consideration – not just efficiency criteria alone. Deeper multidimensional partner networks need to be developed and regulated actively. The role of product manufacturers is evolving into that of service providers, as they provide new products and take back used products as part of a simultaneous process. Product designers need to change their philosophy and not reduce product waste solely in order to reduce costs. Objectives for recycling and return to biological and technical closed cycles raise challenges that are different from those found in classic paradigms. The product development process should address not only the issue of which raw materials and product components should be selected, but also their origin. In many cases local raw materials and components are the better choice as they have the lowest impact overall. In the specific case of packaging materials, for example, natural and locally produced components such as bamboo and rice straw could replace ubiquitous polystyrene packaging. Whereas polystyrene has an environmental impact, alternative packaging is biodegradable. Adapting to local conditions is not only an opportunity to make a product more sustainable, but also to address local practices and consumer requirements. In terms of the value chain, this step entails measuring sustainability to a much higher degree than before. Actions like this are highly disruptive, as the entire value chain needs to be changed – as does the relationship with customers – as soon as the decision is made to incorporate these actions within closed material cycles.

From a sourcing viewpoint, designs need to be assessed to ensure that they are technologically future-proof. Procurement markets are analyzed, price trends evaluated, and possible suppliers qualified. A special kind of expertise is required, combining sourcing and engineering know-how. Sourcing engineering extends far beyond the existing qualification process involved in sourcing or category management. Once product design decisions have been made, it is imperative to tightly synchronize the development and supply chain processes in order to manage the product launch. Any delay in introducing a new product has negative consequences that extend beyond the financial aspects. Delayed product launches can result in high exposure when sourcing raw materials with long lead times, inventory depreciation (often with the associated disposal), or even fines in the worst case scenario. This often happens in the automobile industry.

3.2.3 Building Block 3: Sustainable Supply Chain

The third building block for sustainable value creation is the End-to-End supply chain, that spans from suppliers, through the own operations through to customers and back. Specific objectives and work packages defined in the supply chain strategy need to be applied to day-to-day business. A sustainable product philosophy has many different points of connection with the supply chain right from the design phase: product design has an impact on the choice of components, suppliers, and partners; the design process involves input from operations experts to ensure the development of product designs that ensure more efficient production processes, fewer rejects, and high quality; types of packaging can be selected for improving transport options, reducing transport costs, and involving the purchase of renewable products such as bamboo.

Globalization has brought about major changes to cross-company value chains. The concept of a company's "own 4 walls" has been significantly altered and transformed. The PLAN, SOURCE, MAKE, and DELIVER process elements of the supply chain no longer take place at a small number of company-owned sites near the company's headquarters; these elements have now evolved into a global network, where the various functions are dealt with in different countries. The growth of new markets in emerging countries and increased opportunities for low-cost production in countries with low labor costs have led to the development of supply chains with a global reach. In an initial wave of globalization, the emphasis was often on controlling a company's own value creation by means of overseas subsidiaries and the establishment of own production facilities in emerging countries. Additional opportunities (and levels of complexity) arose for companies through outsourcing. Various waves of outsourcing in recent years and the ensuing competitive pressure have made companies take a close look at their business activities. The resulting separation into core business and noncore business areas has led to the outsourcing of noncore functions, which can then be dealt with more cost-effectively by specialists. Alongside the initial wave of outsourcing primarily focused on relocating production, increasing levels of distribution and product development activity have been relocated in recent years. This has led to less vertical integration and a reduced cost base. On the other hand, such dispersed value creation needs to be managed closely, placing new demands on supply chain managers. Due to the distances, time zones, and cultural differences with internal/external employees, a global supply chain requires not only a special appreciation of varied nuances, but also the ability to deliver the required operational results (capacity, availability, quality, costs).

In addition to managing subsidiaries across time zones and varying cultural environments, there is also a need to be able to "orchestrate" a network of partners and services providers. In order to develop a flexible global supply chain, it is therefore vital to adopt an end-to-end view from customer to supplier and back, whereby it is possible to integrate, render transparent, and accelerate virtually the entire system. For some time now, in order to address the complexity, volatility, and rapidity of global supply chains, there has been a trend for companies to move away from rigid value chain configurations ("one size fits all") toward specific supply chains where individual market- and customer-driven supply chains can run in parallel.

Over the last 20 years, extensive and varied efforts have been made throughout many industries to increase supply chain efficiency, raise delivery service levels and throughput, and to improve response times. Process architectures have been visualized and analyzed around the core processes of the SCOR Model (PLAN, SOURCE, MAKE, DELIVER, and RETURN), waste has been identified, and weak points eliminated. This has resulted in the availability of a wide range of optimization tools that have been developed and further refined by companies and various professional associations such as the Supply Chain Council (SCC), the Association for Operations Management (APICS), and the Council of Supply Chain Management Professionals (CSCMP). APICS and SCC even agreed to combine their respective expertise, to build an even stronger platform for sharing best practice in value chain optimization and announced a merger of both organizations in April 2014. Standards such as Lean, Six Sigma, Operational Equipment Efficiency, Toyota Production System, Good Manufacturing Practices, and Multi-Echelon Inventory Management are just a cross section of the tools and options on offer.

However, to date sustainability has played a subordinate role in the development of these tools, although it has not been forgotten entirely. Corporate sustainability has so far been attached primarily to internal environmental, health, and protection requirements recommended by legislators and even trade associations. In addition to safety and environmental protection, it is also concerned with energy costs savings, CO_2 reduction, saving water, and waste avoidance. Nevertheless, it is important to check the credibility of the sustainability claims made by companies. When companies – motivated by their "green conscience" – are actually claiming environmental cost savings, trading of carbon emissions or selling their waste as "sustainability efforts," most likely these are results of the continuous pressure to reduce operating costs. External factors, i.e., all production elements that the company uses free of charge, but the cost of which is not included in the business accounts, are typically not taken into consideration at all.

When designing a sustainable, flexible, and adaptable supply chain, the first step involves aligning core business processes accordingly:

Core process: PLAN. The integration of planning processes has always been extremely important in the supply chain. The sooner an expected sales forecast can be produced, the more accurately sourcing, production, inventory management, and distribution can be planned. Planning processes not only need continuous improvement but also need to include options for incorporating IT planning systems. Special attention needs to be paid to exchanging planning information with intermediaries and end customers for identifying any changes in demand as early as possible. The ability to react quickly to change provides companies with a significant competitive advantage at the current time. Against a background of increasing uncertainty and high volatility in the global economy, the ability to react quickly is a prerequisite for profitable growth. After the demand and supply planning stage, there are further stages of integration forming part of value chain optimization.

Companies required to adapt to rapidly changing market demands need to combine their supply chain planning with finance planning, as part of their sales and operations planning. The current trend for integrated business planning constitutes the highest level of planning integration in use today. The objective of all planning approaches is the close linking of operation planning (volume, mix) and financial information (forecasting, budgets). With increasing support from planning systems, purely reactive sales planning (based on historical data) can be shifted to proactive planning approaches such as demand sensing and scenario planning. Collaboration with customers, supply planning, and the associated integration of suppliers and partners are increasingly important factors in addition to integration planning. Supply planning is not only critical from the viewpoint of securing the supply. In many instances, most of the environmental effects of a product arise with partners in the upstream value chain stages. The scope of planning activity must be extended in order to ensure sustainable value creation along the chain. While in the past it was enough to restrict the scope of planning activity to one's own company and stages immediately up- and downstream of the supply chain (Tier 1), it is now necessary to include the entire product lifecycle in the planning process (Tier 1-n, including customer use and disposal of the product).

In order to understand and evaluate the effects of value creation, it is also necessary to incorporate a number of upstream supplier and partner stages. Due to their complexity, these requirements also bring certain challenges. However, improved integrated IT systems (big data) provide new opportunities for involving partners from a technical viewpoint. The consideration of external factors in the planning processes presents a new challenge from a planning viewpoint. Due to the consideration of external factors, e.g., the monetary evaluation of "environmental performance," many decisions that are currently useful from a planning viewpoint will certainly turn out differently, and ultimately lead to a relocation of value creation activities.

Core process: SOURCE. Companies are increasingly referring to "sourcing guidelines for sustainable procurement" in order to control sourcing with consideration of sustainability criteria. These guidelines generally supplement codes of conduct aligned with compliance, usually requiring suppliers to follow sustainable working practices in a rather one-sided approach. Codes can stipulate that it is only permitted to work with suppliers operating an environmental management system, for example. With regard to building management and indirect materials such as energy, these sourcing guidelines ensure that companies switch to sustainable energy sources and services more rapidly. The move from nuclear-generated electricity to electricity from renewable energy sources and the use of cleaning services with high environmental standards can be mentioned as an example here. Active supplier management across all sourcing categories has always been within the sourcing remit. Due to the growing complexity of value creation and high volatility even within the supplier base, the importance of close proactive supplier management has increased dramatically. We will explore the role of active management and supplier development separately in Building Block 8 – Supplier &

Partner Networks. While reducing costs is everyday practice in any well organized sourcing department, sustainable sourcing is generally still in the development phase. The right knowledge and methods for including the reuse or design-for-reuse of product components are often not available in sourcing departments. As long as raw materials can still be obtained at "acceptable" prices or additional costs can be passed on to customers, we will see delays in implementing sustainable procurement practices.

Core process: MAKE. The continuous improvement of production processes and cost reduction measures has always been a basis for low production costs and therefore essential to product margins and profitability. Production resource loading (buildings, machinery, and personnel), different production configurations, low exchange costs, and high quality are often critical factors in modern production systems. Until now, most production systems have been run to generate maximum output at minimum cost. This strategy has widely paid off in practice. In recent years, however, production systems have been caught in the tension between a pure emphasis on capacity utilization (push principle) and response to demand (pull principle). Demand-driven management has the advantage that products are only manufactured if customers want them or they are actually needed. This philosophy includes replacing buffer and safety stocks with information, arriving at lower operations costs. The computer manufacturer Dell has gained a global reputation by perfecting the make-to-order (MTO) model. Until now, sustainability in production has mostly revolved around energy and water savings. Driven by rising energy costs, increasing numbers of new models for combining production efficiency and sustainability (saving energy in this case) are being implemented, as the introduction of an environmental management system is a valuable tool for reducing energy, water consumption, and waste. The environmental performance of buildings, plant, and machinery can be measured by KPIs. This means that programs for increasing energy efficiency and consumption can be implemented in a more focused way.

The MAKE process not only consumes energy as part of the production process, it also results in the optimization of packaging, greenhouse gas emissions, solid and fluid waste. The metrics to quantify these results include the energy efficiency of buildings and machinery as a percentage of overall costs/production losses due to noncompliance, CO_2 and greenhouse gases, carbon emissions, air pollutants (COx, NOx, SOx, volatile organics, and particulates), liquid waste, solid waste, and the percentage of recycled waste. In this way, peak capacity can for example be planned for periods when the general demand for energy is low, thus reducing costs. These efforts are measurable by metrics such as energy costs as a percentage of production costs or the proportion of waste as a percentage of products manufactured.

Core process: DELIVER. Along with the underlying issues of transport and warehousing, distribution logistics has a significant impact on a company's flexibility and sustainability. Apart from production, most of the environmental impact and effects of a company's CO_2 footprint are caused in the DELIVER and distribution logistics processes. Customer requirements and the service standards promised by logistics companies have a direct influence on the choice of transportation. The cost and environmental effects of air freight, sea, road, and rail transport vary widely. In this context, sustainability can be achieved by choosing the right form of transportation, load optimization, and proper transport requirement planning. "Cost-to-Serve" analyses that are now being used by leading companies show distribution costs in relation to the requirements and profitability of each customer segment. The consequence of this is that new, differentiated distribution models are being applied which meet the requirements of a specific customer segment while also reducing distribution costs and environmental impact at the same time.

However, also transport companies develop new concepts. One example is the ELVIS logistics network, in which medium-sized European carriers have joined forces to confront growing demands such as transport volumes, increasing volatility, lack of transport infrastructure, and the proliferation of government HGV regulations. The aim of the network and its approximately 20,000 vehicles is to secure an economic business base and also address environmental aspects such as avoiding transport km/miles and empty vehicle runs (ELVIS 2013 – Full Load Network). Innovative ideas from the logistics sector such as the selection of transport service providers according to their CO_2 footprint (based on the technical features of the vehicle fleet) offer interesting approaches. Metrics for measuring sustainability performance, such as fuel, costs as a percentage of distribution costs or the percentage of suppliers who meet sustainability criteria is also used.

Core process: RETURN. In the past, the return of defective products or end-of-life products has either been the remit of a company's service department or part of the value creation process pursuant to statutory regulations. The return process is now being optimized and automated in the same way as other core value creation processes. With the advent of the internet as a customer sales channel, not only have transport volumes for small and medium-sized dispatch items increased, but the number of returned products has also risen. If the customer is not satisfied with a product or wishes to exchange it, the package is simply returned and another one is sent out. This particular way of doing business is convenient for the customer, but ultimately has a greater impact on the environment. Where products are defective or at the end of their lifecycle, it needs to be questioned whether it is actually economical to recycle them, in those cases where they can no longer be repaired. Product recycling has returned to the management agenda as a means to counter the trend of ever diminishing resources for vanity materials. For example, every kilogram of aluminium recycled saves 5–8 kg of bauxite, 4 kg of chemicals, and 14 kW of electricity. This also results in 95 % less air pollution (see UNEP 2009 and Elias-Trostmann 2012). Similarly, there is more gold in a tonne of electronic scrap than in a tonne of iron ore. The number of returned products as a percentage of products sold or the recycling rate for returned products can be used to measure and identify the performance of the return process. IBM is an example of a technology company with a consistent policy of product return at the end of the lifecycle. Whereas the electronics industry faced constant criticism for designing products so that they become defective after a deliberately specified period of time ("planned obsolescence"), IBM has a specialist Global Asset Recovery Services (GARS) department, which has been processing and recycling a wide range of electronic scrap since 1999. IBM GARS recycles 40,000 devices per week worldwide, reusing almost 87 % of resources. Of the 50,000 tonnes of hardware processed every year, <1 % ends up as landfill (IBM 2008). It is interesting to note that GARS was not created out of self-interest but as part of a business model for supplying customers with low IT budgets with cheaper, but refurbished hardware. IBM occupies a unique position in the technology industry in this regard. The European Recycling Platform (ERP), a not-for-profit organization where leading manufacturers of consumer electronics have joined forces, has made similar efforts to reduce waste, increase recycling rates, and return electronic scrap to closed cycles.

A current trend in managing the supply chain throughout the company consists of modeling and operating value creation "configurations" that are customer- and market-specific. The aim of these segmented supply chains is to provide exceptional customer service to each relevant customer category. Customer requirements become the driving factor behind the design of the supply chain, where opportunities for simplifying internal processes and reducing costs naturally play an important role. Nearly all of the companies listed in the Supply Chain Top 25 report, published annually by research company Gartner, put considerable effort into developing and improving customer-specific configurations (Hofman et al. 2013).

3.2.4 Building Block 4: Enterprise Architecture

The ability to adapt to rapidly changing conditions is now a strategic competitive advantage. Clear responsibilities, goals, and time lines need to be identified at the point when value creation strategy is being translated into functional substrategies. The focus should be on continuously questioning the status quo and searching for simple but effective models, whereby the role played by dismantling internal and external barriers is especially important.

The fourth building block for sustainable value creation, is related to the "Enterprise Architecture." The term "Enterprise Architecture" is used to describe the sum of design decisions related to organizational setup, roles, responsibilities, reporting, use of key performance indicators, and the information technology that is supporting the processes within an organization.

Now more than ever, integrated enterprise architectures has become a major prerequisite for flexibility and adaptability. In view of the increasingly volatile levels of demand and mounting competitive pressure faced by multinational companies in today's business environment, challenges are multiple and diverse: Factors such as expansion, concentration, acquisitions, partial buyouts, and changing markets results in a constant requirement to adjust the corporate setup, structures, and responsibilities.

Enterprise architecture is based on modeling a company as a "total system" and defining the required interactions, capabilities, targets, and metrics over time. Enterprise architecture then comes to life and delivers performance through the interaction of people, functions, processes, and supporting information technology. This includes the structural organization, workflow-, process management, as well as incentives and IT systems. All these "components" should be integrated in such a way as to ensure flexibility, adaptability, and sustainability.

Roles and Responsibilities To enable a high performance organization and working towards positive results a clear description of roles, responsibilities, and objectives is paramount. The introduction of a coherent target system, the definition of clear rules for business processes in conjunction with integrated software support solutions are all factors that make an enormous contribution to supporting the necessary operational capabilities.

Although the economic targets of individual departments are often achieved by focusing intensively on the function, this often does not mean to achieve the best result for the company as a whole. In fact, the ability of every single company departments to work toward shared goals is an important lever for capturing potentials in products and the supply chain. This applies to efficiency, cost, and sustainability alike.

Process Management The mapping of essential corporate processes in an integrated process model, which is easy for users to understand, helps not only to provide an improved basis for discussion, but is also often a prerequisite for success. Companies have access to many different methods and tools for achieving this. Users can take their pick from well-tested process models that describe the interaction between product development and the supply chain, such as the SCOR Model and the EFQM approach. A number of companies have seen successful with bringing these aspects together in an integrated representation called a 'process house'. The process house encompasses all relevant corporate processes in a standardized form and describes all roles and responsibilities as well as in- and outgoing information flows.

Information Technology and Systems It is essential to employ modern information technology in order to meet and adapt to the demands of the economic environment. The possibilities afforded by integrated systems such as Enterprise Resource Planning (ERP), Advanced Planning Systems (APS), and Internet technology can be used for developing the relevant core processes and achieving the requisite level of transparency for identifying trends at an early stage. Many different IT solutions are also available as a basis for collaborative product development and supply chain projects. A number of different software vendors have shaped the field of value chain integration and optimization in the last 30 years. The sustainability requirements of global value chains have also been acknowledged and are currently being addressed by market leaders such as SAP and Oracle and a considerable number of niche suppliers. A number of monetary evaluation models that are mapped via software are also emerging for the purpose of evaluating corporate sustainability. Thus, there are some systems already available that allow a monetary evaluation of environmental impacts and even provide transparency with regard to supplier sustainability levels (e.g., the Sedex database). However, many efforts to integrate value chains across several companies often fail in day-to-day business implementation because partners either use outdated systems or are not able to make any major investments in IT. It is for this reason that the integration of sustainable processes via information technology and reporting is becoming a growing challenge for companies. The often manual collection and evaluation of information using calculation tables, surveys, and scorecards has the disadvantage that the growing volume of data is increasingly error-prone. The current plethora of sustainability "standards" and questionnaires does not make the task any easier. Practical solutions need to be applied to ensure that sustainability is incorporated step-by-step and do not fail because the technical requirements for data exchange are excessively high.

Evaluation, Reporting, and Standards Alongside organizational elements, the process model, and the systems in place, it is also useful to have a transparent and easily traceable reporting structure in order to show progress in achieving economic, ecological, and societal objectives. Just a few (but highly informative) KPIs and clear reporting will help sharpen the focus on the relevant issues and manage the change process in an active way. Attempts to incorporate sustainability in the value chain have failed in many companies because the appropriate indicators were not available, not clearly defined, or unmeasurable. Companies need to assess their reporting structure in terms of whether it addresses the scope, content, and orientation of current requirements. Whereas the integration of key operational and financial performance indicators in the value chance is acquiring increasing importance, sustainability KPI reporting needs to be more firmly established.

The Global Reporting Initiative (GRI G3 and G4) standard is undoubtedly the most widely used reporting approach at the current time. The standard was updated and extended by G3.1 in 2011. The new G4 standard and the move towards integrated reporting will definitely bring about harmonisation. In addition to the codes of practice, that in many instances are given their own unique interpretation and corresponding implementation by individual companies, there is also a whole range of standards being adopted by an increasing number of companies. This includes the UN Global Compact, the OECD Principles of Corporate Governance, the ILO Convention, ISO 14001, EMAS, and more. There is also a range of reporting standards such as GRI, designed to render the implementation of sustainable value creation measurable.

A Perspective on Standards and Measurability of Sustainable Value Creation

Companies addressing sustainability issues and seeking to incorporate measurements within the value chain can utilize a whole range of so called "standards." There are various internationally recognized principles, guidelines, industry standards, codes, and voluntary commitments relating to value

(continued)

chain management and sustainability. The choice is extensive and it is not always possible to predict which recommendations will eventually develop into standards in an individual sector. It is notable that there is currently no unified standard for sustainability in the value chain. Efforts at standardization often target subsidiary aspects, all of which are given their own criteria and KPIs. Unfortunately the heightened awareness around sustainability as a "burning issue" has also led to quite a number of new measurement approaches and proposed "standards." **While companies, stakeholders, and society alike would love to make the subject more tangible, a tendency toward increasing standard and reporting fatigue can be observed. The solution to revert this trend, is to cut the amount of existing "standards" by half. Once this is done, the remaining standards and associated reporting requirements should be half as complex – to keep it manageable for all participants**.

As the future adoption of sustainable value creation is tightly related to the acceptance of common, simple, and acceptable "standards" of practice, the following provides an overview of the approaches/standards/labels/measurements developed so far. In addition there is an initial evaluation of their importance to accelerate the adoption of sustainable practices in global value chain.

The following table differentiates standards as follows:

- General guidelines and voluntary commitments
- Finance industry standards
- Standards and norms in product development and the supply chain
- Organizational approaches of interest groups and NGOs
- Reporting for sustainable value creation

Name of standard with description	Assessment
General guidelines and voluntary commitments	
United Nations Global Compact was presented by the UN at the World Economic Forum in Davos in 1999 and has ten basic principles for companies working with the United Nations to ensure that globalization has a greater orientation toward societal and ecological responsibility. Thereby it refers to human rights, employment standards, environmental protection, and anticorruption	Is being ratified by an increasing number of companies to demonstrate their commitment. It brought corporate value creation and adherence to fair working conditions back into the spotlight of public attention. As a selection criterion for critical customers, the public commitment of a company to the UN Global Compact will definitely become a prerequisite
OECD guidelines for multinational enterprises constitutes a code of conduct for global corporate responsibility and represents government recommendations for the economy. They consider behavior	In terms of the value chain, the OECD Guidelines address working conditions, safety, risk management, and the sourcing of raw materials. This makes them more useful than the principles of the UN Global

(continued)

Name of standard with description	Assessment
regarding transparency, employment conditions, the environment, corruption, consumer protection, technology transfer, competition, and taxation. They also relate to international agreements such as the Universal Declaration of Human Rights and the core employment standards of the International Labour Organization (ILO)	Compact for the practical implementation of sustainable value creation. However, they serve more as a base for discussion within a company rather than for holding a dialogue with customers
ISO 26000 provides guidance and direction for the way in which all types of organizations should behave in terms of societal responsibility. It departs from a concept of CSR restricted solely to companies and enlarges the focus to all types of organization. It touches upon transparent and ethical behavior, sustainable development, health and public welfare and thereby takes laws and stakeholder expectations into consideration	It is likely to encourage organizations to discover the significance that sustainability should have within the core business. However, it will not play any role in terms of incorporating sustainability in products and the supply chain as it is on purpose not meant to be measured
Finance industry standards	
UN Principles for Responsible Investment (PRI) they are based on the understanding that, through globalization and trade, the objectives of the UN have converged with those of the private sector and financial markets. Institutional investors act in the long-term interest of their clients. Institutional investors must take into account the major issues of ecology, society, and good governance as part of the overall performance of an investment portfolio	The inclusion of sustainability criteria as a decision-making factor for investors poses a major challenge from the company viewpoint. They have to address the issue of sustainable value creation, in a more focused and credible way. The extension of investment criteria is certain to give new impetus to the implementation of sustainable value creation in products and the supply chain
Dow Jones Sustainability Indices (DJSI) they are a family of stock indices that take ecological, societal, as well as economic criteria into account. This is where DJSIs differ from both traditional stock indices and purely ecology-aligned indices. Since its introduction in 1999, 70 finance companies from 18 different countries have been licensed to work with the DJSI. Inclusion in a DJSI has special significance for companies, as this allows them access to the sustainable investment capital market	Multinational companies in particular have a major incentive for being listed in the DJSI. A high rating gives a company access to significant investment capital. Given its worldwide reputation and strict criteria, companies are also keen to use their position in the DJSI to boost their corporate image

(continued)

Name of standard with description	Assessment
Equator principles[9] the Equator Principles are a voluntary risk management framework to manage environmental, social, and governance (ESG) risks in project finance. The objective is to provide a minimum standard for due diligence to support responsible decision making	The Equator Principles are an important commitment to drive more focus toward sustainable investment behavior. However the idea behind the Equator Principles will only unfold its full potential when being applied to general lending practices as well – ideally when combined with Positive Impact Finance
***KPIs for ESG* of the European Federation of Financial Analysts Societies (EFFAS)** the KPIs for environmental, societal, and governance (ESG) were developed by the German Association for Financial Analyses and Asset Management (DVFA) and have now been adopted by EFFAS, thus enabling their application at European level. They are extra-financial and are aligned with the requirements of the financial sector in that they reveal aspects that could have a material influence on the status and performance of business results. Initially there were 25 cross-sector ESG KPIs; now they include about hundred sector-specific criteria. ESG KPIs are measurable and are intended to render companies comparable. They relate to environmental sustainability, the use of renewable energy sources, product health and safety, employee absence rates, work force changes due to restructuring and corruption, etc. Most of the KPIs relate to risk, as this is an extremely important factor for investors. They are not compulsory, however, and are applicable on a voluntary basis	Given the current European recognition of ESG KPIs, they have become an important reporting element and therefore the basis for access to the capital market
Standards and norms in product development and the supply chain	
Organic guidelines for the production of organic foodstuffs were introduced for the	Product standards and labels help to increase levels of customer trust. Critical

(continued)

[9] See Chapter "Opportunities Through Positive Impact Financing in Banking Value Chains."

Name of standard with description	Assessment
whole of Europe in 1991. Foodstuffs must be produced according to a series of organic criteria in order to carry the organic label	users can be persuaded that not only are products safe in terms of their contents, but that they are also manufactured in line with fair principles. An increasing market demand for sustainable products combined with the increasing focus of advertising messages on sustainable consumption, standards, and labels together are proving the way forward for delivering revenue growth. However, a careful approach is advisable: not all labels deliver on their promises and even established representatives still have their problems. This means that costumer demand needs to drive a small number of reliable standards to the forefront, thereby increasing transparency levels for product constituents. Corporate communications strategies should of necessity promote transparency, i.e., what a label actually means and the extent to which it is applied in industry. "Go beyond the Label" is an important call for action here
Ecolabel it has been successful due to its clear criteria, supporting the requirement for more transparency. The ecolabel can be found in the following products and services: construction and housing; household; garden; office; paper and printing; green energy; green funds; and tourism	
Fair trade this label makes a stand against societal exploitation in developing countries; supporting fairer wages and sustainable management. Due to the growing demand for foodstuffs produced according to fair principles, the Fair Trade label has become increasingly attractive to customers in recent years	
Nordic swan applied in Norway, Sweden, Denmark, Finland, and Iceland, it designates products with a positive impact on the environment. Its criteria relate to environmental factors throughout the entire product lifecycle. It takes into account the consumption of natural resources; energy use; atmospheric, water, and soil emissions; waste generation; and noise pollution	
Bluesign® it is an integrated concept that extends along the entire textile manufacturing chain. bluesign® applies the strictest consumer and environmental protection regulations and thresholds in the world. The bluesign® standard is a kind of insurance policy for retailers and brand manufacturers, providing the guarantee that all certified textiles are as safe as they can be for people and the environment in accordance with the latest scientific standards – from raw materials, chemical constituents, and procedures required for manufacture right up to the end product. The bluesign(r)® standard is applied right at the beginning of the production chain, thus making the route much more efficient, cost-effective and rapid	

(continued)

Name of standard with description	Assessment
Product Stewardship this is a concept relating to the environmental, health, and safety aspects of the product itself. Every individual involved in the product lifecycle needs to take responsibility for reducing any effects on the environment, health, and safety. In practical terms, this means that companies need to plan for product return and bear any costs incurred for disposing of a product at the end of its lifecycle. Product Stewardship is currently a particular issue in the chemical industry, as product component traceability also forms an element of the Product Stewardship concept. It is not only companies that bear responsibility: retailers and end customers also have a responsibility to dispose of products appropriately or return them to collection points. The most familiar system is a deposit on packaging and bottles: a deposit is paid on packaging that is refunded when the item of packaging is returned to a recycling facility. If the packaging is not returned, the deposit is used to finance landfill sites or environmental pollution avoidance measures	
Cradle-to-Cradle this is a design philosophy, whereby products are designed so that they can be returned in their entirety to closed cycles (biological or technical) at the end of their useful life. Cradle-to-Cradle product certification provides end users and customers with an assurance that all product components will be reused as input material at the end of their lifecycle	
Full product transparency this is a philosophy whereby all the essential participants in value creation apply the results of product lifecycle analyses in order to focus on the relevant activities. Internal and external innovation is used to design products to be sustainable right from the start, forming a more stable basis for credible customer communications. Product sustainability is achieved on the basis of reliable facts and figures instead of vague promises (Arratia 2013)	

(continued)

Name of standard with description	Assessment
ISO 14001 as an international environmental management standard, it stipulates globally recognized requirements for corporate environmental management systems. Companies are required to establish an internal environmental policy, environmental objectives, and an environmental program and develop an appropriate management system for achieving the objectives. In order to achieve the desired environmental performance level, an organization has to develop and implement an environmental management system. This family of standards includes many other standards for various areas of environmental management, including eco-balances, environmental KPIs, and environmental performance assessment. They can be applied to both production and service companies. The attraction of ISO 14001 should be increased, especially for small and medium-size businesses, by stating requirements even more simply and clearly; Special consideration should be given to the eco-balance sheet and value chains for identifying and evaluating the environmental impact that products have. The challenge should be accepted of establishing an external communications strategy that includes communications objectives, identifies interested parties, and describes the reporting requirements (DIN 2012)	It is the most important corporate environmental management standard. It has achieved the status of an established and broadly accepted standard within corporate value chains. Companies qualify their supplier base in terms of whether or not suppliers have an ISO 14001 certified environmental management system. They will usually not be accepted if they do not meet the requirements
SA 8000 this is the first auditable standard in the field of corporate social responsibility. It is based on International Labour Organization (ILO) conventions, the Universal Declaration of Human Rights, and the UN Convention on the Rights of the Child. Its basic requirements relate to the important issues of child labor, forced labor, health and safety, the freedom of association, wages, and the management system. The number of SA 8000 certified organizations is constantly increasing and was at a total of 3,083 sites in 65 countries throughout the world in June 2012 (DQS 2013)	It is growing in popularity in terms of its application within companies, as the standard gives access to the difficult issue of human rights in the value chain. Given that the standard is certifiable and subject to continuous monitoring by experts, it can be assumed that more companies will make the commitment to adopt this standard in their value chains

(continued)

Name of standard with description	Assessment
ISO 28000:2007 it is a specification for global supply chain security management systems. In a world of global networks, the security management for international supply chains has taken on a bigger role. The requirements of ISO 28000:2007 include checking all major aspects in order to increase security levels along the entire supply chain. The affected areas include finance, production, information management, infrastructure and packaging, warehousing, and the transportation of goods between various transport systems and delivery points	Due to the increasing danger of environmental disaster and terrorist attack, the issue of security in the supply chain has increased in significance. It can therefore be assumed that ISO 28000 will become a basic prerequisite for collaboration within the supply chain. However, this will only relate to intercompany relationships
ISO 50001:2011 it is a voluntary energy management standard published by the International Organization for Standardization (ISO). It describes the requirements for an energy management system, the standard being suitable for all sizes of organization. The aim of the standard is to cut global energy consumption by up to 60 % (ISO 2013b)	Due to an increasing global population and the associated demand for products, professional energy management has become essential to the value chain. Companies with high energy consumption are already benefiting from the application of ISO 50001[10]
Good Manufacturing Practices (GMP) this describes a series of production and testing procedures that help to ensure product quality. The manufacture of pharmaceutical and medical products in particular is regulated in many countries. The application of the GMP standard is compulsory in these industries, the ultimate intention being to ensure patient health and product quality. The application of GMP is compulsory in the pharmaceuticals industry. The basic elements of GMP are the clear definition of production processes, the evaluation and possible reassessment of any changes to the processes, description of the processes using clear language, training of equipment operators so that they can carry out and document the processes, record-keeping at each individual process stage as part of a seamless process and provision of these records for inspection, organization of packaging and distribution to prevent any impact on product quality, and establishment of a return system for production batches	Good Manufacturing Practices and the associated requirement for transparency form an important aspect of product security in the value chain. Although most patients are unfamiliar with the concept of GMP, the standard is an important basic factor in drug safety. In a market where drug product fraud is unfortunately on the rise, transparency has become an important counter-measure

(continued)

[10] See the chapter "HENKEL: sustainability in the Value Chain – From Philosophy to Practice."

Name of standard with description	Assessment
Eco Management and Audit Scheme (EMAS) this is a voluntary instrument of the European Union. EMAS is designed to provide support for companies and organizations of all sizes in the continuous improvement of their environmental performance. The number of organizations registering with EMAS rose from 3,300 in 2006 to around 4,600 as at 30.06.2012 (EU Commission 2012: EMAS)	EMAS and the associated auditing process are enjoying increasing levels of popularity in terms of implementation within companies. For many companies, EMAS certification is the "label" of choice as part of their public image as environmentally responsible and this certification is bound to prevail as a European standard
Standards defined by interest groups and NGOs	
Carbon Disclosure Project (CDP) this is a nonprofit organization and has the aim of encouraging companies and communities to publish their environmental data. Once a year, the CDP collects data and information for investors using a standard voluntary questionnaire relating to corporate CO_2 emissions, climate risk, reduction targets, and strategies. The CDP now manages the biggest database of its kind in the world, representing major companies, listed of stock exchanges. The CDP is independent, financing itself through a wide range of sponsors and member contributions via special projects and partnerships. Data provided by companies and annual CDP reports are made available to all interested parties on the CDP website free of charge	The Carbon Disclosure Project is an established standard for the evaluation of corporate environmental impact. Reporting on climate objectives and monitoring of progress via the CDP are part of a any credible reporting effort
Labor standards of the International Labour Organization (ILO) the ILO is responsible for the formulation and implementation of international employment and societal standards. Global minimum standards are designed to protect employment rights, ensuring decent working conditions for all throughout the world (ILO 2013). The main area of activity with relevance for the value chain is adherence to employment, safety, and societal standards. Its core employment standards and Declaration on Fundamental Principles and Rights at Work have laid the foundations for all major CSR initiatives	Labor standards of the International Labour Organization are part of the basic framework of corporate sustainability. Against a background of increasing public awareness and criticism of employment conditions in global value chains, companies can apply the ILO principles to demand the necessary standards from their own value chain and from their suppliers
Code of Conduct of the Business Social Compliance Initiative (BSCI) BSCI is a business-powered platform originating in Europe for improving social standards in global value chains. It provides commercial enterprises with a systematic monitoring and qualification system for improving	Like the EMAS environmental management system, the BSCI has seen a sharp rise in the number of companies joining as members over the last 5 years. The question remains whether it will ultimately be possible to incorporate sustainability

(continued)

Name of standard with description	Assessment
employee working conditions. The number of BSCI member rose from 69 in 2007 to over 700 in 2011 (EU Commission 2011)	within value creation through a large number of voluntary codes of practice
The World Business Council for Sustainable Development (WBCSD) this is an organization run by company boards of directors that is concerned solely with the issue of "Business and Sustainable Development." It sees itself as an intermediary for disseminating to companies the concept of sustainability and helping them to change the way they do business. An important element of the WBCSD philosophy is that companies can only operate in a sustainable way if they can still earn money. The WBCSD represents business interests on the international political stage in terms of sustainable development and CSR. The Business Council for Sustainable Development and the World Industry Council merged to become the World Business Council for Sustainable Development (WBCSD) in 1995	Based on collaboration between companies, the work of the WBCSD has provided much impetus for establishing business practices with a greater focus on sustainability. Under the banner of "eco-efficiency," the well-known approach of "less is more" is what characterizes the sustainability efforts of most companies. This should be welcomed and more companies need to adopt this philosophy. However, there is some criticism of the concept of eco-efficiency, namely that the process of resources consumption is merely slowed down and fundamentally new design alternatives are not being developed. By focusing on profitable sustainability, the work of the WBCSD is to help create a meaningful link between economy and ecology
Sustainability Code of the German Council for Sustainable Development the sustainability code was developed as part of an interdisciplinary project run by the Rat für Nachhaltige Entwicklung, also known as the German Council for Sustainable Development. The sustainability Code was recommended for voluntary application with a focus on increasing levels of transparency, commitment, and comparability for corporate sustainability performance. It can be applied at international level to companies and organizations of all sizes and legal forms. In order to comply with the sustainability code, companies enter into a Declaration of Compliance, documenting how sustainability is implemented in the value chain. It has 20 criteria, each with up to two KPIs, covering Environment, Social, Governance (ESG)	Because the sustainability code arose from a cross-functional and interdisciplinary process, it can be assumed that it also covers the major requirements from a business perspective. The sustainability code is increasing in popularity with companies in Germany. Given that Germany serves as a role model both in Europe and worldwide in terms of its approach to sustainability, it can be assumed that in future the sustainability code will also be applied by other companies at European and global level
SCOR Model of the Supply Chain Council the Supply Chain Operations Reference model (SCOR) introduced by the Supply Chain Council in 1996 provides a standardized process model for value chains. In addition, the SCOR model describes not only each individual task as well as the support processes and the input/output parameters, the model also specifies	Due to its versatile application and the option it provides for benchmarking, the SCOR framework is a valuable tool for improving value chains, while at the same time measuring economic and environmental improvements in a standardized way. Sustainability has been added to the model with GreenSCOR in 2008. This is a good starting point; however, the full

(continued)

Name of standard with description	Assessment
the KPIs that are relevant for evaluating a process. The objective of the SCOR model is to help companies improve their operational performance rapidly and to a significant degree. The SCOR model is currently one of the most widely accepted value chain standards, as it is applicable to simple and complex structures alike	potential of the integrated model will be available once sustainability is fully embedded in a value chain processes (including product development)
OMBOK of the Association of Operations Management (APICS) APICS is one of the most influential organizations in the field of supply chain management both within in the USA and abroad. The Operations Management Body of Knowledge (OMBOK) is a valuable resource that sets standards for processes in production, inventory management, materials management, sourcing, and logistics. It defines sustainability in the context of operations management: the support of good business practices through the control of input and output flows in the transformation process, supporting sustainability as a business practice. In this context, sustainability means the alignment of operations processes, products, and services in a way that is societally, environmentally, and economically responsible (APICS 2012)	Through its comprehensive programme of education and further training, APICS has helped to disseminate approaches for optimizing corporate value creation. Although sustainability has been acknowledged as an issue, it will probably be some time before a comprehensive training programme is available for enabling the current and next generation of supply chain managers to implement sustainability.
The EFQM Model of the European Foundation for Quality Management (EFQM) this simple model consists of a three-pillar strategy: people, processes, and results. All employees are involved in a continuous process of improvement with the aim of achieving excellent results in terms of sustainability. Information on current status, continuous improvement, and future trends is accessed and processed via the continuous monitoring of all processes. The EFQM model is a tool providing support for the creation and continuous development of a comprehensive management system. It is designed to identify internal strengths, weaknesses, and potential for improvement, aligning corporate strategy appropriately. The basic concepts and the eight basic principles include sustainability as a basic principle: Achieving Balanced Results – Adding	Thanks to its system-based approach, EFQM provides a solid foundation for harmonising organizational improvement across economic, environmental, and societal aspects at the same time. However, the EFQM has so far restricted its remit to the corporate context. Customers are not actively involved in the process. Likewise, EFQM is not a standard that can be used to convince customers of a company's sustainable value creation. Therefore, the adoption of EFQM principles is more of a management decision than a standard that is geared toward consumers

(continued)

Name of standard with description	Assessment
Value for Customers – Leading with Vision, Inspiration & Integrity – Managing by Process – Succeeding through People – Nurturing Creativity & Innovation – Building Partnerships – Taking Responsibility for a Sustainable Future (EFQM 2013)	
The Supply Chain Management Standards of the Council of Supply Chain Management Professionals (CSCMP) this is a trade association that is committed to the improvement and dissemination of SCM research and expertise. The Supply Chain Management Standards booklets were published for the first time in 2004, providing tools for improving Supply Chain Performance. The CSCMP pursues a philosophy similar to that of the Supply Chain Council: it provides companies with the opportunity to significantly improve their Supply Chain Performance and gives them access to benchmarking	It has now established itself as the second largest practitioner community after the Supply Chain Council. The benefits of the applying the CSCMP recommendations occur across the value chain, with focus on operational improvements. It has also been concerned with sustainability in the value chain for some time now, but it is a little too soon to start talking about a standard
Reporting progress on sustainable value creation	
Global Reporting Initiative (GRI) these guidelines for sustainability performance reporting are recommendations only, enabling companies to voluntarily provide a basis for contrasting and comparing the economic, environmental, and societal aspects of their products and services. Indicators are designed to identify a company's societal and environmental performance more clearly. The GRI develops guidelines for the preparation of sustainability reports for large companies, small and medium enterprises (SMEs), governments, and NGOs as part of a participative process. The Global Reporting Initiative can be seen as a continuous international dialogue involving many different stakeholders. The basis of GRI-compliant reporting is transparency and its objectives are standardisation and comparability. The GRI operates throughout the world, with the active participation of companies, human rights groups, environmental groups, labor organizations, state organizations and other stakeholders	The Global Reporting Initiative should be seen as the number one sustainability reporting standard at the current time. Large companies in particular refer to the standard; many of them also have an active role in developing the standard. GRI sustainability reporting is now being applied by an increasing number of companies. It should be assumed that the number of companies applying GRI reporting will increase sharply in future

(continued)

Name of standard with description	Assessment
Integrated reporting framework the International Integrated Reporting Council (IIRC) is the initiative of international companies, investor and analyst representatives, NGOs, public sector representatives, as well as standard-setters such as the International Accounting Standards Board (IASB) and the Financial Accounting Standards Board (FASB). It aims to develop a jointly accepted framework to put sustainability on the balance sheet, whereby financial, environmental, societal, and government information can be brought together in a single "integrated" format. Integrated reporting should be designed to provide recipients of company reports – primarily investors – with a comprehensive insight into a company's ability to generate short-, medium-, and long-term value. The objective of the Integrated Reporting Framework is to add more KPIs to existing company reporting, in order to improve the disclosure of integrated value creation	It provides today's standard for the reporting of the future. Given that this framework marks the departure from the separate reporting of sustainability data toward an integrated component of business reporting, we can expect to see a wide-scale switch towards integrated reporting sooner or later
Environmental Profit & Loss Statement (E P&L) of PUMA the sports goods manufacturer PUMA has set itself the corporate strategy objective of operating in a more environmentally friendly way and becoming the most sustainable sports goods manufacturer by 2015. An environmental profit and loss statement was prepared for the very first time – a unique approach at that point in time. The approach is based on two fundamental prerequisites for sustainable operations: extension of corporate responsibility to include the total value chain and acknowledgement within the company that something can only be changed if it is measured and included in the target system. This environmental profit and loss statement includes environmental costs along the total value chain, from raw materials to the customer. In the first step, greenhouse gases, water consumption, land use, air pollution, and waste along the value chain were taken into consideration and evaluated	Thanks to its environmental profit and loss statement, PUMA has reached an important milestone in terms of embedding sustainability in the core business and in company reporting. This approach has been warmly welcomed by financial analysts and other companies. PUMA's parent company Kering (formerly PPR) has taken the decision to roll out this successful model across all of its brands in the next few years
"Measuring shared value" approach of Michael Porter and Mark Kramer this is also aimed at the measurability of results in all aspects of value creation, as is the case with Integrated Reporting (IIR) and PUMA's environmental profit and loss	It is a new and interesting approach, taking the economy, ecology, and society into consideration from the outset. It focuses on creating opportunities as well as linking corporate success with environmental and societal well-being. Shared Value extends

(continued)

Name of standard with description	Assessment
statement (EP&L). In contrast to IIR and an EP&L however, the Shared Value Concept is aimed at making societal value creation measurable. The measurability of shared value is currently being tested by leading companies such as Nestlé and Coca-Cola. In addition to identifying the societal problems that companies need to address, there is a requirement to draw up a business case (objectives, activities, costs, benefits, value-added), monitor progress, and measure results in order to combine economic and societal value creation. For its Coletivo pilot project in Brazil, Coca-Cola developed an approach to improve employment opportunities for young people from low-income families by including them in their business model for commercial distribution, brand loyalty, and increasing local sales. Young people are trained by local NGOs in the areas of business development, business, and entrepreneurship. When they have completed their training, they go on to work with retailers, who then generate strong brand loyalty for Coca-Cola through increased sales. Progress monitoring is based on relevant parameters such as the number of young people trained, self-reliance, increased revenue, and retailer participation in order to create a conscious connection between societal metrics and core business. Investors will not see the full contribution made by societal value creation if no connection is made between the societal dimension and business metrics such as revenue growth, cost reduction, and profitability	beyond the eco-efficiency approach and aims at the measurability of factors that were previously "soft." Given that the results of societal initiatives do not necessarily fall within a certain reporting period, Measuring Shared Value still involves a large variety of challenges. According to Porter, if there are no metrics available for investors to understand and accept, it will be difficult to establish societal responsibility within core business (Porter et al. 2013 – Measuring Shared Value). However, despite the challenges, Measuring Shared Value is a concept that addresses sustainable value creation in terms of creating business opportunities
The "Sustainable Value Measurement" of the ADVANCE Research Project this is a measurement and comparison approach that resulted from the ADVANCE research project sponsored by the European Union. It was developed to close former gaps in the measurability of sustainable value creation. The approach is based on the opportunity cost approach and is aimed at achieving comparability for corporate sustainability performance. It adopts the logic of the	Sustainable Value Measurement is a new approach that transfers the logic of financial analysis from the business sector over to the field of sustainability. Because the Sustainable Value Measurement approach is value-based, it can definitely develop into another alternative for the existing evaluation of commercial success. The question remains as to how far this approach can map the aspects of product development and the supply chain

(continued)

Name of standard with description	Assessment
capital market, which does not reward corporate impact or risk, but rather the ability to deliver value. The idea behind Sustainable Value Measurement is to extend this value-based system to include the environment, society, and corporate value creation. This expresses sustainability performance in monetary terms. According to this method, a company that generates more income per tonne of emitted CO_2 performs comparatively better. Several hundred companies have tested and validated Sustainable Value measurement (see Figge 2012 and German Council 2012)	

Critical Appraisal of the Standards in Relation to Sustainable Value Creation
The issue surrounding those standards that already exist or are in development relates to how relevant these standards are and whether their relevance is even important. The listing in an index or the achievement of the standards for reporting are certainly good opportunities for corporate marketing departments to demonstrate how they have implemented sustainability.

However, do all of these guidelines, standards, and indices equate with quality? At least the Dow Jones sustainability Index, the Carbon Disclosure Project, and the Global Reporting Initiative rate highly within the corporate sector. According to Dan Bross, Senior Director at Microsoft, rankings help companies to assess their efforts in terms of benchmarking, giving investors the opportunity to evaluate companies against certain criteria.

Inclusion in a sustainability index from the DJSI family is a highly attractive proposition for listed companies, as investors managing sustainable investments (amounting to around USD 6 billion in the case of the DJSI in 2012) are generally guided by the index. However, the necessity of a company that exploits oil and coal appearing in a sustainability index is questionable to say the least (and in the case of the Canadian company Enbridge that appeared in 2012, a company that is also responsible for massive environmental pollution in the USA).

The EU Commission has called on all European companies to sign up by 2014 to at least one of the following standards in their CSR strategies: OECD Guidelines for Multinational Enterprises, the UN Global Compact or ISO 26000 Societal Responsibility. The EU has also called on all multinational enterprises based in Europe to sign up by 2014 to the International Labour Office (ILO) Tripartite Declaration on Multinational Enterprises and Societal Policy (EU Commission 2011).

However, standards for measuring sustainable value creation have still not seen success in companies across the board. It is widely the norm that decisions affecting financial parameters have to be based on an investment proposal or business cases.

One reason for the current delayed implementation of comprehensive reporting and measuring systems is that there is currently no "common denominator" specifically for measuring economic, environmental, and societal aspects (Figge 2012). Given that, as yet, sustainable value creation is only rendered measurable by a small number of approaches, the arguments in favor of sustainability will continue to be weak until the effects can be tangibly shown and accountability for these effects can be assumed. It is undisputed that public debate has led to the development of a considerable number of measurability initiatives and proposals. To date, the standards that are most commonly used to work toward sustainability in the value chain are the UN Global Compact, the Carbon Disclosure Project, ISO 14001, the Dow Jones Sustainability Index, and the Global Reporting Initiative. GRI is supplemented by one or more of the above standards as and when required. The Initiative Integrated Reporting (IIR) initiative will henceforth mean significant progress in terms of unifying various approaches for true integrated reporting. Benefits of sustainable value creation

3.2.5 Building Block 5: Stakeholder Collaboration

A study carried out by APICS in 2011 revealed that Management, employees, and customers are – in that specific order – as the most important stakeholders on terms of sustainability (APICS 2011). This result underlines yet again the importance of the role played by companies as drivers of sustainable value creation. Sustainability must be driven from inside-out by companies and practised by the entire organization and its staff on a daily basis. Only then we will see a credible dialogue with employees, customers, governments, NGOs, and other stakeholders emerging. Transparency and the desire of all participants to forge new paths for collaboration are the prerequisites for success. A systematic analysis of the risks and opportunities associated with open collaboration can provide the basis to identify shared values and benefit for all involved parties. This is where everyone directly or indirectly affected by a company's products and supply chain can make a contribution to its workings.

Close collaboration with customers is gaining in importance, both from the product development and supply chain management viewpoint. There is also an increased willingness on the part of customers to help companies by giving them important indicators for improving products; this applies particularly, but not only in terms of sustainability. The Internet and Social Media have helped to establish new communication channels here. This information, which is gained in this manner, can be a major impetus for a company in redesigning products or developing new ones. For instance, products can be improved by removing harmful constituents and new information on previously unknown requirements can help completely new products to come into being. A considerable number of product innovations have already been developed from using customer innovation potential. Successful shared product development approaches have also resulted from this collaboration (e.g., the so called "co-creation"). Food manufacturer Procter & Gamble (P&G) has been using this "external innovation" for many years now for finding solutions to difficult development challenges. P&G publishes these "challenges" to actively seek out a dialogue with external innovators. Through this process external innovators subsequently start to act as stakeholders and often drive hugely successful products and business models. This has caused intermediaries to become established in this field, for example the service provider Innocentive, which encourages innovation for companies by connecting them to people who are capable of solving their respective problems.

Henceforth one can see new cooperative ventures between companies, stakeholders, government representatives, and entrepreneurs coming into being throughout the world. All of which have the aim of promoting sustainability in the value chain and solving societal problems on a grand scale. The "B Team" initiative launched by Sir Richard Branson and former PUMA CEO Jochen Zeitz is such a venture. Its mission is to "catalyze a better way of doing business for the wellbeing of people and the planet." The "B Team" works collaboratively with companies, thought leaders and entrepreneurs to discuss new business avenues and implement them on a large scale.

However, until now, only few companies have given customers an insight into the suppliers they work with or the conditions which are prevailing in a supplier's value chain. In the past, customers have usually had no understanding of how or where products or their components were manufactured. Now, they increasingly demand transparency, wanting to know for example where their wood has come from, where and under what circumstances their metals were extracted, or where their coffee was grown. A philosophy of openness toward customers, investors, employees, suppliers, and stakeholders is a prerequisite for company success. Such openness does not relate to the transparency required by compliance. Rather, this openness is much more a matter of proactively seeking opportunities to bring valuable information and creative ideas into a company "from the outside." Expectations, requirements, and ideas can relate to products, the supply chain, or both. A company's strong inward focus is gradually replaced by closer collaboration with the relevant stakeholders. This gives them the opportunity to become part of a company's overall system and influence its decisions: unions and third sector organizations identify problems and exert pressure in order to achieve improvements; they can also work with companies constructively to find solutions. Through their sourcing, purchasing, and investment decisions, consumers and investors can jointly ensure that socially responsible companies are rewarded by the market. The media can draw attention to both the positive and negative consequences of corporate activity. Authorities and other stakeholders should also act in a socially responsible way when dealing with companies (EU Commission 2011).

Transparency One important aspect and requirement for successful stakeholder collaboration is transparency. To increase the transparency in the value chain, more and more companies are providing insights into their practices. According to the chain of custody concept, traceability of products along the extended value chain becomes a necessary prerequisite to ensure that internal sustainable business commitments remain transparent (Lopez 2012). The world's biggest food company Nestlé has already banned logging from its supply chain and ensures compliance via traceability in the value chain. This approach was then extended to palm oil, fiber, and paper and ten other goods categories. According to Nestlé' S.A.'s Executive Vice President of Operations and GLOBE, José Lopez, "The challenge here is to engage with consumers in a way that is honest, fact-based and looks at the sustainability performance of the entire lifecycle of a product rather than exaggerating individual aspects, which will sooner or later be exposed as 'greenwashing'" (Lopez 2012).

3.2.6 Building Block 6: Sustainable Business Planning

The special role of integrated and efficient processes covering the entire corporate "system" has already been addressed in Building Block 4. Now the focus should be on the processes that need to provide the required transparency to deliver products to customers in a fast and efficient manner – value chain planning processes.

The ability to react to changing demands at an early stage is vital for product innovation, sourcing, production, and distribution in such a way that the right products are available in the right place and at the right time, without waste and with ideally zero negative impact to environment and society. Changes in consumer behavior will affect the extended value chain – also having a knock-on effect on product launches in certain circumstances. In addition to meeting customer requirements, environmental protection is also an important aspect, as products that are not needed should not be produced at all. In a value chain context one will find planning, execution, and enabling processes. While planning processes have always been an important prerequisite for economic activity, there is a growing trend for planning to include stronger integration with internal and external constituents of organizations and value chains. Today's economic environment and volatility calls for transparency across a links and nodes of a value chain. Integration should include on the one hand the provision of information to value chain partners and on the other hand volume and finance planning activities.

Even before the financial crisis, there has been a growing trend for companies to improve the speed and quality of their planning activity. Following on from the role-based planning approaches of the past, Sales & Operations Planning is now available as an important tool for companies to use. Sales & Operations Planning (S&OP) brings together the available information from the market and balances it with the supply capabilities of the organization. As part of the process S&OP takes different planning horizons (long term, mid-term, and short term) into account. The granularity of information required to support decision making, is higher in the short term horizon and at aggregate level in mid- and long term horizons.

A good S&OP process also takes promotions and new product launches into account. This input is vital to ensure that products are available at scale for the market launch of a product. While S&OP has emerged as a key tool, to counter demand uncertainties and volatility, the aspect of sustainability has been less in focus so far. However, the inclusion of sustainability factors such as environmental and societal impact in the decision-making model, will provide an additional dimension of value creation into the planning process. As many companies have started to understand and manage their environmental footprints, the S&OP process and its decision meeting are the right place, to make the trade-offs between environmental impact and profitability. The sustainability strategy of the company can be perfectly aligned with the different planning horizons of S&OP and thus ensures integrity of decision making and the right balance of profits and sustainability. The benefits of including sustainability aspects as decision criteria in the S&OP process, are not only limited to short-term gains like waste reduction, energy cost reduction, and avoidance of obsolete products. It also helps to increase long-term customer loyalty.

Integrated Business Planning (IBP), is a process where all operational parameters are planned and monitored in parallel with financial parameters, representing the next qualitative stage of maturity. IBP is a process that spans marketing, R&D, operations, and finance functions in order to make the best planning decisions. Differentiated planning models are employed for individual value chains strongly aligned with actual customer demand (pull principle) in order to forge the link between various planning and forecasting approaches (differentiated by product, customer, and market dynamics) and the value chain. Various calculation models are used to simulate the effects on the balance of accounts, profitability, and liquidity. "What if" scenarios are simulated for the value chain in order to support the optimum allocation of resources. The need for integrated business planning comes from the requirement to be able to respond appropriately to many different fluctuations in demand. Integrating sustainability considerations into the IBP process, give management and operational teams the chance to truly understand the full opportunities and impacts of their value chain decisions. Getting the visibility upfront on the impacts of decisions to outsource production, to optimize the distribution network, to source from a different country/region, to build a new plant in a resource constrained area, or to schedule plants to capture lower energy pricing, is a key asset in making better and more sustainable decisions. It have been proven, that companies who take sustainability into consideration in a planning and decision making processes are more successful than others.

3.2.7 Building Block 7: Sustainable Cost Reduction

The implementation of sustainable value creation will often only have any chance of success if the competitiveness of products or services can be maintained or improved. Nowadays, most products and services are up against global competition and under permanent cost pressure. In order to secure profitability, it is essential to establish a continuous process for identifying and implementing any potential for reducing costs in products and the supply chain. This is not about achieving short-term quarterly targets – rather it is a matter of maintaining an effective and competitive product portfolio by means of creatively examining value drivers, cost drivers, functions, constituents, etc. This is where cost reduction develops into value analysis (Fig. 15).

In order to identify the potential for cost savings, a structured process needs to be conducted across all functions, whereby a deliberate "creative destruction and restructuring" reveals the main drivers for value, cost, and sustainability. The process should include a critical evaluation of whether customer requirements have changed or if customers are prepared to pay for certain product functions (or for certain constituents) at all. It is only with this knowledge that it is possible to weigh up functionality against cost. A regular, structured, and moderated cross-functional cost reduction process usually results in the identification of significant potential for making savings. This not only helps to keep the product competitive but also leads to a timely restructuring of the supply chain.

Product designs are often influenced by the designer's love and ambition of what is technically feasible or frameworks provided by a set of cost parameters – in many cases, both extremes lead to product specifications that do not focus on what actually delivers value for the customer. Weighing up cost considerations against product safety and environmental impact is a particularly important part of this process. Many "product scandals" could have been avoided if sustainability considerations would have been taken care of at the outset of the development process. Many companies now configure their "sustainable and cost-efficient" product portfolios and supply chains according to the "less bad" principle – i.e., fewer

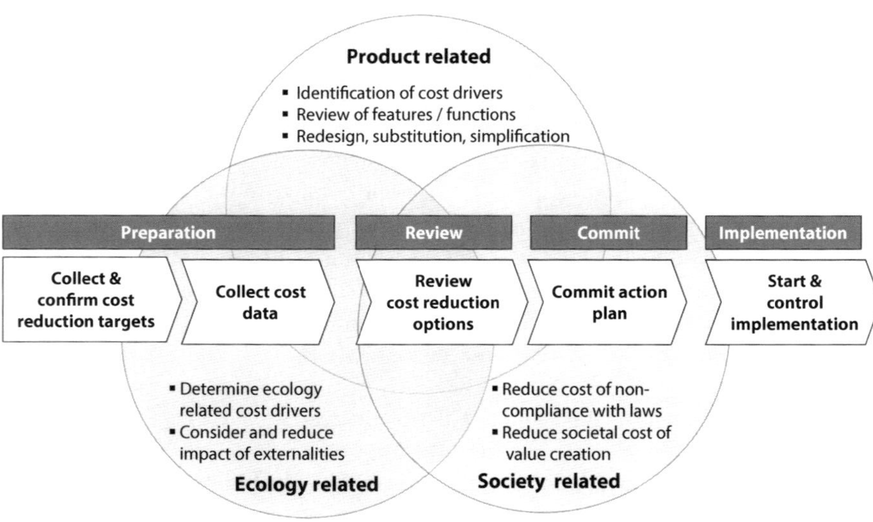

Fig. 15 Cross-functional product cost reduction/value engineering process (shared.value.chain 2013)

harmful effects should be generated. This begs the question whether toxic and other harmful substances could not be completely replaced or even avoided altogether. Why should it not be possible to replace metal product components with renewable raw materials such as bamboo, thereby meeting customer requirements and taking advantage of cost benefits at the same time? (Dell 2012). Companies who have already incorporated principles such as Cradle-to-Cradle in their product philosophy are now designing according to the principle that products can be recycled completely.

In addition to identifying the potential for reducing costs, a disciplined approach to realizing this potential throughout the entire product lifecycle is essential to the success of any continuous cost reduction process. Reporting systems for tracking the maturity level of an individual idea (from inception to implementation) are proven tools, making it possible to put a company's creative potential into practice.

3.2.8 Building Block 8: Supplier and Partner Networks

The role of suppliers in a value chain has changed significantly in recent years. Suppliers are not seen merely as a transaction partner whose own value creation occurs in some kind of "black box," which is only subjected to closer examination during annual negotiations. The selection of efficient suppliers has become increasingly important during the various waves of outsourcing experienced in recent years. Diminished vertical integration within companies has made it necessary to work with suppliers more closely and frequently and to treat them as partners. Working with suppliers does not relate solely to joint business planning, joint development partnerships, and joint operations planning (e.g., for critical components with long lead times). Suppliers must have sustainable and fair practices as "entry criteria" for long-term collaboration. It should be noted that this often involves a balancing act between product quality, delivery capacity, and working conditions.

The role of sourcing is also changing in this regard. Excellent product knowledge and the requisite buying skills are often no longer sufficient to meet today's differentiated demands. Over and above knowledge of markets and technology, suppliers need to be segmented strategically and relationships need to be managed on a continuous basis in terms of operations, with the right degree of escalation in the event of any problems with performance. Sustainability issues present new challenges for buyers. Each suppliers might operate as provider of innovation to a companies product development department and as supplier of goods withing a global production network. The integration of several of these suppliers can only be successful if it is coordinated by experts with an in-depth understanding of sourcing, supply chain and sustainability. Due to the many different interpretations of sustainability, companies still have some catching up to do on this particular point. This involves developing the necessary skills for meeting customer and company requirements. Until now, sustainability in terms of suppliers and partners has only been an issue when there was a need to comply with legislation and communicate expected sustainability standards via compliance regulations. Agreeing a code of conduct is now the method companies adopt for defining expectations in terms of supplier ethics and working practices.

Choosing the right partners, integrating them into your own value creation system and managing supply risk are important factors for flexible value creation and ultimately for the success of a company. Flexibility and sustainability can only be implemented credibly and for the long-term where suppliers and partners are truly embedded in processes and procedure, acting according to the same or similar value creation principles. Company stakeholders need to be provided with an assurance that suppliers operate on the same principles as the company itself. Most codes of conduct implemented focus on the following:

- Adherence to laws
- Ethical integrity
- Clean business practices
- Open communication
- Anti-corruption
- Respect for human rights
- Safe working conditions
- Fair pay
- Environmental protection
- Societal commitment.

Most codes of conduct are still too generic and put all the responsibility one-sidedly on the supplier. Apart from unannounced audits by independent experts, there are usually no further measures in place. This makes the application of the code of conduct difficult in legal terms. Most codes of conduct are recommendations only and are not binding.

For example, although the German "Zentralverband der Elektroindustrie (Z.V.E.I.)" – the German Electrical and Electronic Manufacturers' Association, describes in its code of conduct what sustainability means "in terms of working conditions, societal and environmental compatibility, transparency, trustful cooperation and dialogue" (Z.V.E.I. 2012), nevertheless application of the code is only a recommendation. The Z.V.E.I. sees its role as an association as purely a support function via which members can "respond to the various global market conditions [...] and face the challenges of societal expectations [...] arising from increasingly networked collaboration within the value chain. A shared basic understanding of societally responsible corporate management provides the basis of the Code of Conduct. For the signatory company, this means taking responsibility by considering the consequences of its business decisions and activities in economic, technological, societal and ecological terms alike and achieving a reasonable balance between various interests. Within the scope of the available possibilities and its activity, the signatory companies make a voluntary contribution to the prosperity and sustainable development of global society at its operational locations. It aligns itself with universal ethical values and principles, particularly integrity and respect for human dignity. The signatory companies also undertake to promote compliance with the code of conduct in its dealings with its suppliers and throughout the rest of the value chain within the scope of the available options" (Z.V.E.I. 2012).

It is questionable, however, if sustainability can really be embedded in a suppliers core business on such a nonspecific and voluntary basis. The lack of controls and recommendations for transparent reporting, means that many codes are well-intended but not very well executed. Selective and unannounced audits by independent experts are often put into practice and provide a good start for ensuring compliance with a code of conduct. But compliance is not equal to commitment. Proactive and consistent implementation in the form of sustainable sourcing practices will definitely have greater effect.

Compliance is not what is called for here, but commitment and active cooperation. Nestlé has now extended its Responsible Sourcing Guidelines to the 10 most important sourcing categories. Nestlé viewed and analyzed all the relevant supply chains in order to gain a better understanding of the relevant sourcing approaches and supply chains. Nestlé UK & Ireland has been monitoring the supply chains of its sourcing groups closely since 2012 and began sourcing 100 % sustainably produced palm oil in the same year (Guardian 2013).

Capability Building in the Supply Base Transparency and trust are prerequisites for achieving a shared value network with suppliers and partners. Whereas sustainable sourcing guidelines often relate only to transactions, within the networks themselves a proactive approach is characterized by actively helping suppliers to develop new skills. Customers can only be offered a choice of sustainable products where there is collaboration with the right suppliers. Working with suppliers and strategic partners as part of Supplier & Partner Networks is an important element for embedding flexibility and sustainability in the supply chain. Given that very few companies are now able to operate with complete independence, partner quality and performance capability are important factors for success. Responsibility for value creation does not end with the company itself. It is therefore in a company's interest to take responsibility for value creation end-to-end. This does not mean the authority to dictate to the parties involved. A transparent way of working aligned with shared benefits will ensure that problems can be identified and resolved at an early stage.

Risk Management However, costs and proactive default risk management continue to be the main concern of current approaches to working with suppliers. There may be talk of sustainability, but the responsibility for complying with standards is put solely on the shoulders of suppliers.

Apart from the cost aspects of product and service sourcing, operational security of supply continues to play an important role in the choice and structuring of supplier relationships. Efforts to reduce lead times based on collaboration are to be welcomed but usually only focus on economic factors (e.g., lower stocks while ensuring delivery capacity at the same time). Financial crises and environmental disasters have forced companies to implement new approaches to risk management and to manage their supplier bases more closely than in the past. Corporate risk is identified under the concept of Supplier Risk Management or Business Continuity and minimized via a range of risk management levers. Both product factors and depth of supplier integration are reviewed to evaluate any effect and costs associated with supplier default. Sustainability is an issue in terms of Supplier Risk Management insofar as product quality and constituents can have a material effect on production conditions.

Companies have access to a growing number of supplier databases for integrating sustainability aspects within Supplier Risk Management. A good example of this is the SEDEX database, which contains data for around 27,000 production sites around the world (SEDEX 2013). SEDEX is a "not-for-profit" member organization that strives for ethical improvements in global value chains. A questionnaire is used to establish societal responsibility status and analyzed as a tool for risk management. International standards and recommendations on human rights, employee protection (SA 8000), environmental protection (ISO 14001), and health protection (OHSAS 18001) are taken into consideration. The issues addressed in the questionnaire relate to working guidelines, health and safety in the workplace, business practices, and environmental management.

Risk always has a potential for opportunity and should be of primary consideration when working with suppliers:

- Which partners do we need to work with in order to be able to work faster and with greater flexibility?
- Which partners can help to improve products?
- How can shared infrastructures be used?
- How are risks and opportunities divided?
- Which partners can help their third-party suppliers with sustainable value creation?

Genuine collaboration with a supplier can be defined as a relationship between two parties that is built on trust and mutual commitment to each other. The role of suppliers is becoming increasingly important, as they not only help with the achievement of objectives, but also often have to bear the costs of investments in sustainability or carry out the work involved in audits. Transparency and collaboration are prerequisites for the successful broad-base implementation of initiatives in order to achieve greater sustainability in the value chain. The Zero Discharge of Hazardous Chemicals (ZDHC) program is a good example of collaboration between companies that are in the same industry and share the same supplier base. Adopted by Nike, Adidas Group, C&A, H&M, Li Ning, and PUMA, the aim of the program is the total elimination of dangerous chemicals from products and the value chain by 2020. The group's shared road map provides a way for collaborating with the relevant suppliers in order to achieve this goal. This road map paves the way for extending the scope of individual value chains via pilot stages (starting with dying plants, which have the greatest impact) (Nike 2011). However, even greater opportunities will be afforded by close collaboration with the suppliers themselves. Suppliers usually have specialist expertise and creativity. A product lifecycle analysis will highlight requirements that can serve as an incentive to suppliers to offer totally new and sometimes radical solutions for reducing the environmental impact of their products. This is a valuable type of collaboration and delivers greater value than any survey containing hundreds of questions on compliance and codes of conduct ever could (Arratia 2013).

4 Benefits of Sustainable Value Creation

The reasons for addressing sustainable value creation are as many and varied as the economic, environmental, and societal benefits that can be achieved. Any assessment of the benefits that can be derived from sustainable value creation will therefore depend on the industry sector, market environment, and the maturity of individual companies.

In terms of the benefits of sustainability, we can identify increased revenues due to the willingness of customers to pay premium prices for "sustainable" products, lower risks and liabilities due to sound environmental management (and therefore lower insurance premiums), better access to capital, fewer days of absence, and an improved attitude toward work due to pleasant family-friendly working conditions. A company's reputation can also be improved. Business managers can sometimes obtain personal benefits through the public recognition of green or societally committed corporate leadership (through honors, awards, and articles in the specialist media, etc.) (Wirl 2012: 7).

While the debate on the benefits of sustainability is well underway in societal, scientific, and industry circles, many clearly measurable benefits have been achieved in the supply chain through long years of experience with optimization. Clearly measurable monetary benefits can be measured via standardized financial and operational KPIs. Given that measurability is a basic prerequisite for integrated management, it is no wonder that the sustainability KPIs currently applied are aimed at areas where measurability is relatively inexpensive. This usually involves setting targets for reducing CO_2 emissions, energy, and water consumption and the production of waste.

However, by seeing sustainability as an integral component of value creation and including societal value, we can see even more benefits being generated from the business model, from the products and from the value chain. Figure 16 shows a number of these benefits, although the list is not exhaustive.

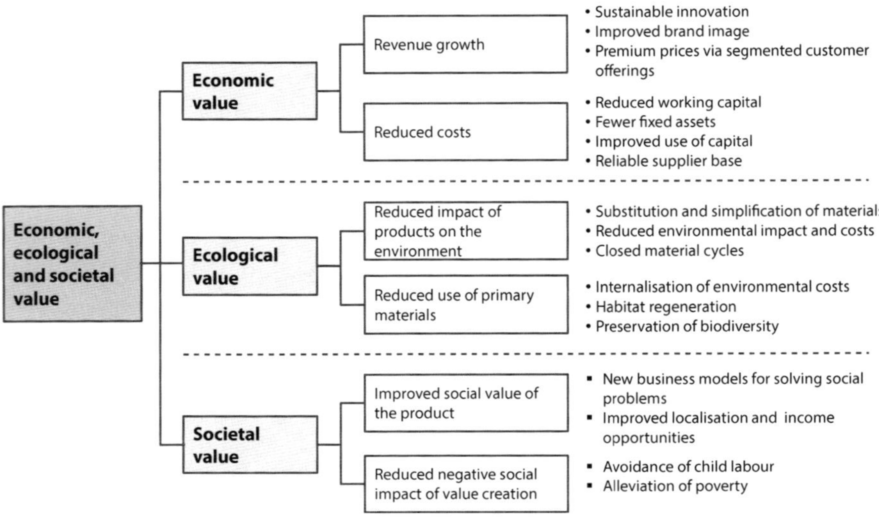

Fig. 16 Benefits of sustainable value creation (examples) (shared.value.chain 2013)

4.1 Economic Value

According to management guru Peter Drucker, companies do not exist merely for their own benefit, but to fulfil certain societal functions and a certain societal need arising from groups or individuals. They are not the end, but the means to that end. Companies find their reason for existence or "purpose" by looking beyond themselves and making themselves useful to customers and society through their performance. However, the majority of companies currently sees corporate purpose as acquiring as many customers as possible for their products and services and ensuring that these customers remain loyal to the brand. Profitability expectations are often excessive and detrimental to long-term corporate survival. Given that a customer's appreciation of a product or brand is always subjective and that competitors court the same customers, it is necessary to achieve revenue growth from product innovation that addresses new requirements in order to survive in the market. With more information available to them, customers do not consider solely economic factors, i.e., price. They also take other factors into account, such as the ecological and societal value of a product when making their purchase decision.

The economic benefit and advantages of sustainable value creation derive from a combination of market opportunities and the potential for optimization along the entire corporate value chain. The following examples illustrate these opportunities:

Developing and Maintaining Competitive Advantage A culture of innovation coupled with the potential to explore new avenues are both factors that help companies to take up market opportunities sooner, thus securing competitive advantage.

Revenue Growth Through Sustainable Product Innovation The development of products based on sustainability criteria makes products more attractive to customers, can mitigate societal issues, increase sales and customer loyalty. Economic and ecological values are created at the same time. Kraft Foods achieved double-digit revenue growth by introducing the Fair Trade Label and voluntarily committing itself to 100 % sustainable sourcing for the European coffee market. Thanks to sustainable sourcing practices, a 100 % increase in sales was achieved for instant coffee products sold in Sweden (Wirl 2012: 7).

Revenue Growth Through Products and New Customer Groups Most of the future population growth will occur in emerging and developing countries. The desire for prosperity in these particular countries will lead to increased product demand. Given that incomes are far lower, companies can take advantage of the opportunity to increase local market growth and revenue through product innovation. The application of an "inclusive" business model has its advantages, as not only will resources from developing countries be drawn into the value creation system, but local buyer categories (often from the so-called 'Bottom of the Pyramid') can be tapped to create new consumers.

Improving Productivity Motivated employees make for improved productivity. They make better use of their individual potential for innovation, accelerating, simplifying, and integrating product development and/or value creation processes. Employees who understand the relationship between their daily work and the value it generates (economic, ecological, and societal) are an important factor for competitiveness. It is also generally the case that fewer absences are observed for these employees.

Reducing Operating Costs Along the Extended Supply Chain Design concepts that incorporate sustainability and value creation factors can help to reduce operating costs. Approaches such as Design for Manufacturing and Design for Reuse/Sustainability lead to lower sourcing cost and reduced environmental impact. In the Cradle-to-Cradle approach, products are designed so that they can be returned in their entirety to closed cycles at the end of their lifecycle, thus avoiding the cost of the extraction and manufacture of primary raw materials. As a rule, 60–90 % of a company's costs are incurred in the supply chain. The Supply Chain Council analyzed the cost structure of large US enterprises in 2006, arriving at the following percentages for value chain costs: GM 94 %, Ford 93 %, Conoco 90 %, Wal-Mart 90 %, IBM 77 %, GE 63 %. Due to this specific cost structure within a company, most optimization activity in the supply chain is aimed at reducing operating costs and accelerating throughput. According to a study conducted by the Supply Chain Council, costs can be reduced by around 30–35 % through supply chain optimization, freeing up circulating capital and delivering 3–5 % revenue growth (Supply Chain Council 2006). Integrated planning identifies changes in consumer behavior at an early stage, thus eliminating unnecessary orders for raw materials and redundant process stages, as there is no real consumer requirement for them. The Bullwhip Effect is reduced significantly in the supply chain when planning information is filtered across a number of stages working in collaboration (Leitl 2009).

Particularly during the financial crisis, considerable attention was paid to having the "right" stocks to ensure optimum capital commitment and liquidity. Inventory management can reduce stocks significantly, while still maintaining or even improving the reliability of delivery. The cost of noncompliance can be reduced by choosing the right suppliers. Fixed assets can be reduced via horizontal integration and the use of shared infrastructures, increasing capacity at the same time. Examples of this include transport infrastructure and distribution warehouses. By collecting and reprocessing used products, sourcing, energy, and environmental costs can be reduced and important raw materials for production recovered, such as the reprocessing of rare earths in the lighting industry.

A More Stable Supplier Base A stable supplier base makes it possible to maintain business operations continuously, even in times of crisis. Risks and costs incurred through supply failures, production delays, and customer delivery delays can be reduced through the creation and development of strategic partnerships. This also makes the entire system more resilient to external impacts.

4.2 Ecological Value

Accounting for Externalities Ecology is a pillar of sustainable value creation. Besides the positive effects generated by eco-efficiency efforts (reduction of inputs of raw materials, soil, water, air, etc.), there is an even bigger opportunity to generate ecological value: The consideration of "externalities" into a companies financial system and evaluation.

By factoring in the "services provided by the environment," a change is brought about whereby ecology is not merely viewed as an unlimited resource. This gives rise to a new understanding of nature's balance sheet, resulting in an increased number of programs for environmental protection and the renewal of natural resources together with a corresponding monetarization of scarce natural resources. This ultimately means that alternatives to scarce resources will need to be identified and value creation activities carried out with significantly reduced recourse to natural resources. PC manufacturer Dell is a good example of how both economic and ecological value can be generated: it was able to access important information on how to change its packaging model by including customers in the process. Whereas customers wanted robust packaging, Dell wanted to make it lighter to reduce costs. Based on customer feedback, Dell took the decision to combine both requirements in an intelligent way. Due to the new opportunities available for rapidly renewable raw materials (in this case, bamboo from accredited cultivation), 70 % of the packaging has been changed so far (MIT 2013).

Reduced Use of Primary Raw Materials and Other Natural Resources Sustainable value creation gives new meaning to the services provided by the environment, whereby the use of limited natural resources such as water can be re-evaluated. The agriculture and food industries use a large percentage of available drinking water. The goal of minimizing water consumption is leading to the innovation of new irrigation systems and processing practices. The Campbell Soup Company has introduced a considerable number of initiatives for reducing water consumption in the supply chain: partnerships with farmers for new irrigation systems, systems for recovering steam from the production process, and improved vegetable washing systems have been introduced. These projects deliver a 15–20 % internal "return on investment" for the company and reduce water use significantly (Kaye 2013).

Reducing the Negative Impact of Products on the Environment Sustainable value creation pays particular attention to the value contributed by the environment. Procedures aimed at reducing the CO_2 footprint and energy consumption, saving energy and avoiding waste lead to a reduced impact on the environment. In addition to these advantages, which are based on a philosophy of eco-efficiency, other benefits are also derived from new product philosophies such as Cradle-to-Cradle. Depending on their features, products that have been designed according to the Cradle-to-Cradle concept can be returned to biological cycles at the end of their useful life, helping to protect deposits of primary raw materials and reduce landfill volumes. For example, the Pure Origin product range from the textile manufacturer Triumph has set an unprecedented quality standard. Manufactured according to the Cradle-to-Cradle concept, these products are characterized by their economic efficiency, low or zero environmental impact and high level of user-friendliness. They involve a paradigm shift in terms of the industrial manufacture of products. The innovative design not only redefines shape, functionality, and components; products that have been designed Cradle-to-Cradle also achieve a new dimension in terms of quality and safety within endless cycles (EPEA 2010). These efforts also lead to reduced environmental costs for the company. The substitution of toxic product components not only makes products safer to use, but also reduces harmful effects at the end of the lifecycle.

Those companies that for some time have made efforts to avoid environmental impact and assess their products and supply chains for "eco-efficiency" can report considerable success. Thus, 3M has saved over USD 750 million in environmental costs since the introduction of its 3P Program (Pollution Prevention Pays) in 1986 (3M 1997). British retailer Marks & Spencers initiated Plan A in 2007 with the objective of ensuring that all its traded products would have certain attributes of sustainability by 2015; after breaking even in just 2 years in 2009, Plan A has made a double-digit million contribution to company earnings every year since then (Marks and Spencer 2013).

4.3 Societal Value

Solving Societal Problems by Reconceiving Products An assessment of the product portfolio in terms of sustainable value creation will lead to the discovery of new opportunities for increasing the societal value of products. Likewise, the traditional way of using products can be reconceiving in order to solve societal problems. The German technology manufacturer Nanogate worked with the British adaptive spectacle technology manufacturer Adlens to provide spectacles for use in developing countries such as Ruanda. In this way, corporate growth goes hand in hand with solving societal problems.[11]

Increasing the Revenue of Local Producers Through Business Model Innovation The conscious consideration of how products can generate societal as well as economic value also constitutes a driver for product innovation and new business models. SAP used its StarShea Project to establish a new societal business model for providing female shea nut farmers in Ghana with improved and more secure income opportunities. The pilot project gave female farmers in Ghana direct access to trading venues with the support of SAP technology and intermediaries. This connection is generating improved levels of income in the region and customers are seeing higher quality products at the same time.[12]

Likewise, the food manufacturer Nestlé has also been able to improve the income levels of a further 4,500 families of smallholders in the Ivory Coast by extending its cocoa sourcing policy to include Fair Trade criteria. In addition to improving income levels, a premium is also paid, which is used to sponsor the construction of schools and health centres (Guardian 2013).

Eliminating Societal Injustice and Corruption Because sustainable value creation takes into account the extended supply chain and entire product lifecycle, it not only provides transparency but also supports the efforts made to eliminate child labor and inhumane working conditions via new forms of collaboration with suppliers. Questions about the supply chain for a smart phone and the conditions in which raw materials are extracted help consumers to identify issues such as conflict minerals, child labor, and corruption. After all, the buying behavior of customers and the company standards expected by customers will ultimately lead to a U-turn in common practices. Alongside customer buying decisions, new legislative frameworks means that there is now greater transparency, particularly in the supply chain, aimed at eliminating inhumane working conditions, corruption and even the financing of wars. An example of this is the requirement of the US Dodd Frank Act, which stipulates that companies listed on US Stock Exchanges must ensure transparency in terms of the use of conflict minerals (tantalum, tin, tungsten and gold – 3TG) in their products and supply chains. Such transparency is intended to help reduce the forced child labor and the financing of war in the Democratic Republic of Congo. Given that these minerals are mainly used in the electronics industry and that the requirements apply to all listed companies in the USA, this can

---

[11] See also contribution by Ralf Zastrau in the chapter "Nanogate AG: Sustainable Value Creation in Technology Companies."

[12] See also contribution by Heino Kantimm on the StarShea Project in the chapter "SAP AG & StarShea Limited (Ghana): Sustainable Value Creation Through Collaboration with Companies, NGOs and Intermediaries."

only lead to increased transparency. This kind of regulation will have a direct impact on product design, sourcing decisions, and future corporate investment. Working conditions will improve and corruption will diminish as a result.

4.4 Soft Benefits

In addition to the abovementioned benefits created in the economic, ecological, and societal dimensions, there is also a whole series of "soft" benefits or additional advantages that are tracked only partially via KPIs, if at all. Although customers have an intuitive awareness of these benefits, these benefits are often excluded from any fact-based evaluation due to an absence of clear definitions. However, it is just as important to deliver these "soft" benefits as all the other measurable corporate objectives, until we ultimately have measuring systems and KPIs for soft benefits too.

Improved Company Reputation/Brand Repositioning A risk to the reputation of a listed company can prove to have significant effects within a short space of time. CSR and sustainable value creation activity is therefore carried out in order to increase loyalty of the customer to the company and its brand. A company does not necessarily have to offer sustainable products. The US mobile operator Sprint Nextel does not have any sustainable products, but it targets indirect benefits. It introduces initiatives that have a positive impact on the environment. The company then publicizes them via reputation rankings in order to position the brand itself as "sustainable."

The benefits of a holistic approach to sustainability can also be seen indirectly. After launching its Sustainable Living Plan in 2010, the consumer goods manufacturer Unilever embarked on a series of sustainability initiatives affecting extensive areas of its portfolio. While the company is not ready to link this shift toward sustainability to performance, the correlation is definitely there: sales went up 6.5 % between 2010 and 2011; and then jumped 6.9 % in 2012. Meanwhile the company's stock has been on an impressive upward trajectory (Kaye 2013).

Greater Flexibility and Adaptability Response times can be reduced and market opportunities used more efficiently through the intelligent networking along the intersection points along the supply chain and collaboration based on mutual trust. Value creation networks, where partners are all working towards similar objectives and providing each other with the relevant incentives, deliver strategic and operational competitive advantage.

Meeting/Exceeding Standards By meeting standards and even exceeding them voluntarily, not only is an economic, ecological, and societal contribution made, but collaboration with the authorities and auditors is also made easier. Customers are certainly going to consider a company's compliance with accepted standards when making their sourcing decisions.

Building Customer, Supplier, Stakeholder, and Shareholder Trust Credibility and integrity in setting targets for implementing sustainable value creation is the basis for long-term relationships and profitable growth. Where a company succeeds in differentiating itself through a collaborative approach to product development and the supply chain, working with societal and political stakeholders can make a major contribution toward developing new forms of collaboration and improved framework conditions.

In terms of the aforementioned benefits, the financial KPIs are at least measurable and comparable due to reporting requirements. Comparability is achievable for the supply chain via operational KPIs. Due to recent efforts made to render environmental impact measurable, there are now KPIs available for the emission of CO_2 and other gases that can also be verified by independent auditors. Whereas financial and operational KPIs are widely disseminated across corporate reporting, the KPIs expected by stakeholders in relation to nonfinancial performance are less clearly defined and not as widely accepted. While internal goals are aimed primarily at profitability and value contribution for the company, the emphasis outside the company is much more on investment potential and public opinion. Despite all the current problems, it makes sense to render sustainability measurable, as this is the first step toward embedding it within core business. There are already a number of tools available to assist with data collection and analysis – but they relate to a wide range of different areas and remits. Lifecycle assessments are very popular with manufacturing companies and private organizations. The demand for lifecycle assessments has led to improved data, methods, measuring systems, and comparability. In the medium term, this has led to reduced expenditure on conducting analyses and provided opportunities for including various stakeholders in the process. An interesting trend can also currently be observed in the USA. There is great interest in an organizational form called a Benefits Corporation (B Corp.). A certified Benefits Corporation is required to apply sustainability as the basis for its corporate decision-making process and render this transparent. Irrespective of the issue of the specific measurability of sustainable value creation, the Massachusetts Institute of Technology (MIT) has reported a pleasing trend. According to a study conducted by the MIT Sloan Review, at least 37 % of the companies surveyed are confident that they benefit either directly or indirectly from sustainability (MIT 2013).

5 From Idea to Action

When it comes to the practical implementation of sustainability, managers currently see themselves faced with a wide range of different issues and demands. They need to work with the Senior Management Team to decide which aspects of sustainability a company should get involved in, how sustainability commitments can be made transparent, and the benefits a company can obtain from its commitment to sustainability. In addition, Industry specific challenges are numerous and vary greatly.

While employee motivation, fair working conditions, and corporate societal commitment are to a considerable degree part of daily business operations, issues such as the impact of demographic trends, product responsibility, and the involvement of external partners are multi-layered and cannot always be easily linked to the company's core business. Given that there is a diversity of issues coupled with only a tenuous link between product creation and supply chain, the result is that every situation requires a highly individual interpretation and customized approach

Despite the challenges, a number of companies have already started implementing sustainable value creation, with even initial activities showing already success. Companies experiencing the greatest success are driven by (often visionary) management teams, essentially driving the organization to embed sustainability as part of the core business across all functions and across all levels of the hierarchy. Also known as eco-entrepreneurs, these entrepreneurs stand out for their "personal drive and the ability [...] to tap into ecologically and societally important market opportunities in the role of First Mover" (Brix et al. 2006).

These corporate pioneers are currently the main drivers of change – although not all of them act from totally altruistic motives. Successful and profitable growth remains a primary goal: the company must make money. The difference in terms of the approach taken toward the strategy for implementation lies in addressing risks and dealing with issues in a proactive manner. This stance goes way beyond the widespread approach to compliance and may be interpreted as purposeful and open commitment.

The timing of a company's decision to move toward sustainable value creation depends on many factors. Company stakeholder expectations, competitive landscape, product portfolio, supply chain structure, and internal business goals (such as profit orientation and personal monetary goals) are important parameters in this context. As with individual people, individual companies have different trigger points, when one is no longer satisfied with the status quo and recognizes "that we have reached or passed the limits of our current economic model of consumer-driven material economic growth." This is the point where we decide to switch from a situation of "knowing and not acting" to "active engagement" (Gilding 2012: 5).

To move from the idea of sustainable value creation to concrete action, six basic principles are emerging, that leaders in the field apply to drive their efforts. They impact the corporate strategy, product designs, collaboration internally as well as external to the organization.

5.1 Principle 1: Embed Sustainable Value Creation in the Corporate Strategy

The Senior Management Team has to see the importance to position sustainability within the company's core business, i.e., sustainability ought to be embedded as part of corporate strategy and a commitment must exist to assess the extent to which sustainability forms part of the company's purpose. Value creation aimed at sustainability and generating value for all those involved, is not to be interpreted as a response to legal regulations or adherence to codes of conduct (compliance). It is about agreeing on what sustainability means in the context of corporate core business and giving a clear commitment. It can also mean switching from the philosophy of increased efficiency ("do more with less") to an approach geared toward producing only what is really necessary and delivers ecological, societal, as well as economic value.

Senior managers need to communicate the company's voluntary commitment to sustainable value creation in an open and transparent way in order to give the necessary signal for change. This voluntary commitment should involve a conscious decision to adopt a new paradigm, rather than simply improve the old model bit-by-bit (Braungart and McDonough 2002: 182). This voluntary commitment needs to be documented in the form of a sustainability agenda that draws attention to the company's public support of sustainable value creation. Ensuring that this voluntary commitment is followed at all management levels is the first step toward a true commitment to sustainability.

Set objectives for "healthy" growth and persuade stakeholders to be part of the journey: Voluntary commitment and supply chain analysis should deliver a clear system of objectives and an operational framework focused on "healthy" growth, i.e., achieving the appropriate balance between economic, ecological, and societal aspects and making good use of all potential opportunities arising. Transparency and active communication are required in order to attract stakeholders for new practices and ensure the successful implementation of initial projects. In this context, the ability to motivate your employees so that they see the connection between their daily work and the effects on the environment and society has also now become a core competency (Hofman et al. 2013). Including the objective of sustainable value creation as part of the strategy, requires to change from a one-dimensional context to a systemic view.

The growing public interest in sustainability drives an increasing numbers of studies that are looking into the features and behaviors that characterize successful companies, operating sustainable supply chains, and also generating profit with this approach. Successful companies that take up the opportunities offered by sustainable value creation share the following characteristics:

- Sustainability has attention and full commitment of the Management Board & the Senior Management Team.

- Sustainability is seen as an opportunity to position the company for the future and develop new business models.
- A systematic approach serves as a basis for acquiring a clear understanding of risk, value for customers and society, the opportunities for exerting influence as well as associated dependencies.
- The review of opportunities and risks, deliberately also include cost of externalities (e.g., natural capital) in the development of financial models
- The Senior Management Team is proactive and makes decisions on sustainability even without a clear business case.
- Customers are actively included in drafting the sustainability agenda and the appropriate information is made easily accessible.
- Sustainability is embedded in the core business in a consistent manner (i.e., both in the products and the supply chains).
- The company has the courage to be transparent – a transparent and collaborative way of working together is practised both internally and externally.
- Statements are precise and not misleading. Information can be checked and verified. Clear and simple language is used.

The Senior Management Team needs to have the courage and vision to set ambitious goals to achieve this, but this is not enough to incorporate it end-to-end. Breaking down the targets set forth in the value creation strategy into manageable work packages, defined timelines for implementation and consistent follow-up are becoming important drivers for internal corporate change (even without a clear business case in some instances). Corporate culture, collaboration, and communication play a decisive role here. In many cases, embarking on the path to sustainable value creation also means introducing cultural change within the company. The necessity of "tearing down the internal walls and obstacles" paves the way for credible and robust collaboration. Communication with internal stakeholders needs to be clear and transparent.

The often poor implementation of sustainability within the core business is partly due to well-intended but badly executed strategies. There is a need for action to agree on concrete operational objectives, responsibilities, and the required management support to get there. Implementation can only be a possibility where sustainability is firmly embedded in the management system and is a clear criterion for operational and financial decisions.

The consideration of external factors ("externalities") plays a special role in the strategy development process. As a rule, external factors are not taken into account when determining optimum market demand volumes and pricing structures. In concrete terms, this means that a company does not bear all the societal costs incurred in the form of environmental pollution, noise pollution, and habitat degradation. Including these external factors explicitly in the corporate decision-making process is an important step toward valuing nature as a supplier of resources, whose services are used by companies. Widening perspectives and considering "natural capital" as an important variable in business accounting affords the possibility of reassessing value creation and providing an incentive for change. The costs/

opportunities associated with external factors are then no longer left to politics and society. The particular challenge of integrating sustainability in the core business, consists of combining a profit-oriented business approach, acceptance of personal responsibility at various levels of the organization, with short-, medium-, and long-term requirements to keep the organization viable. Short-term profit orientation often neglects the requirement to maintain a viable foundation of the business, i.e., not only in financial terms. Despite increasing efforts by companies to position themselves with regard to sustainability, it is clear that many activities are still carried out on an extremely superficial level and that sustainability is not actually embedded within the core business. In many instances, the links between sustainability, products, and supply chains are very weak, yet companies only admit to this behind closed doors. In many companies, sustainability is still predominantly managed by Public Relations, Governmental Affairs, and communications departments, which – as part of their specific remit – focus on communications with customers and stakeholders. Regular and structured exchange between product development and supply chain managers is rarely an established part of any corporate system. It is particularly important to develop a clear understanding of roles and responsibilities as part of the strategy definition process. This involves reassessing the previous understanding of roles and defining new adapted objectives. Equally important is regular cross-functional operational exchange for the joint discussion of progress and eliminating any obstacles to achieving objectives.

5.2 *Principle 2: Honestly Assess Your Starting Position, Determine What Matters, and Where the Organization Can Generate Positive Impact*

A company's current position should be determined through a structured analysis of the potential for sustainable value creation and benchmarking of the major KPIs. Opportunities for sustainable value creation can be discussed in cross-functional workshops and meaningful objectives can be set for the company. A materiality analysis can be used as a tool to identify not only those areas where the company should make a commitment, but also ascertain the willingness of customers to pay for sustainability in terms of products, processes and commitment. Writing up the stakeholder expectations and the importance for the company in the form of a "materiality matrix" has proved to be a helpful tool for the internal and external communication of relevant issues and goals. The materiality matrix brings stakeholder requirements in line with corporate objectives. The materiality matrix is also useful for communicating corporate self-regulation in a simple and transparent form (Fig. 17).

A more comprehensive approach is used by the "Shared Value Opportunity Analysis" that analyzes both product development and supply chain opportunities by mapping value creation points and value streams. All stages of product

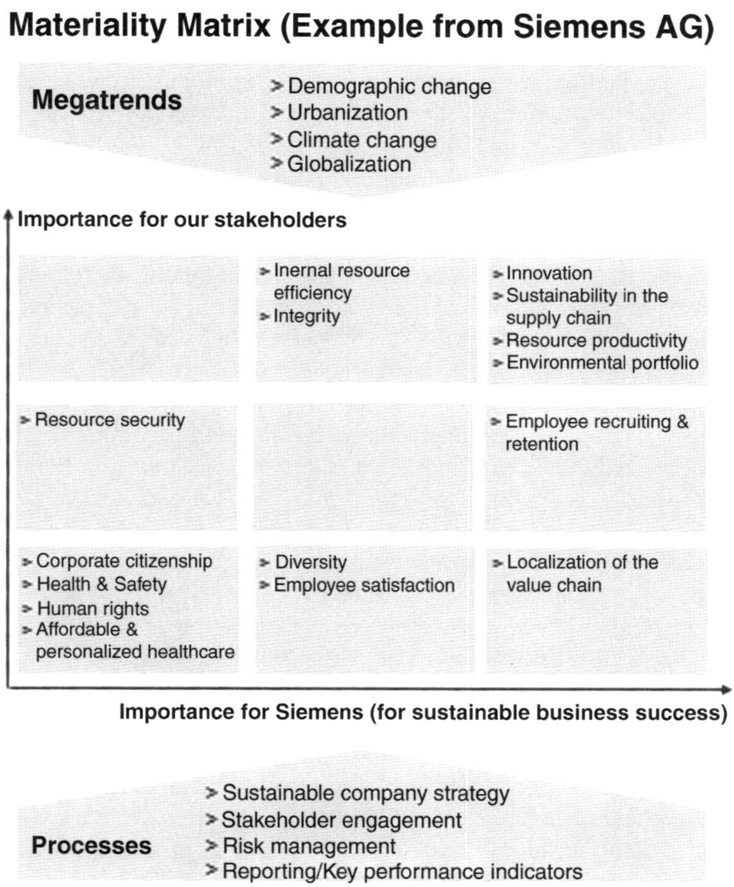

Fig. 17 Example of a materiality matrix: Siemens AG (2013)

development and the product lifecycle, stakeholders, suppliers, partners, and production sites should be included. Risks and opportunities should be identified for each value creation point in terms of their potential for ecological, economic, and societal value creation.

5.3 *Principle 3: Consider Sustainability Criteria Throughout the Entire Product Lifecycle*

Decisions on product/service designs, cost, and lifetime are not the only decisions to be made during the product creation process. Particularly at the beginning of the lifecycle (product idea and design), a major proportion of the economic, ecological, and societal consequences of a product will be determined by design decisions as well as the selection of supplier and components – even before the start of production. The sourcing, manufacturing, and distribution of products has economic, ecologic, and societal consequences. Even while the customer is using a product, effects on the environment and society will continue to arise – right up until the final act of "disposal" in landfill at the end of the product's lifecycle, which, according to the current philosophy for product use, unfortunately means: take, make, use, throw away.

This limited view is not workable for the future, as it represents a senseless waste of resources on the one hand and creates environmental problems for subsequent generations on the other. There is an urgent need for a change in perspective here. A consideration of the entire lifecycle from the sustainability angle also needs to be part of the new product development and product management repertoire. Not only the roles and responsibilities of product developers will have to change. Product managers and marketing teams also have to understand, that responsibility for a product does not and should not end when it is sold to the customer. Until now, most departments in any company have failed to recognize their accountability or responsibility for the entire lifecycle within their sphere of influence. Product support is indeed provided during the utilization phase in the form of instructions and support, service, and warranty agreements. However, in most cases there is ultimately no obligation with regard to return and disposal, as these processes are usually very costly for the company. The first sensible step may be to move toward "eco-efficiency" in order to manufacture products with fewer resources and less environmental impact. However, "eco-efficiency" on its own does not go far enough, as it fails to solve the basic design problem. A move toward "eco-effectiveness" is therefore preferable. Effectiveness is concerned with "doing the right things" and not optimizing the "wrong" things. Companies need to consider all of the relevant parameters in the product decision-making process as part of their product philosophy, i.e., the economic, ecological, and societal aspects. Ideally, negative effects and waste are to be avoided from the outset – or at least kept to a minimum, where this is not possible. Where possible, there should be an option to return products (or at least product components) to a closed biological or technical cycle in order to protect natural resources. Customer feedback and the sales history of existing products should be considered in addition to product lifecycle decisions. In effect, those products that do not even make it to the manufacturing stage because there is no demand or only low demand for them are the products with the least impact on the economy, ecology, and the environment. Profit and margin can be optimized if business resources can be diverted into products showing good sales performance. The Spanish fashion company ZARA is a pioneer of this model. ZARA only offers products in limited quantities and for limited periods. Using modern point-of-sale technology, ZARA is able to gather data on a daily basis and divert production capacity away from products that are not selling to products that are selling well.

5.4 *Principle 4: Take Responsibility Across the Extended Supply Chain*

For many companies, sustainable supply chains have to date only ever been an issue if these have become a topic of management discussion due to pressure from stakeholders (customers, investors, employees, legislators, NGOs) or if there has been a scandal relating to a product or supply chain. Studies dating back to 2006 show that companies see their greatest challenges in the issues of sustainability regulations, incentive systems, supplier management, and integrated supply chain management (Seuring and Müller 2008: 464). For many companies, the commitment to sustainability in the supply chain has so far been a response to legislative requirements and articulated customer preferences (Geissbauer and D'heur 2008).

As a rule, improvements to the supply chain are made within internal operational areas of responsibility and are aimed at being strictly functional. Current optimization approaches focus on the economic aspect, accelerated delivery capabilities, and flexibility: reducing the cost to run the supply chain, increasing efficiency, delivering higher profitability levels.

Savings in these areas have long been considered a necessity and are part of process optimization for achieving lower operations costs. These optimization activities make an essential contribution to the profitability of production or service delivery. If, in addition to improved capacity, accelerated processing and throughput times, optimization also leads to reduced water, waste, and energy costs, the opportunity to position the savings "as a contribution to sustainability" is eagerly embraced by many companies. An inspection of company sustainability reports reveals frequent references to savings in these areas.

The societal dimension of value creation has so far only been a topic of discussion when there has been a need for action due to a scandal, e.g., poor working conditions. The reactions of companies where scandals occur upstream in the supply chain (e.g., the textile and electronics industries, mining) range from the total denial of responsibility, the introduction of codes of conduct and public acceptance of responsibility. If there are any problems with the working conditions of their suppliers, companies are quick in citing either their own or also standard industry codes of conducts without actually working at solving the underlying problem in a proactive way. However, the consideration of all three dimensions – economy, ecology, and society – right from the design phase involves taking active responsibility for the supply chain and thus extends beyond the bounds of the company itself.

Despite their interest in the issue of sustainability, companies are slow to act, arguing that opportunities for influence or control are beyond their own corporate remit. Thankfully, examples of leaders adopting a different course of action are becoming increasingly frequent. Driven by his personal vision for taking a new approach to responsibility across the supply chain, PUMA CEO Jochen Zeitz redefined the concept of responsibility for product, supply chain, environment, and society. PUMA's strategic goal is to become the most desirable and sustainable sports lifestyle brand. The core task involves taking responsibility for the entire

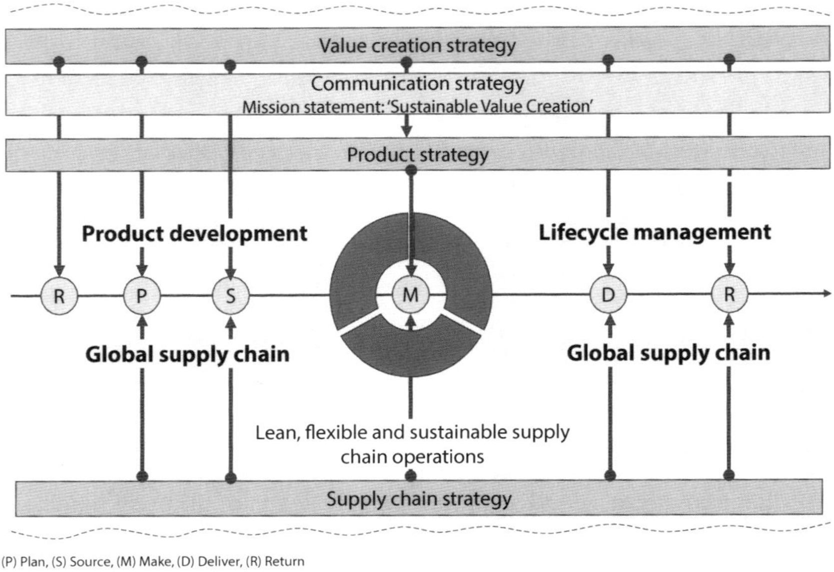

Fig. 18 Integration of sustainability in the core business (shared.value.chain 2013)

supply chain from the purchase of raw material to their return at the end of the lifecycle. PUMA customers should be able to purchase a product and rest assured that the corporation has taken into account all parties and resource suppliers involved in the value creation process based on fair and equitable principles (Fig. 18).

In this way, the interpretation of the supply chain is extended to a concept of responsibility, leading to a network of partners working together for the benefit of customers, the environment, and society.

5.5 *Principle 5: Drive Change*

The implementation of sustainable value creation based on fundamental principles will succeed, if applied to an organization's entire system as part of a structured and collaborative process. The type of organization is not important. The important factor is to design the practical implementation of sustainable value creation in such a way that, in addition to bringing the relevant stakeholders on board at the outset, keeping close track of the implementation process is pursued as an equally important objective.

The ultimate measure of implementing sustainable value creation is how many of the relevant stakeholders adopt the concept (buy-in) and then change their actual behaviour (stay-in)

Given that the route to sustainable value creation varies greatly depending on an organization's current position, maturity, and interests, it makes sense to proceed in several stages to ensure that any communication on voluntary commitment is consistent with the reality.

Anyone wishing to implement sustainable value creation within an organization needs to be fully aware that this is a multi-stage process requiring clear objectives, patience, and commitment. As a rule, changing a product development philosophy and coming up with innovations does not happen instantaneously, but is part of a creative (and often long-term) process. Value chain optimization has always been a process composed of a number of different stages. This means that the implementation of sustainable value creation is not a "stand-alone project" by the end of which all goals have to be achieved. Rather, it is part of a continuous process in which the improvement of individual components of the overall system and each learning outcome serve as base for further improvements.

At this point, one should consider the results of basic research into the diffusion of innovation in societal systems. The Innovation Diffusion Theory developed by Everett Rogers is concerned with processes whereby innovation is introduced and diffused in a societal system, e.g., a company. Innovation includes all ideas, processes, and objects that a societal group subjectively perceives to be new. The theory of adoption is a helpful starting point, describing on an individual level the factors that lead to the adoption or rejection of an innovation. Diffusion curves can be extrapolated from the aggregation of individual adoption processes. They show the number of people who have already adopted an innovation.

A distinction is usually made between the various stages of the adoption process:

1. Knowledge – finding out about an innovation;
2. Persuasion – being persuaded of an innovation to a positive/negative degree;
3. Decision – deciding for or against an innovation;
4. Implementation – implementing the innovation;
5. Confirmation – confirming the decision to implement the innovation and applying it in the future or retrospectively.

The implementation phase is only reached if an innovation proves to have features that are appropriate for adoption, including high relative benefit, low complexity, high compatibility, and high observability (Rogers 2003).

Even if an innovation brings obvious benefits, implementing it within an organization can be extremely difficult. Implementing new ideas in the context of a global supply chain is a special challenge, as the departments and companies involved often have different values and objectives. Moreover, operational processes are not always closely linked, even in the case of regular trade partners (So et al. 2012). Many innovations have been around for years before they have actually been adopted and implemented across the board. Sustainable value creation is no exception here. In recent years, much effort has been invested in the public debate around and validation of the management approach "Triple Bottom Line" – People, Profit, and Planet. Thankfully, companies are increasingly subscribing to this management approach. However, the timing aspect of implementing the Triple Bottom Line is not addressed in the majority of discussions (Wells 2010).

Sustainability is not an issue that can be solved within a company itself. Sustainability is much more the case of an inter-organizational issue, whereby supply chains – which by their very nature map the operational relationships existing between companies – need to be harmonized so that sustainability goals can be achieved (So et al. 2012). The key question is which process to pursue for a company's commitment to sustainable practices in the supply chain or how this commitment can be accelerated (So et al. 2012).

Figure 19 proposes a model based on the work of Parker and So, that uses the Innovation Diffusion Theory as the basis. The model describes the implementation of sustainable value creation in the context of products and supply chains.

The following stages of the adoption process apply to sustainable value creation:

1. **Knowledge** – What knowledge and experience does the company already have of sustainability (in the sense of innovation)?
2. **Persuasion** – What are the perceived benefits of sustainability for the company and individuals? ('What's in it for me?')
3. **Decision** – Sustainability within the company is either adopted or rejected.
4. **Implementation** – Sustainability with all its associated successes and problems related to operational activity is implemented.
5. **Confirmation** – Decision-makers acknowledge the benefits and value of sustainability and incorporate sustainable value creation as part of the core business (long-term commitment).

The time before the decision phase is particularly important, as it involves persuading a company's management team of the value of sustainability and managing employee attitudes. It is a case of clarifying existing misunderstandings and concerns by means of transparency and communication (So et al. 2012). As the implementation phase of sustainable value creation will depend on the stating

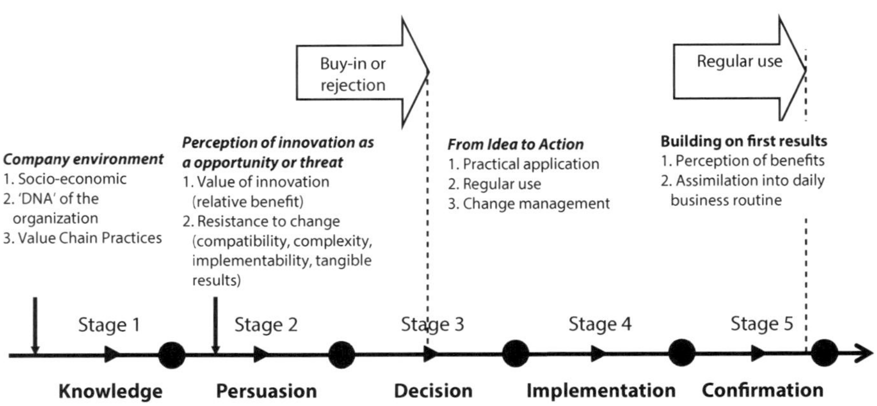

Fig. 19 Implementation of a sustainable value creation over time (shared.value.chain with reference to Parker and So, 2012)

position of a company and industry context, it is important to uncover doubts and objections early in the process. Transparency of information is the key here. The greater the willingness on the part of the entire organization to adopt sustainability, the faster it can be implemented in the supply chain (So et al. 2012). It is possible to reach a point where innovation becomes part of the organization and a routine activity. Innovation loses its "innovation tag," whereby the innovation process comes to a standstill. For some people, the follow-on confirmation phase is still necessary. Empirical studies show that the decision for or against innovation is often revised. When the decision has been made in favour of innovation, a considerable number of people also consult further information in order to confirm their decision (Rogers 2003).

5.6 *Principle 6: Collaborate, Deliver Tangible Results, Inspire Others*

Sustainable value creation occurs in a complex environment. Experts, industrial organizations, NGOs, and governmental groups should therefore be called on for support. This is particularly true in the case of implementing sustainable value creation, as a considerable number of people involved in corporate processes need to be persuaded to move away from their established practices and adopt new ways of working. This means consistent implementation based on the "Go beyond the Label" approach, support for implementation from concept stage to the point when behavioral change is achieved and at the same time, measuring progress, and eliminating obstacles along the way.

It is essential, that managers that are engaged in implementing sustainable value creation, take the lead and challenge the status quo: In order to overcome the challenges of the future, we need more companies to take the lead in solving economic, ecological, and societal problems. There are plenty of different ways of doing this, giving companies room to change the status quo in their particular industry.

To inspire others it is also important to share experiences and make adjustments to the deployment approach, where necessary: Given that companies and their stakeholders are on a learning curve throughout the implementation of sustainable value creation, it is vital to exchange information both internally and externally. Recourse to expert groups and communication platforms is the way forward to find case studies and inject fresh momentum. Open and transparent communication becomes an essential accelerator for the implementation of sustainable value creation.

No implementation of sustainable value creation will last long without tangible results. In this context, it is necessary to establish the required KPIs for demonstrating the value of sustainability so that adoption and implementation can be accelerated. By doing this, a positive correlation arises between the perceived value of sustainability and the speed of implementation in the supply chain (So et al. 2012).

One example for a set of metrics that is driving sustainable value creation, is the sustainability supply chain scorecard the global electronic components distributor AVNET is using (Fig. 20). As long-term member of the Supply Chain Council, AVNET has also helped to refine the SCOR Model and associated metrics. Because a major part of AVNET's value creation involves the acquisition of electronic components for customers, speed and service are the determining factors for its business and therefore for competitive advantage. The AVNET Supply Chain Scorecard was previously structured according to traditional SCOR model metrics, with operational performance as main focus area.

Costs and fixed assets were used as the economic parameters, namely the following: indicators and comparators: cost of goods sold, total supply chain management cost, total cost of ownership (TCOO), PPV (purchase price variance) and selling, general, and administrative expenses (SG&A).

The metrics applied for fixed assets are as follows: cash-to-cash cycle time (time between the investment to acquire goods and receipt of payment from the customer); return on assets; return on working capital; and inventory turns.

The focus is on three main areas in terms of customer service: reliability, responsiveness, and risk management. All three areas are important, as the electronic components trade lives and dies by the ability to provide the customer with the right components and the right specifications every time. Due to advances in technology, electronic components are exposed to short lifecycles and sharp falls in market prices.

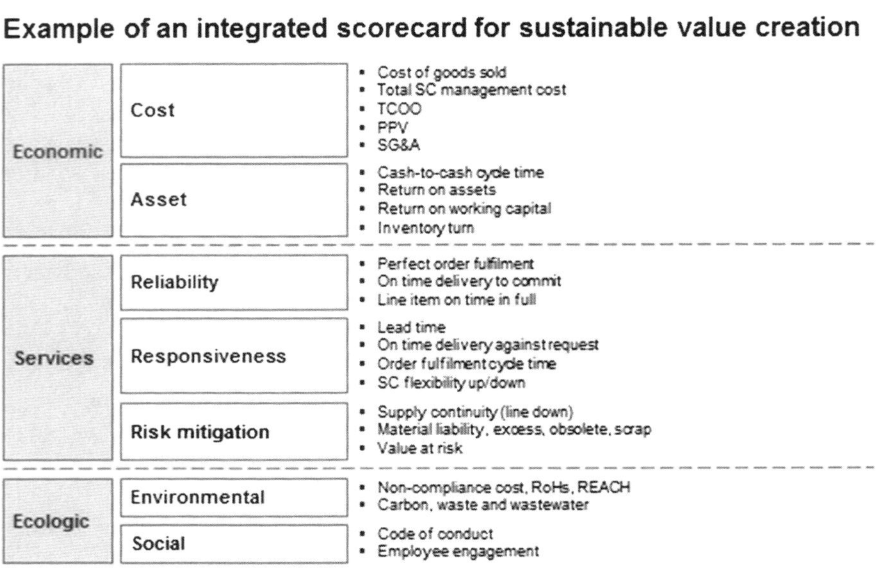

Fig. 20 Example of an integrated scorecard (Supply Chain Council 2012)

The following delivery performance metrics are extremely important in terms of meeting customer requirements:

- Perfect order fulfilment
- On time delivery to commit
- Line item on time in full

Supply chain responsiveness to market changes is measured via the following metrics:

- Lead time
- On time delivery against request
- Order fulfilment cycle time
- Supply chain flexibility up/down

Given that increasing numbers of suppliers are defaulting due to the financial crisis and natural disasters, there has been an increased focus on risk management. Risk management strategies have been applied relating to the following metrics:

- Supply continuity – line down
- Excess, obsolete, scrap
- Value at risk

In view of the regulations prevailing in the electronics sector, AVNET has also included the ecological dimension on its scorecard. This relates to sustainability and the environment. The following metrics are applied in terms of sustainability and environmental KPIs:

- Cost of noncompliance
- Adherence to legal requirements (such as ROHS and REACH)
- CO_2, waste, and wastewater reduction
- Adherence to code of conduct
- Employee engagement

The AVNET Sustainable Supply Chain Scorecard is one of the few approaches rendering the economic, environmental, and societal dimensions of value creation measurable via a standardized system. AVNET has only just started to address the environmental dimension, but hopes to become a pioneer in the sector.

A whole range of other KPIs are available for ensuring that the results and performance of corporate sustainable value creation are measurable, some examples of which are given below:

- Number of new products on the market,
- Green portfolio,
- Delivery service level (e.g., 95 % + for A customers),
- Stocks (raw materials, semimanufactured goods, end products),
- Flexibility/response time,
- Security,

- Percentage of sustainable/certified raw materials (e.g., 15 % increase in the percentage of sustainable raw materials),
- Adherence to code of conduct (companies, suppliers – e.g., number of positive audits completed),
- Production efficiency,
- Waste (reduction of solid waste volumes from production, reprocessing of x % of "used" products in order to reuse raw materials, avoidance of landfill costs),
- Reduction of harmful emissions (CO_2, SO_2, NO_x, . . .),
- Reduction of absolute emissions of greenhouse gases (scope 1&2) in product manufacturing,
- Reduction of harmful emissions/water consumption,
- Energy consumption (15 % improvement in energy efficiency of outsourced energy),
- Workplace (public reporting of commitment to training, health and infrastructure in communities where company is active),

6 Are You Ready?

6.1 The Core Business: Driver for Sustainable Value Creation

Bridging the gap between today's way of doing business and sustainable value creation will happen at the intersection points between corporations and society: in corporate product development departments, local, regional and global corporate supply chains, the public sector, and NGOs.

Currently addressed on a rather isolated basis, the areas of sustainability, product development, and global supply chain management are gradually merging to form an integrated approach. Economic, ecological, and societal value creation will be integrated via products, services, and supply chains. A stable basis for future growth can only be achieved by embedding sustainability in the core business.

The general public and the corporate sector are becoming increasingly aware that the current linear model of the economy is not sustainable to deliver growth, prosperity, and a liveable environment. In the past, campaigning for sustainability was mainly an activist's choice, most of which could not imagine working with companies to find solutions. Despite progress made in the past, there are still major challenges to overcome before sustainable value creation can be applied end-to-end. Many companies in the EU still have not incorporated societal and ecological concerns fully within their operational management and strategy. A small minority of European companies still stand accused of infringing human rights and failing to meet basic employment standards. Only 15 of the 27 EU member states have a strategic national framework for promoting sustainability (EU Commission 2011).

Sustainable value creation provides a framework that companies can use to make systematic preparations for the future. The implementation of sustainable value creation will be largely determined by a company's individual starting point and its level of maturity, simply because these starting points vary greatly according to industry, company history and competitive situation.

Public debate refers continuously to maturity models and rankings, whereby companies are evaluated according to various CSR and sustainability criteria. This mainly involves market research agencies, consultants, NGOs, and foundations, which look at companies in ever-changing ways in order to rank their sustainability performance. When assessing the amount of progress companies have made in implementing sustainable value creation, it is important to note that the companies whose products are closest to the end user (e.g., food, clothing, cars) invest more in sustainability than the others. Sustainable production in the preliminary stages of the chemicals industry or working conditions in the manufacture of electronic components have not yet become criteria affecting the end-user's decision-making process.

By considering the economic, ecological, and societal aspects of products and supply chains, shared.value.chain has developed a maturity model to indicate how companies measure up along the sustainability and supply chain dimensions (Fig. 21). The model indicates where companies still have room for improvement (the basis for classification is the implementation of practices across the 8 building blocks of sustainable value creation).

As illustrated in Fig. 21, companies in the consumer goods, food, and textile industries are the most assiduous in terms of implementing sustainable value creation. Even major retailers such as Marks & Spencer are establishing sustainability within their core business. All of these companies have complex global

Fig. 21 Pioneering companies drive the implementation of sustainable value creation (shared.value.chain 2013)

supply chains – but none of them shies away from addressing the task proactively. Companies such as Monsanto and BP occupy the dark zone of greenwashers and toeing the line of compliance. They do make the effort to acquire a greener image via sustainability, but both their products and their supply chains are far from meeting the requirements of sustainable value creation.

6.2 The Time Is Right, Leaders Show the Way

Sustainable value creation is the backbone of any viable industrial economy – now and in the future. The evolutionary process is already underway, leading to a greater number of new approaches. There should not be competition between different approaches and schools of thought: the public dialogue on the issue of sustainable value creation allows more than just one approach. In an increasingly networked global economy it is important to interpret sustainable value creation as an open source concept, thus paving the way for joint discussion and development.

The importance of and interest in sustainable economic activity cannot be disputed. The growing frequency of events on the subject and increased public awareness are part of an active dialogue between various stakeholders and new forms of collaboration. However, the practical implementation of sustainable value creation is still significantly driven forward by corporate and NGO visionaries. Areas of interest and legal frameworks differ so widely that it will definitely take some time before sustainable value creation is implemented end-to-end. Nevertheless, the growing debate on the necessity of sustainable value creation occurring in think tanks, political circles, and cooperative efforts between NGOs and companies is to be welcomed. The greater the number of companies following the good example of the visionaries that forge ahead as drivers of innovation, the sooner end-to-end implementation will begin to happen on a large scale.

The following four factors will help the adoption of sustainable value creation:

- **Customers** and the communities where companies create value are no longer satisfied in terms of the product alone, but want to know how and under what conditions products are manufactured, what the impact is on people and the environment as well as how the products can be returned to a closed cycle. Corporate transparency and stakeholder dialogue will accelerate improvement.
- **Entrepreneurs** that understand that business and profit are only viable where economic, ecological, and societal aspects are combined – and where these aspects are demonstrated credibly to the outside world as being linked to the core issue of corporate purpose.
- **NGOs** see the value in collaborating with companies, instead of refusing to work together. This new level of collaboration will help to jointly identify solutions to societal problems.

- **Governments** are no longer able to pass on the cost of uncontrolled industrial activity (factors external to companies) to the taxpayer. The appropriate legislation will soon apply across the board.

All these factors rely on mutual trust and the measurability of sustainable value creation. Reference models for global supply chains that are already long-established should be supplemented by the aspect of sustainability along the supply chain from end to end; from raw materials to customer and back again. Sustainability is not a new issue and continues to be affected by fluctuations due to economic cycles and various scandals. Embedding sustainability in the corporate operating model will ultimately be the key to accomplish a breakthrough.

The future belongs to sustainable value creation that is fully embedded within the core business – defined and implemented from inside out – as part of an integrated corporate management system, enabling partners in the supply chain and in society to work on equal terms. Sustainable value creation is a proactive concept, giving access to market opportunities via sustainable products, processes, and practices. The value generated in this way is therefore not only the next stage in the optimization of corporate processes and global value chains – it will be a fundamental customer prerequisite. The Licence to Operate ultimately becomes the Licence to Lead (Edelman 2013).

I am convinced that the time has come for a new way of doing business and that the best is yet to come.

Are *you* ready for sustainable value creation?

Bibliography

3M. (1997). *Pollution prevention pays.* http://www.3m.com/about3m/environment/policies_about3P.jhtml. Accessed June 11, 2013.

Aachener Stiftung Kathy Beys. (Ed.). (2013a). *Lexikon der Nachhaltigkeit*, Keyword: Greenwashing. http://www.nachhaltigkeit.info/artikel/greenwashing_1710.htm. Accessed June 11, 2013.

Aachener Stiftung Kathy Beys. (Ed.). (2013b). *Lexikon der Nachhaltigkeit*, Keyword: ESG-KPI. http://www.nachhaltigkeit.info/artikel/dvfa_schluesselkriterien_zur_nachhaltigkeit_esg_k_1630.htm. Accessed June 11, 2013.

APICS. (2011). *APICS operations management body of knowledge* (3rd ed.). Chicago, IL: APICS.

APICS. (2012). *APICS supply chain sustainability survey*. Chicago, IL: APICS.

Arratia, R. (2013). How full product transparency can embed sustainability at the core of your Business. In *GreenBiz*. http://www.greenbiz.com/blog/2013/04/25/how-full-product-transparency-can-embed-sustainability-core-your-business?page=0%2C1. Accessed June 11, 2013

Ayres, R., & Neese, A. V. (1989). Externalities: Economics and thermodynamics. In F. Archibugi & P. Nijkamp (Eds.), *Economy and ecology: Towards sustainable development* (p. 93). Netherlands: Kluwer.

Bain & Company. (2010). *Rückkehr zu neuem Wachstum*. http://www.bain.de/Images/Results%2002_2010.pdf. Accessed June 11, 2013.

BBMG. (2013). *Delivering trust through shared value*. Accessed December 11, 2013 from http://de.slideshare.net/sustainablebrands/delivering-trust-through-shared-value

Belz, F.-M., & Peattie, K. (2009). *Sustainability marketing: A global perspective*. Hoboken, NJ: Wiley.
BITKOM. (2014). *Erstmals mehr als 100 Millionen Alt-Handys zu Hause*. http://www.bitkom.org/files/documents/BITKOM_Presseinfo_Althandys_22_01_2014.pdf. Accessed May 31, 2014.
Bluesign Tech. (2013). *Bluesign standard*. http://www.bluesign-tech.com/index.php?id=151. Accessed June 11, 2013.
Braungart, M., & McDonough, W. (2002). *Cradle to cradle, remaking the way we make things* (1st ed.). New York: North Point Press.
Brix, K., Bromma, B., & Jänisch, J. (2006). *Nachhaltiges Unternehmertum*. Centre for sustainability Management (CSM) e. V., Universität Lüneburg.
Bundesregierung. (2012). *10 Jahre Nachhaltigkeit "made in Germany": Die Nationale Strategie für eine nachhaltige Entwicklung*. http://www.bundesregierung.de/Content/DE/_Anlagen/Nachhaltigkeit-wiederhergestellt/2012-05-09-kurzpapier-zum-fortschrittsbericht-2012-barrierefrei.pdf?__blob=publicationFile&v=3. Accessed June 28, 2014.
Carbon Disclosure Project. (2013). *CDP supply chain program*. https://www.cdproject.net/en-us/programmes/pages/cdp-supply-chain.aspx. Accessed June 11, 2013.
Christensen. (2012, November 3). A capitalist's dilemma, whoever wins on tuesday. *The New York Times*. http://www.nytimes.com/2012/11/04/business/a-capitalists-dilemma-whoever-becomes-president.html?ref=business&_r=1&. Accessed May 31, 2014.
Climategreenwash. (2009). *Climate Greenwash Awards 2009*. http://www.climategreenwash.org/. Accessed June 11, 2013.
Crane, A., Matten, D., & Spence, L. J. (Eds.). (2008). *Corporate social responsibility: Readings and cases in a global context*. London: Routledge.
Dell. (2012). *Bamboo – Natures eco-friendly packaging solution*. http://www.dell.com/Learn/us/en/uscorp1/corp-comm/bamboo-packaging. Accessed June 11, 2013.
DIN. (2012). ISO/TC 207 startet Revisionsarbeiten zur ISO 14001. In *DIN Normenausschuss Grundlagen des Umweltschutzes*. http://www.nagus.din.de/cmd;jsessionid=AD41131C7BE3DC915AB4A5E147DE13E0.1?cmsrubid=47243&menurubricid=47243&level=tpl-artikel&menuid=47224&languageid=de+&bcrumblevel=1&cmstextid=158414&cmsareaid=47224. Accessed June 11, 2013.
Drucker, P. (1986). *The effective executive*. New York: Harper Business.
DQS. (2013). SA8000 – Wertschätzung steigern durch soziale Verantwortung. In *DQS GmbH Deutsche Gesellschaft zur Zertifizierung von Managementsystemen*. https://de.dqs-ul.com/standards/nachhaltigkeit/soziale-verantwortung/sa8000.html. Accessed June 11, 2013.
Eban. (2013). Dirty medicine. In *Forbes*. http://features.blogs.fortune.cnn.com/2013/05/15/ranbaxy-fraud-lipitor/. Accessed June 11, 2013.
Edelman. (2010). *Edelman goodpurpose® Study 2010. Fourth annual global consumer survey*. http://ppqty.com/GoodPurpose2010globalPPT_WEBversion%20%281%29.pdf. Accessed June 11, 2013.
Edelman. (2012). *Edelman goodpurpose® Study 2012. Fifth annual global consumer survey*. http://de.scribd.com/doc/90411623/Executive-Summary-2012-Edelman-goodpurpose%C2%AEStudy. Accessed June 11, 2013.
Edelman. (2013). *Edelman trust barometer 2013*. http://de.scribd.com/doc/121501475/Executive-Summary-2013-Edelman-Trust-Barometer. Accessed June 11, 2013.
EFQM. (2013). *Fundamental concepts* http://www.efqm.org/en/Home/aboutEFQM/TheEFQMExcellenceModel/FundamentalConcepts/tabid/169/Default.aspx. Accessed June 11, 2013.
Elias-Trostmann, K. (2012). Taking it back to the original source. In *6 Heads*. http://6-heads.com/2012/07/12/1100/?goback=.gde_2101817_member_134630074. Accessed May 3, 2013.
ELVIS. (2013). *Elvis – Full load network – vision und mission*. http://www.elvis-ag.com/de/ELVISAG-Da-wollen-wir-hin-208,239.html. Accessed June 11, 2013.
EPEA. (2010). High Tech-Innovationen mit Cradle to Cradle® weckt Emotionen: Pure Origin. In *EPEA Switzerland*. http://www.epeaswitzerland.com/case-studies/case-study-1/. Accessed June 11, 2013.

European Commission. (2011). Eine neue EU-Strategie (2011-14) für die soziale Verantwortung der Unternehmen (CSR), KOM (2011) 681 endgültig, Brüssel.
Europäische Kommission. (2012). EMAS. http://ec.europa.eu/environment/emas/pictures/Stats/2012-06_Overview_of_the_take-up_of_EMAS_across_the_years.jpg. Accessed June 1, 2013.
Figge, F. (2012). *Sustainable value – The concept*. http://www.sustainablevalue.com/theconcept/index.html. Accessed June 11, 2013.
Gabler Verlag. (Ed.). (2013). *Gabler Wirtschaftslexikon*, Keyword: Wertschöpfungskette. http://wirtschaftslexikon.gabler.de/Archiv/145581/wertschoepfungskette-v6.html. Accessed June 11, 2013.
Gang, J. (2012). *Don't be fooled 2012 – The worst greenwashers*. Report April 2012, The Green Life.
Hofman, D., Aronow, S., & Nilles, K. (2013). Supply chain top 25. In *Gartner, 2013*. http://www.gartner.com/id=2493115. Accessed June 11, 2013.
Geissbauer, R., & D'heur, M. (2008). Erste Schritte in Richtung einer nachhaltigen Supply Chain. *Logistik für Unternehmen, 10*.
Gilding, P. (2012). *The great disruption: Why the climate crisis will bring on the end of shopping and the birth of a new world*. New York: Bloomsbury Press.
Goldratt, E. (1999). *Theory of constraints*. Great Barrington, MA: North River Press.
Guardian. (2013). Sustainability case studies. In *Guardian sustainable business*. http://www.guardian.co.uk/sustainable-business/series/sustainability-case-studies. Accessed June 11, 2013.
Günther, E. (2008). *Ökologieorientiertes Management: Um(weltorientiert)-denken in der BWL*. Stuttgart.
Harting, D. (1994). Wertschöpfung auf neuen Wegen. In *Beschaffung aktuell 7/1994*.
Hillmer, H.-J. (2013). *Rahmenwerk zum Integrated Reporting*. http://www.compliancedigital.de/ce/rahmenwerk-zum-integrated-reporting/detail.html. Accessed June 11, 2013.
Hopp, W., & Spearman, M. (2000). *Factory physics: Foundations of manufacturing management* (2.Aufl.). McGraw-Hill Higher Education.
IBM. (2008). *Gute Gründe für eine IT-Finanzierung*. http://www-304.ibm.com/businesscenter/cpe/download0/181058/GFS03057DEDE.pdf. Accessed June 11, 2013.
ILO. (2013). *Die Grundprinzipien der ILO*. http://www.ilo.org/berlin/arbeits-und-standards/kernarbeitsnormen/lang–de/index.htm. Accessed June 11, 2013.
ISO. (2013a). *ISO 26000 – Leitfaden zur gesellschaftlichen Verantwortung*. https://www.iso.org/obp/ui/#iso:std:iso:26000:ed-1:v1:en. Accessed June 11, 2013.
ISO. (2013b). *ISO 50001:2011, ISO*. http://www.iso.org/iso/iso_50001_energy.pdf. Accessed June 11, 2013.
Kaye, L. (2013). The business case for sustainability is becoming easier to make. In *Triple Pundit*. http://www.triplepundit.com/2013/03/business-case-for-sustainability/. Accessed June 11, 2013.
Lee, H. L. (2010). Don't tweak your supply chain – Rethink it end to end. *Harvard Business Review, 88*(1), 62–69.
Leitl. (2009). *Supply chain management – Was ist das ... ?* http://www.harvardbusinessmanager.de/heft/artikel/a-621606.html. Accessed June 11, 2013.
Lin-Hi, N. (2013). License to operate. In Gabler Verlag (Ed.). *Gabler Wirtschaftslexikon, Keyword: Licence to operate*. http://wirtschaftslexikon.gabler.de/Archiv/18118/licence-to-operatev6.html. Accessed June 11, 2013.
Lopez, R. (2012). *Keeping nature's balance sheet in balance*. University of Cambridge – Programme for sustainability leadership. http://www.cpsl.cam.ac.uk/~/media/Files/Resources/SoSL/SoSL_2012/Jose_Lopez_Keeping_natures_balance_sheet_in_balance_SoSL12_low.ashx. Accessed June 11, 2013.
Marks & Spencer. (2013). *How we do business report 2012, Marks & Spencer's*. http://corporate.marksandspencer.com/file.axd?pointerid=24f35ecfc08e4eb1992603107c4ec51a. Accessed June 11, 2013.

MIT. (2013). Sustainability: The new business model opportunities. In *Massachussetts Institute of Technology Sloan Management Review*. http://sloanreview.mit.edu/article/video-sustainabilitythe-new-business-model-opportunities/. Accessed June 11, 2013.
Mitchell, S. (2012). Walmart's Greenwash. In *The Institute of Local Self-Reliance, March 2012*. http://www.newrules.org/sites/newrules.org/files/walmart-greenwash-report.pdf
Nike. (2011). *Joint roadmap: Toward zero discharge of hazardous chemicals*. Nike Inc. http://nikeinc.com/system/assets/5408/111118_JointRoadmap_original.pdf?1321567189. Accessed June 11, 2013.
Nölke, A. (2012). Banken. In *Bundeszentrale für politische Bildung*. http://www.bpb.de/politik/wirtschaft/finanzmaerkte/51718/banken. Accessed June 11, 2013.
Porter, M. (1989). *Competitive advantage*. New York, NY: Free Press.
Porter, M., & Kramer, R. (2011). *Creating shared value*. http://hbr.org/2011/01/the-big-idea-creating-shared-value. Accessed June 11, 2013.
Porter, M., & Kramer, R. (2012). Shared Value: Die Brücke von Corporate Social Responsibility zu Corporate Strategy. In Schneider, A. & Schmidpeter, R. (Ed.). *Corporate social responsibility: Verantwortungsvolle Unternehmensführung in Theorie und Praxis*. Berlin
Porter, M. E., Hills, G., Pfitzer, M., Patscheke, S., & Hawkins, E. (2013). *Measuring shared value*. http://www.fsg.org/Portals/0/Uploads/Documents/PDF/Measuring_Shared_Value.pdf. Accessed June 11, 2013.
PRTM. (2008). Global supply chain trends 2008–2010.
PRTM. (2010). Global supply chain trends 2010–2012.
Randers, J. (2013). *The 2052 Forecast* (Pestel Institut). http://www.2052.info/presentation.php. Accessed June 11, 2013.
Reuters. (2013). *Pollution 'worst on record' in Beijing: Greenpeace*. http://www.reuters.com/article/2013/01/13/us-china-pollution-idUSBRE90C01Q20130113. Accessed May 31, 2014.
RNE. (2012). Nachhaltige Unternehmensführung: Kosten kennen – Nutzen erschließen. In *Rat für nachhaltige Entwicklung*. http://www.nachhaltigkeitsrat.de/ideenwettbewerb. Accessed June 11, 2013.
Riess, B. (Ed.). (2012). *CSR WeltWeit – Ein Branchenvergleich*; Bertelsmann Stiftung. http://www.csr-weltweit.de/fileadmin/inhalte/Publikationen/CSR_Weltweit_Ein_Branchenvergleich.pdf. Accessed June 11, 2013.
Rogers, E. (2003). *Diffusion of innovations* (5th ed.). New York, NY: Free Press.
shared.value.chain. (2012, 2013). *Company profile*. Accessed June 11, 2013 from https://sharedvaluechain.com/about-us
Schmidpeter, R. (2013). Das Gegensatzdenken proaktiv überwinden. In *Verantwortung Zukunft, Ausgabe 02*. http://www.verantwortungzukunft.com/sites/verantwortungzukunft.de/files/images/VZ_Magazin_Ausgabe_2_2013_Titelstory.pdf. Accessed June 11, 2013.
Schneider, A. (2012). Reifegradmodell CSR – eine Begriffserklärung und –abgrenzung. In Schneider, A. & Schmidpeter, R. (Ed.). *Corporate Social Responsibility: Verantwortungsvolle Unternehmensführung in Theorie und Praxis* (pp. 19–24). Berlin.
Schneider, A., & Schmidpeter, R. (Ed.) (2012). *Corporate Social Responsibility: Verantwortungsvolle Unternehmensführung in Theorie und Praxis*. Berlin.
Schürmann, H. (2011). Puma will ökologischen Fußabdruck verkleinern. In *Ingenieur.de*. http://www.ingenieur.de/Politik-Wirtschaft/Unternehmen/Puma-oekologischen-Fussabdruckverkleinern. Accessed June 11, 2013.
SEDEX. (2013). *About Sedex*. http://www.sedexglobal.com/about-sedex/. Accessed June 11, 2013.
Seuring, S., & Müller, M. (2008). Core issues in sustainable supply chain management – A Delphi study. *Business Strategy and the Environment, 17*, 455–466.
Shanahan. (2013). Setting bold sustainability goals. In *Why bold sustainability targets make good business sense; The EcoInnovator*. http://corporateecoforum.com/ecoinnovator/?p=7612. Accessed June 11, 2013.

Siemens AG. (2010). *One Siemens – unser Weg zur nachhaltigen Wertsteigerung*. http://www.siemens.com/about/pool/de/vision/one_siemens_broschuere_d.pdf. Accessed June 11, 2013.

Siemens AG. (2013). *Materiality as a guiding principle*. Accessed June 11, 2013 from http://www.siemens.com/about/sustainability/en/sustainability-at-siemens/materiality.htm

Slade, G. (2007). *Made to break – technology and obsolescence in America*. Cambridge, MA: Harvard University Press.

So, S., Parker, D., & Xu, H. (2012). A conceptual framework for adopting sustainability in the supply chain. In *ANZAM operations, supply chain and services management symposium*, Conference paper (pp. 397–413). University of Queensland.

Supply Chain Council. (2006). *Supply chain council executive briefing*. http://supplychain.org/f/2010/SCC-Executive-Overview-final.pdf. Accessed June 11, 2013.

Supply Chain Council. (2012). *Capitalizing on complexity: Reducing multi-tier supply chain risk*. Accessed June 11, 2013 from https://supply-chain.org/f/Pascal%20Fernandez%20-%20EMEA%20Avnet%20Velocity.pdf

The Economist. (2011). *Milton Friedman goes on tour, 27.01.2011*. http://www.economist.com/node/18010553. Accessed June 11, 2013.

United Nations. (1987). *Report on the World Commission on Environment and Development, "Our Common Future" unter dem Vorsitz von Gro Harlem Brundtland*. http://www.undocuments.net/ocf-cf.htm. Accessed June 11, 2013.

United Nations Environment Programme. (2009). *Recycling – From e-waste to resources*. http://www.unep.org/PDF/PressReleases/E-Waste_publication_screen_FINALVERSION-sml.pdf. Accessed June 11, 2013.

United Nations. (2013). *United Nations Global Compact – Die zehn Prinzipien*. http://www.unglobalcompact.org/Languages/german/die_zehn_prinzipien.html. Accessed June 11, 2013.

WBCSD. (2010). *Vision 2050 – The new agenda for business*. http://www.wbcsd.org/pages/edocument/edocumentdetails.aspx?id=219&nosearchcontextkey=true. Accessed June 11, 2013.

Wells, P. E. (2010). *The automotive industry in an era of eco-austerity: Creating an industry as if the planet mattered*. Cheltenham: Edward Elgar.

Wilkerson, T. (2008). GreenSCOR: Integrating green supply chain practices into the SCOR mode. In *Supply Chain Council*. http://supply-chain.org/node/3989. Accessed June 11, 2013.

Wirl, F. (2012). Individual firm and market dynamics of CSR activities. In Wirl, F., Feichtinger, G. & Kort, P. M. (Eds.). *Research report*, Institute of Mathematical Methods in Economics, Vienna University of Technology.

Z.V.E.I. (2012). *ZVEI Code of Conduct – Unternehmen bekennen sich zu gesellschaftlicher Verantwortung*.http://www.zvei.org/Themen/GesellschaftlundUmwelt/Seiten/ZVEI-Code-of-Conduct.aspx. Accessed June 11, 2013.

Beiersdorf: Generating Joint Added Value Through Collaboration, Planning, and Evaluation

Daniel Weber and Dorle Bahr

1 CSR and Sustainability at Beiersdorf

1.1 Beiersdorf AG: 130 Years of Corporate Responsibility

The cosmetics company Beiersdorf AG is registered in Hamburg and has 16,000 employees worldwide. In 2012, it achieved a turnover of EUR 6,040 billion. It has been listed on the German share index DAX since December 2008 and its brand Nivea is the leading skincare brand in the world (cf. Beiersdorf 2013a; Euromonitor 2011). Its internationally successful portfolio also includes Eucerin, La Prairie, Labello, 8x4, and Hansaplast. Its affiliate tesa SE is one of the leading manufacturers of self-adhesive products and system solutions for industry, commerce and consumers in the world. Beiersdorf has over 130 years of experience in skincare and is renowned for its innovative and high-quality products (Beiersdorf 2013a).

Beiersdorf has a long history of acting in a sustainable fashion. It has notched up notable success in terms of taking up corporate responsibility for the society and its employees. For example, at the start of the last century Dr. Oscar Troplowitz gradually introduced a 48-h week with no loss of payment, paid vacation, Christmas bonus and offered employees a free lunch if they agreed to reduce their lunch break from two-and-a-half to two hours. He also founded TROMA, a retirement and survivors' pension fund, which still exists today, albeit in a different form. Besides this, in 1938 Beiersdorf was one of the first companies in Germany to open a kindergarten (see Fig. 1).

D. Weber (✉) • D. Bahr
Beiersdorf AG, Unnastrasse 48, 20245 Hamburg, Germany
e-mail: daniel.j.weber@t-online.de; dorle.bahr@beiersdorf.com

M. D'heur (ed.), *Sustainable Value Chain Management*, CSR, Sustainability, Ethics & Governance, DOI 10.1007/978-3-319-12142-0_2

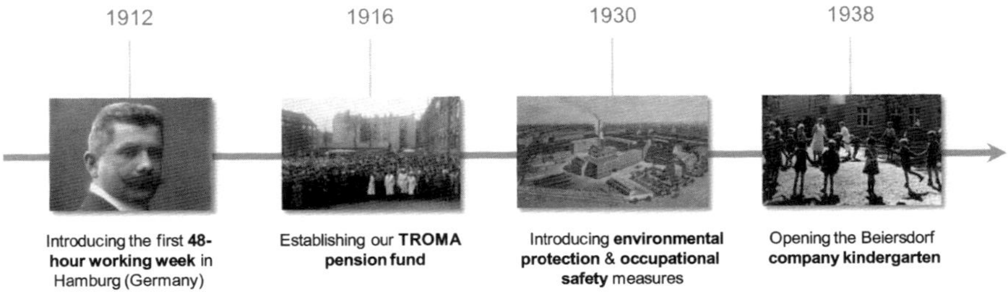

Fig. 1 Beiersdorf: first sustainability milestones. *Source*: Beiersdorf (2013c)

1.2 Sustainability Management: The Beginning

The first recorded environmental, health, and safety measures date back to 1930. The pledge to dispense with animal testing and the engagement during the introduction of the "Green dot" – the German collection and recycling scheme for consumer packaging waste – provide evidence that economic efficiency, environmental protection, and social responsibility are deeply ingrained in Beiersdorf's corporate culture.

The foundations for sustainability management were laid in 2003; a sustainability reporting system and global environmental protection and health and safety management system were introduced based on OHSAS 18001 and ISO 14001 and have since been consistently expanded. Simultaneously, the first corporate citizenship strategy was developed as the first overarching structure for social commitments in 2006 and a Sustainability Advisory Board was established from various central functions and a number of subsidiaries to act as a steering committee. At this time, responsibility for sustainability was only laid out in the mission statement for environmental protection, health and safety and for Corporate Citizenship, which was later renamed Corporate Social Responsibility. A comprehensive strategy with specific objectives across the entire value chain did not exist at the time.

1.3 "We Care" as Core Value of Our Sustainability Culture

When the Corporate Sustainability function had been introduced in 2010, which contains a direct reporting line to the CEO, an interdisciplinary, international project team developed a new global sustainability strategy in 2011.

Beiersdorf's approach to sustainability focuses mainly on improving the company's resilience and adaptability. The name of the new sustainability strategy ("We care") ties in with our brand values and reflects the core value of our corporate culture. "We care" is based on six focal points in three sectors of particular relevance to the company. The classic "3P" of sustainability (profit, planet, people)

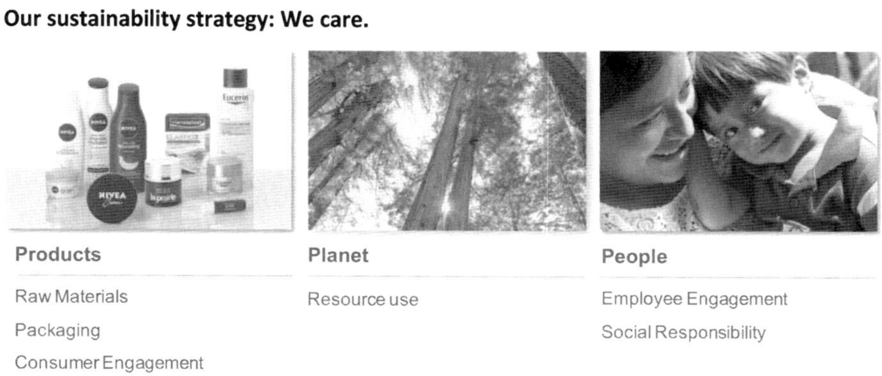

Fig. 2 Cornerstones of the Beiersdorf sustainability strategy. *Source*: Beiersdorf (2013c)

is reflected in the strategy sectors "Products" (as the source of our profit), "Planet," and "People" (Fig. 2).

1.3.1 Our Sustainability Strategy: We Care

In the "Products" sector, Beiersdorf focuses on sustainable use of raw materials and conservation of resources, which is done by minimizing packaging and developing sustainable packaging solutions, and also on active support for sustainable customer behavior. The "Planet" sector includes responsible use of scarce resources in production and sales. The "People" sector is on the one side about motivating Beiersdorf's employees to take an active part in sustainable action. On the other side it supports local as well as global social initiatives.

The main challenges in implementing a sustainability strategy are based in the overall complexity of the issue, which means that differentiated criteria must be used in order to evaluate the sustainability of a process or product. Besides this, possibly also due to this complexity, sustainability is often seen as an issue for experts and thus not as an integral component of core processes within the value chain. However, integrating sustainability into core processes is fundamental for strengthening the company's adaptability.

Sustainable action has been introduced as an integral component of our business by setting clear, long-term targets for each of the three "We care" sectors (Fig. 3).

1.3.2 Ambitious Commitments for 2020

In order to achieve our core targets, we have developed a series of indicators to systematically measure our performance in each of the (six) focal areas. Based on these indicators, we shall develop a system to operationalize our core targets.

Fig. 3 Targets for sustainable added value. *Source*: Beiersdorf (2013c)

1.4 New Impetus for Collaboration

The focal areas listed above extend across the entire value chain and are therefore directed at everyone involved in internal and external processes (Fig. 4).

It will be impossible to achieve our long-term targets without intensive collaboration with and between internal and external partners along the entire value chain.

Intensive collaboration with all our stakeholders is required if we are to increase the sustainability of our product portfolio. The stakeholders include innovation partners, such as suppliers and scientific organizations, as well as nongovernmental organizations and service providers at the end of the value chain, such as disposal service providers.

Intensive collaboration with partners throughout the entire life cycle is also needed in order to establish life cycle assessments to analyze and improve the environmental impact of our products. Close collaboration with suppliers and service providers is necessary to collate specific primary data and improve life cycle models. Increasingly optimized life cycle products and processes will be developed in the future within the framework of development partnerships and open innovation processes, implying that, here too, more collaboration will be required outside the company.

1.5 Social and Ecological Responsibility in the Supply Chain

In times of value added partnerships across value networks, a company's responsibility does no longer end at its factory gates (Lee 2010). Serious penetration of individual supply chains is necessary in order to safeguard minimum social standards and sustainable handling of raw materials. Moreover, when defining and applying sustainability criteria and standards close collaboration between everyone

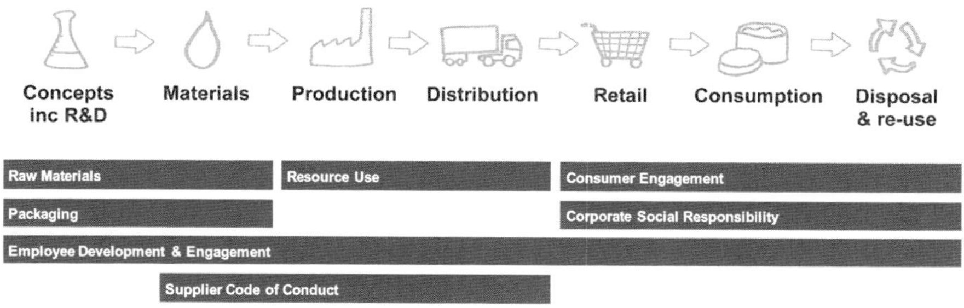

Fig. 4 Driving forces behind collaborative action. *Source*: Beiersdorf (internal)

involved in the value chain is crucial. For example, the process of extracting and producing raw materials is highly complex as supply relationships extend beyond industries. To define lean management systems which guarantee sustainable working of renewable raw materials, without jeopardizing the efficiency of downstream raw material production processes, multi-stakeholder dialogue has to take place.

Our aim of reducing our carbon footprint includes internal and external measures. However, using alternative resources in various processes also plays an important part. Both internal and external partners need to help us to achieve this aim for all processes along the value chain. For example, internal process optimization is needed beside collaboration with transport service providers, material suppliers and finished goods suppliers. Cooperation along the value chain is not a new idea; it has been seen in the consumer goods industry since the mid-1990s in various Efficient Consumer Response (ECR 2013) projects. The overall focus on the major global challenges posed by climate change and scarce resources offers a chance to overcome the limits set on previous collaborations by discussing the distribution of costs and benefits.

2 Beiersdorf Supply Chain

2.1 *From Functional Focus to Close Collaboration*

Beiersdorf's supply chain has changed from functionally orientated, nominally connected, and inward focused functions to a modern, demand-driven supply chain (Gartner 2012). At the same time, supply chain requirements in the consumer goods industry have tightened in recent years. Market cycles have shortened and the demand of large trade groups is increasing continuously. In addition, consumers have changed. While a few years ago, the trend towards large, global brands seemed unrelenting, local brands are now growing again and product ranges have to satisfy highly differentiated markets. At the same time, legal requirements are becoming increasingly stringent and sustainability gains importance. This calls for an increased

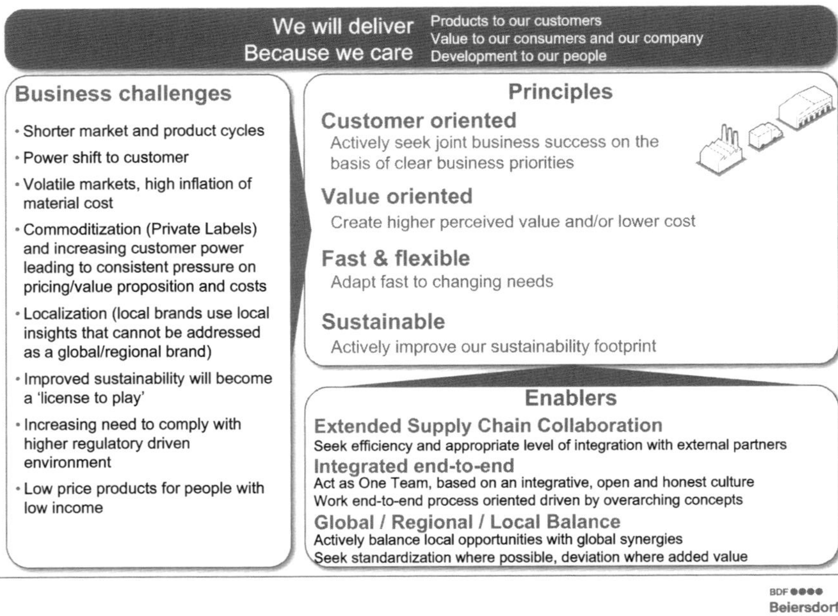

Fig. 5 Beiersdorf AG supply chain objectives. *Source*: Beiersdorf (internal)

level of flexibility in the supply chain, to respond promptly and comprehensively to wide-ranging demands.

It was in this light that we introduced our Blue Supply Chain Agenda (Fig. 5).

The self-imposed commitment of the supply chain is clear in the very first sentence of our mission statement: "We will deliver. Because we care."

First, the English word "care" is an important emotive reference to our brands, which care for skin. Second, we want to set a statement that "care" also refers to our consumers, customers, staff, and Beiersdorf as company, i.e., it is a far-reaching commitment to honor our responsibility towards our various stakeholders and, in the broadest sense, a commitment towards sustainable action. In order to keep developing in this direction, we have laid down four principles. Without going into the underlying strategy in detail here, it is nonetheless obvious that all four are important in terms of long-term, sustainable growth.

For example, "Fast & Flexible" includes production close to markets, which ensures that we are agile and resilient, while at the same time improving our sustainability by reducing transport and increasing local added value. We have also committed to three "enabler" principles. The very first promotes close integration and collaboration in the extended supply chain, i.e., with external business partners. We also took inspiration for the strategy from papers by leading academics, such as Martin Christopher and Hau Lee (cf. Christopher 2010; Lee 2010).

Overall, the strategy illustrates that sustainability is deeply embedded in our thoughts and deeds. As stated above, the job of cultivating and developing brands such as Nivea cannot be achieved with a short-term focus and demarcated management style.

We are convinced that efficiency and sustainability are not mutually exclusive but go hand in hand as they stand. On the one hand, very short-term efficiency measures often also improve sustainability (for example when transportation is reduced through optimized planning, better collaboration, and joint use of means of transport). On the other hand, our long-term approach forces us to concentrate our scarce resources on solutions that promise success in the long term. Mostly, these are solutions that do not inflict obvious disadvantages in terms of sustainability in the short run. Based on this, at times we debate in this article how to improve the efficiency of a process, which will simultaneously become a more sustainable process.

The most important core process in Beiersdorf's supply chain is the "Perfect Order Fulfilment" process as an end-to-end process.

Individual process steps are defined based on the Supply Chain Council SCOR Model (SCC 2011).

Figure 6 illustrates the flow of information (planning data) from the customer to the supplier as well as the physical flow of materials to factories, of products to warehouses, and to the customer. In this figure, the Source process relates to the call-off process; in the remaining article the term "Source" refers to the strategic procurement process.

Even though this article does not discuss purely internal measures to increase sustainability, for Beiersdorf to implement the fourth core principle of "sustainability" in our supply chain strategy (Fig. 5), various measures have been set-up, for example to reduce energy and water consumption.

The planning process is an important driving force behind any successful and sustainable supply chain. The planning procedure and its development are presented here briefly. We will subsequently focus on joint added value generated

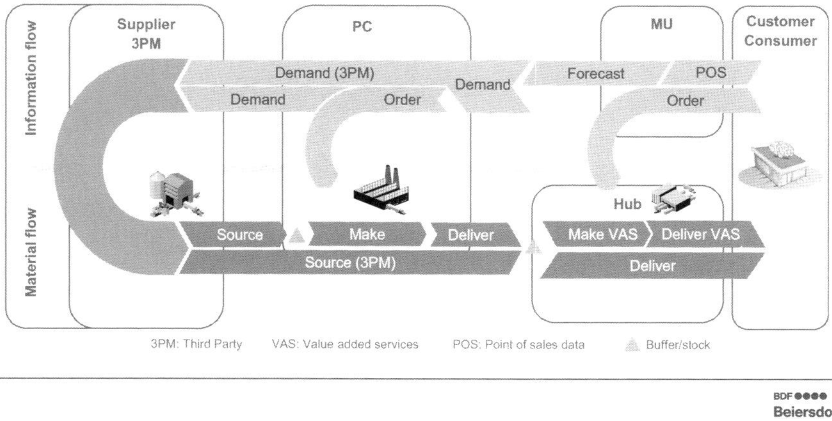

Fig. 6 Core processes in Beiersdorf supply chain. *Source*: Beiersdorf (internal)

with our partners in the extended supply chain; customers (downstream), suppliers (upstream) to service partners.

2.2 Planning as Core Instrument for Steering a Supply Chain

The model shown above illustrates an important element of our supply chain: we are a strict make-to-stock manufacturer. Normally, it takes 48–72 h from the receipt of an order to the delivery of goods to the customer's warehouse. These lead times can only be achieved at a tenable cost if the goods are stocked close to the customer and dispatched directly from there.

Accurate planning is crucial in order to supply our customers reliably and, at the same time, guarantee a highly efficient and sustainable supply chain. This becomes apparent from the following considerations:

- Products and materials which are produced according to an overoptimistic sales planning and which are not sold must be destroyed at a later date. This causes ecological and financial wastage.
- Raw materials, which are purchased but not processed as a result of discrepancies between procurement and production planning, are also ecological and financial wastage. This equally holds true when raw materials remain as a result of preproduction or production errors at suppliers. (and must be destroyed)

These considerations are trivial in themselves; however, the result is complex:

Tens of thousands raw and packaging materials have to be ordered and received from hundreds of suppliers. Several thousand finished products have to be planned and manufactured in various factories for them to be delivered to numerous customers worldwide. Proper planning is vital in order to meet these challenges every day.

The purpose of planning the sales volume is to collate all the information available to make predictions about the future. Traditionally this has been done using statistical prediction models, complemented by the experience of the planners in charge.

In the first step towards optimization, the quality of information for our volume planning was increased by improving internal collaboration. This was done through intensive and systematic collaboration between sales planners and the marketing and sales divisions. Consequently, we developed from an independent volume planning within the supply chain to an integrated business planning approach called Sales & Operations Planning. In order to make planning more accurate further customer information and consumer trends are used, for example from social media platforms. Expanding the information platform is challenging and has only been partially completed so far. The accuracy of sales planning is continually measured using performance indicators (KPI, BPI, PPI). Different drivers, such as promotions or new product launches and various prediction horizons, in the form of M1, M3, and M18 planning are analyzed (i.e., next month, in 3 months, in 18 months). These detailed measurements help everyone involved to continually improve the quality of forecasts.

The production plan is prepared based on the sales plan, which in return is used as basis for procuring raw materials and packaging. This is a well-known process, which initially runs only within the supply chain and is therefore relatively easy to organize, but rather complex in detail. Historically, suppliers and service partners were instructed at short notice based on master contracts and must then supply the goods and services ordered.

2.3 Added Value Through Collaboration with Customers

The challenges named in the strategy (shorter product cycles, greater volatility, pressure on costs) require a highly developed collaboration with customers towards joint planning processes. Inventory is optimized via the various value added steps in terms of associated costs, security, and agility.

Beiersdorf's supply chain was based historically on a very transactional understanding of collaboration with customers. Order management was part of the Sales Organization and Process. Where necessary the sales department discussed issues with the customer and dealt with delivery problems. The supply chain was responsible for the execution of supply. Even in our customers' firms, the supply chain organization was not always optimally represented by their buyers, who often have only basic understanding of supply chains.

However, from a supply chain perspective, our customers are the beginning and end of our supply chain. Close collaboration between the customer's supply chain and our own supply chain is therefore a basic driving force behind improving overall efficiency. Subsequently, order management now belongs to the supply chain organization. Step by step we are strategically converting order management into more broadly focused supply chain customer management. This means reducing the cost of order management by improving processes and systems, the resources which this frees up are used for supply chain customer management. One element of these improved processes and systems is the electronic product and supply data exchange established with numerous customers. This reduces errors in order management and deliveries and hence the number of complaints, allowing standard transactions as a whole to be highly automated.

Today we have already significantly reduced order management resources in the first countries. For major key accounts dedicated supply chain resources collaborate closely with the customer's supply chain counterparts. Meetings on optimum stock levels, delivery times, or orders are held for the purpose of collaborative optimization of the supply chain. This close cooperation increases the delivery service, sometimes even at reduced costs.

The information generated during the course of this cooperation is used directly for sales planning at customer level. Besides this, detailed customer data, for example from till scanners is also being used to identify trends early on. Precise information about the flow of goods in commerce allows sales planning to be more accurate.

Key performance indicators, which are used, differ considerably, depending on whether they are measured by manufacturers or trade partners. Measuring the "Service level," for example, may differ by 10 % or more. These discrepancies are mainly the result of definitions, for instance when orders could not be delivered because they were processed with wrong master data. Indicators need to be set and defined clearly, as this allows the two supply chain teams to focus on identifying and resolving weaknesses in the chain.

One challenge in achieving closer collaboration with customers lies on the technical side, in the procedures and systems used to process mass data. This, however, is an issue which can be dealt with through suitable project management.

A more serious challenge lies in achieving collaboration beyond functional and corporate boundaries. Internal functions (marketing, sales) expect the supply chain to have an excellent delivery service at low cost, but they sometimes have difficulties in recognizing their own responsibility within the supply chain and to seriously work on it. In external collaboration with customers (win/win), this willingness is sometimes even more limited, though a basic commitment to partnership exists. Joint added value often requires an initial investment in personnel and systems, from which benefits are then generated over time. Both parties need to go beyond the limits of traditional relationship, which is often marked by stiff competition (win/lose). Employees sometimes need to change their approach in order to identify potential added value opportunities.

Collaboration to generate joint added value with customers is therefore under way and good examples of functioning collaboration exist. Nevertheless, despite intensive discussions at conferences and in associations, day to day collaboration is still very often dominated by competition and demarcation.

2.4 Collaboration with Suppliers

As global manufacturer of various groups of products we have numerous and diverse suppliers, which makes collaboration with our partners versatile. As a company, which has always placed importance on a partnership-driven approach, we have established relationships with numerous suppliers since many decades. These relations are often marked by trust and collaboration. However, the nature of more extensive collaboration with suppliers depends on a range of factors. Somewhat simplified, our suppliers are evaluated in a matrix with strategic importance on one axis and maturity on the other.

Two particular cases can be highlighted here: global, specialized suppliers with a high degree of maturity and small, local suppliers with very simple structures.

Regarding global, specialized, and highly-developed suppliers, we rely heavily on extensive integration of processes and systems. There are just a few of these suppliers within a product group and they are usually open to extended collaboration. At the same time, however, long-term partnerships need to be checked regularly regarding their profitability. Procurement takes up a leading position in the supplier relationship management. Cross-functional purchasing teams define the requirements for a specific product group, including logistical and transactional requirements, such as quantity, quality performance indicators, supply forms and frequencies, lead times, and quantitative flexibility. Goods are put out to tender on this basis with strategic partners, where necessary with the addition of new suppliers. Contracts are executed either for a specific term or for the entire life cycle of the product.

The transactional process is then moved to an electronic platform to exchange master data, orders, order confirmations, and delivery advices. With strategic suppliers, we gradually switch from a make-to-order process to a make-to-stock or vendor managed inventory system. This involves providing suppliers with volume estimates via the same integrated platform, so that they can independently plan their production process even earlier and synchronize it with our requirements. The most progressive suppliers integrate their own systems with the platform, meaning that data exchange is efficient and reliable. Here again, the resources, which are freed up (compared with manual data exchange via fax and email) can be used to identify ways of improving the joint supply chain process. With more extensive integration and more coordinated supply chain processes (including financial processes such as payments), the overall efficiency of the partnership improves and so does its longevity. The product development process is not discussed in this paper; but it ought to be mentioned that these suppliers are increasingly involved in the innovation process of Beiersdorf and are therefore important partners for product development.

Collaboration is regularly measured using key performance indicators (KPI), based on a supplier evaluation, for which the procurement department is responsible. Suppliers are evaluated according to various qualitative parameters such as reliability, quality, and cost. Several internal contacts, for example the factories in which most contact takes place on a day-to-day basis, are asked to provide structured feedback. Central quality management is another source of information. This gives us the opportunity to establish how a supplier deals with his own errors and issues, how well he can resolve them in a sustainable and quick fashion. Moreover, it informs us about the level of support a supplier offers us in exceptional situations, for example when we have additional demand outside the agreed lead-times or when we have to cut back demand. This evaluation is discussed openly and intensively with suppliers and impacts on the position of the supplier in our network in the medium term.

Collaboration with small, less developed partners is very different. This may be collaboration with individual farmers in Europe who cultivate a particular natural raw material for us, collaboration with an agricultural cooperative in South Africa from which we buy ecologically certified Argan oil or collaboration with small industrial suppliers, e.g., of glass or plastic jars in developing regions. These partnerships are based on support in basic business processes rather than extensive integration of processes and systems. Where necessary, these partners are trained in the quality processes. Sometimes they are trained in manufacturing processes but more often in international business procedures in general. As a result, we therefore help these partners to become more professional and to expand their business in the long term. For Beiersdorf this means that we have reliable and loyal suppliers in emerging markets that we need for strong growth in these regions.

Therefore, in an ideal case, a symbiotic relationship is established in which close integration gives rise to a certain mutual dependence and increased switching costs for both partners. However, mutual benefits also improve, in the form of improved costs, high service levels, and simplified business processes based on well-established teams and processes as well as joint crisis management. Contracts are good, trust is better.

2.5 Collaboration with Service Providers

In addition to close collaboration with customers and suppliers at both ends of the supply chain, we also rely on collaboration with various service providers. Many of the processes described for suppliers also apply for service providers in the supply chain. As with suppliers, we evaluate service providers in terms of their strategic importance to us and subsequently decide on integration and collaboration models.

There are different types of collaboration of which two examples are given here. One group of service providers are container shippers. Even though we are producing increasingly close to markets, we still need to ship products and materials by container regularly. For shipping companies, our demand is very small. Container shipping offers a highly standardized and regulated product: shipment of a container from port A to port B. As a small customer on the market, we do not pursue close collaboration with individual shipping companies; instead we try to achieve good costs through tendering procedures or by buying services based on daily rates (spot buying).

A different situation exists with service providers, which operate warehouses for us (third party logistics or 3PL). With 3PLs we need a very closely integrated system in order to exchange data on a daily basis. Logistics partners must be familiar with our processes and products if they are to serve the strict requirements of our customers. These services are also put out to tender, usually every 3–5 years. In addition to logistics services (warehouse and dispatch), 3PL increasingly perform other services, such as manufacture of promotional packaging, which increase both our integration and our collaboration. Our strategic objective is to partner with 3PL providers, which can bundle several manufacturers (including direct competitors) with similar customers. By bundling orders of several manufacturers in one warehouse, logistics partners can deploy staff across several manufacturers and thus straighten out order peaks. When supplying the same customers, they can bundle different manufacturers' orders in the same consignment to the customer. This reduces costs as well as the number of trucks on our streets. Collaboration extends beyond 3PL here, as different manufacturers must synchronize their delivery windows with the customer for the purpose of bundling consignments.

3 Conclusion

Modern supply chains collaborate in complex, global supply chains with numerous partners. In fact, the term value network is nowadays a better description than supply chain. There are two fundamentally different approaches when it comes to adding value at the various interfaces between in-house and third-party services: short-term transactions and longer-term partnerships. In most companies, these approaches operate simultaneously, as no one can generate sustainable joint added value via extended collaboration with all suppliers and partners at the same time. In some areas, there is no identifiable added value, which improves on the simple exchange of goods for money.

However, current developments confirm what we have always believed:

- Collaboration generates considerable joint added value.
- Long-term collaboration reduces internal staff and management costs compared with knee-jerk action.
- Strong networks, based on solid relations, are more resilient, as all partners have a joint, vested interest in overcoming crises.
- Staff satisfaction improves, as staff feel they can deal legitimately with partners and the workload caused by short-term crisis management declines.
- Even good partnerships must be evaluated regularly (through KPI) and tested (e.g., tendering procedures) in order to ensure that partnerships are objectively assessed and are not based on mere trust or habit.

Bibliography

Beiersdorf. (2013a). *Data & facts; Beiersdorf AG*. http://www.beiersdorf.de/Investoren/Daten_Fakten/Aktuelle_Daten.html. Accessed 11 June 2013.

Beiersdorf. (2013b). *Press release by BDF AG dated 26 March 2013, Beiersdorf AG*. http://www.beiersdorf.de/Presse/Pressemitteilungen_News/Personell_Ver%C3%Aenderungen_im_Beiersdorf_vorstand/html. Accessed 11 June 2013.

Beiersdorf. (2013c). *Sustainability reporting and Beiersdorf AG, Beiersdorf AG*. http://www.beiersdorf/de/Nachhaltigkeit.html. Accessed 11 June 2013.

Christopher, M. (2010). *Logistics and supply chain management* (4th ed.). Financial Times/Prentice Hall.

ECR. (2013). *What is ECR?* http://ecr-all.org/about-ecr-europe/. Accessed 11 June 2013.

Euromonitor. (2011). *Skincare brands. Sales turnover 2011*.

Gartner. (2012). *The Gartner framework for aligning global logistics with a demand-driven supply chain strategy*. http://www.gartner/com/DisplayDocument?doc_cd=239631. Accessed 11 June 2013.

Lee, H. L. (2010). Don't tweak your supply chain – Rethink it end to end. *Harvard Business Review, 88*(1), 62–69.

SCC. (2011). *Supply Chain Operations Reference (SCOR) Model*. In Supply Chain Council. http://supply-chain.org.scor. Accessed 11 June 2013.

Fairphone: Sustainability from the Inside-Out and Outside-In

Tessa Wernink and Carina Strahl

1 The Human Cost of Technology

1.1 Sustainability in Technology Products: A Complex Issue

The last few years have seen a growing demand in fair trade, organic and sustainable products, which can now be found in nearly every supermarket in the developed world. Events such as mad cow disease in the 1990s or the recent European horse meat scandal caused a drastic increase in public awareness about food production. The devastating collapse of a Bangladesh garment factory in 2013, which resulted in more than 1,000 injuries (CNN 2013), and wide-ranging criticism of sweatshops had a similar effect on the fashion industry. These events, as well as the recent economic crisis, have increased and accelerated public awareness and made consumers more conscious about their purchases.

While fair trade or organic food is now quite common, this trend has not carried over into the technology industry. The way consumer electronics are sourced, produced, and assembled, and whether these steps incorporate "fair" principles remains outside people's perception. This remains true, despite the fact that consumer technology (including mobile phones) plays a significant role in our everyday lives. In 2013, more people owned a mobile phone than had access to a working toilet (Time 2013). In 2014, it has been predicted that the number of mobile phone subscriptions will surpass the total number of people on the planet (BBC 2013).

T. Wernink (✉)
Fairphone, Piet Heinkade 181A, 1019 HC Amsterdam, The Netherlands
e-mail: tessa@fairphone.com

C. Strahl
shared.value.chain, Adams-Lehmann-Str. 56, 80797, Munich, Germany
e-mail: cstrahl@sharedvaluechain.com

M. D'heur (ed.), *Sustainable Value Chain Management*, CSR, Sustainability, Ethics & Governance, DOI 10.1007/978-3-319-12142-0_3

There is of course a significant difference between these two "commodities." Most consumers view their electronics as a single product. In reality, however, a smartphone includes about 40 different minerals, such as tantalum, tungsten, copper, iron, nickel, aluminum, lead, tin, silver, chromium, gold, cobalt, and palladium. Each mineral performs a different role that is essential to the phone's functionality. For example, every modern smartphone uses tungsten in the vibration mechanism; tantalum is required to make the capacitors smaller; cobalt is used in rechargeable batteries.

Besides the processes involved in production, all the components, parts, and minerals have different origins and a unique supply chain. Tracing where, how, and under which conditions each of these different elements are produced is a challenging exercise, not made any easier by the global nature of supply chains and the multitude of players. The process – and therefore the product itself – is a collection of supply chains that stretch across the globe. These supply chains are opaque to buyers, and even manufacturers rarely have a complete overview of all the connections in their supply chains.

An experiment by designer Thomas Thwaites illustrates the huge complexity of consumer products and their materials, minerals, metals, and components (TED 2011). In an effort to build a toaster by himself, Thwaites disassembled a basic toaster and laid each component side-by-side. He demonstrated that even the cheapest, simplest toaster consists of around 400 different parts, all of which are manufactured separately. While Thwaites failed to actually build a working toaster, his attempt provided insight into the convoluted nature of manufacturing processes. His experiment also showed that consumers know very little about the circumstances under which materials and components were sourced or manufactured. Given the low retail price of Thwaites' original toaster, one can also assume that not many people benefitted financially from its production.

1.2 The Effects of Mass Technology Production

The fact that consumer technology is now seen as essential has made most people turn a blind eye to its human and environmental costs. Conflict minerals, environmental problems, poor production conditions, and increasing amounts of electronic waste (e-waste) are just some of the results of technology production.

Conflict Minerals A considerable percentage of the world's minerals come from developing countries. In states where the federal government is weak, these natural resources, which are often critical to local economies, may lead to conflicts between state and non-state actors over control of the mines and their related revenue. Conflict minerals are therefore defined as minerals "*whose control, exploitation, trade, taxation, or protection contribute to, or benefit from the context of, armed conflict*" (Le Billon 2003, p. 216) as rebel groups operating in mineral-rich areas may finance their struggle through the revenues from mining (Bryceson and Jonsson 2014).

The best-known example of this is the prevailing situation in the Democratic Republic of Congo (DRC), which the United Nations described to be one of the worst humanitarian crises in the world with an estimated six million Congolese being killed since 1998 (Jeffrey 2012). Here the sale of minerals, especially coltan, a source of tantalum, fuels fighting (Melcher et al. 2008). With 7 or 8 % of the world's coltan supply, this mineral is essential for the economy of the DRC (Seay 2012). Public awareness of conflict minerals has in part contributed to the "OECD Due Diligence Guidance for Responsible Supply Chains of Minerals from Conflict-Affected and High-Risk Areas" and the "Dodd–Frank Wall Street Reform and Consumer Protection Act" (Dodd–Frank Act). The former, adopted in 2011, provides detailed guidance for companies to respect human rights and to avoid support of conflicts through their mineral purchases (OECD 2014). The latter, passed in the United States in 2010, requires – according to section 1502 – publically listed companies to report on their use of conflict minerals from the DRC and its adjoining countries. It was predicted that this legislation would have an effect on around 6,000 companies (Narine 2013). However, by limiting the definition of conflict minerals to only gold, tantalum, tungsten, and tin sourced from a single region, the Dodd–Frank Act leaves much room for improvement. An integrated, voluntary approach to stop profits from mineral trade being used to finance armed conflicts was recently proposed by the European Union (EU) (see also Jeffrey 2012). To strive towards transparent supply chains the focus will be set on tin, tantalum, tungsten, and gold coming from all over the world (EUROPA 2014; Financial Times 2014).

Environmental and Health Effects of Mining The effects of mining on the environment and human health are also significant, especially in countries that do not have government regulation regarding mining practices. Fifteen million people, for example, work as artisanal small-scale gold miners (ASGM) and use mostly mercury during gold extraction as this is a cheap and simple method (Gibb and O'Leary 2014). Using mercury can result in serious health problems for the entire region, polluting local water sources as well as the local food supply (Gibb and O'Leary 2014; MIT 2014). The effects of ingesting mercury are extremely serious: It can cause tremors, insomnia, memory loss, neuromuscular effects, and cognitive dysfunction (WHO 2013). As mining is often done by women, mercury poses a particular threat to those who are of childbearing age or pregnant (Schmidt 2012).

E-Waste The growing number of people using electronic devices combined with rapid product innovation and replacement cycles is causing e-waste to become a severe issue (StEP 2014). Most developed countries have established e-waste recycling programs in place. This is not the case in developing countries. E-waste contains numerous substances such as heavy metals. These become dangerous when not handled properly, as is often the case in the informal sector of developing countries (Schluep et al. 2013; Akormedi et al. 2013). To obtain steel, copper, or aluminum e-waste is frequently burnt in the open, which results in high levels of toxins in air, water, and soil (Schluep et al. 2013). Similarly, primitive e-waste recycling exposes children to high levels of lead, which can cause adult osteoporosis (Yang et al. 2012). It is not only the lack of recycling facilities in developing countries which contributes to e-waste problems, but also the fact that companies from the developed world actually export used or broken electronic goods to these countries (Guardian 2013b). Since 1996 trade in e-waste has more than doubled and now surpasses one billion kilograms; the US, for example exports 50 to 80 % of its e-waste rather than recycling it domestically (Lepawsky 2014). Through the Waste Electrical and Electronic Equipment (WEEE) Directive, the European Union (EU) has adopted clear guidelines to help national port authorities to tell the difference between secondhand goods and e-waste (EUROPA 2010).

1.3 Business Model Choices Are the Drivers

When evaluating leading technology manufacturers and their respective business models, it becomes clear that their core business and values greatly differ. Some companies focus on design leadership and innovation and others on affordable pricing. The different models that drive decision making and internal processes have a major effect on the products these companies bring to market. While their products are different, manufacturers have common goals: to be commercially successful, increase market share and shareholder value.

Today most companies are addressing sustainability in one way or another, but it is often an "add-on" to current business practices. Besides a few corporate social responsibility initiatives, it is mostly "business as usual." Within large corporations, sustainability is often understood in terms of money. For example, decreasing energy consumption can be understood internally as a way to cut costs and improve branding while also reducing their environmental impact. Sustainability is frequently connected to the marketing department, which promotes products (not company practices) as "green" or "eco-conscious," making it difficult for consumers to distinguish between real, lasting changes in practices and efforts that simply support sales targets. However, more companies are now incorporating sustainability into their corporate principles – realizing that technology production brings with it great, long-term, and systemic problems. Many companies now understand that solving these issues takes more than just the production of a few goods in a different manner.

Fairphone is a social enterprise that has made fairer production its mission. The following sections outline Fairphone's objectives, how it operates, and what drives its choices. Fairphone aims to encourage consumers to think about and develop a new relationship with their products and works together with the industry to create systemic change through the production of a smartphone.

2 Fairphone: Fairness as Core Belief

2.1 *Starting Fairphone*

In 2010, Waag Society, a Dutch organization that utilizes technology for social innovation (Waag Society 2014), was approached by ActionAid Netherlands, an NGO campaigning for the fair distribution of mineral wealth in sub-Saharan Africa, and Schrijf-Schrijf, a communications company. They proposed to jointly draw attention to the connection between conflict minerals and consumer electronics. As a result, Bas van Abel, Creative Director at Waag Society, started Fairphone with two colleagues in March 2010 as a nonprofit campaign within Waag Society. With a background in Interaction Design, van Abel hoped to raise awareness for conflict minerals by engaging designers, creatives, and technical experts in the process of making a phone based on fair principles. The idea was an open design platform that applied the principles of open source software to hardware design. He saw Fairphone as a design challenge and a campaign, with the phone serving as a storytelling artifact to connect both makers and users to the processes behind production.

In 2010, a Danish documentary film claimed that the mobile phone industry relies on conflict minerals from the DRC, yet isn't taking responsibility for the environmental degradation and human suffering caused by mining practices. Called "Blood in the Mobile," this documentary amplified the conversation already in full swing in the US through the Enough Project and accelerated Fairphone's presence in the media. Van Abel started to appear in the Dutch press and took part in debates about conflict minerals. In 2011, Fairphone also went on a fact-finding mission to the DRC and produced its own documentary about it. As awareness for Fairphone grew, an active community began following its progress on the Fairphone website and through social media.

Remaining independent from investors and staying true to its principles has been essential to Fairphone's vision since its foundation. From the beginning, Fairphone's goal was to make an actual phone. In 2012, Fairphone was awarded a place in the Bethnal Green Ventures accelerator program in London and formulated its business strategy. A few months later, it received its initial funding and became a social enterprise[1]. With an initial investment of EUR 400,000, Fairphone had sufficient funding to set up and operate the company for approximately 5 months. To move forward as an independent company, Fairphone chose to focus on crowdfunding, asking supporters to prepay for the new phone as a means to raise funds to start production.

Within this framework, Fairphone began selling its first (virtual) product with an online campaign. People could preorder Fairphones at EUR 325 each through the Fairphone website, and be part of bringing the project into its next phase. To raise

[1] A social enterprise is a company for which sustainability of income and a social mission are the ultimate objectives and this should be evident in every decision taken (Guardian 2014b).

money for the down payment to the manufacturing partner, and to prove that there was demand for the product, Fairphone aimed to sell 5,000 phones within the first month. By the end of the crowdfunding campaign, it sold 10,186 phones – more than doubling the target. This provided 3 million euros in funding, which was more than enough to begin production. The trend continued and the demand for Fairphones grew. Over the following 6 months, a total of 25,000 phones were sold. Fairphone sold out of its first batch, even before the first phone had been produced. This allowed the company to remain independent and showed that consumers could actively be part of the change.

Between December 2013 and January 2014, 7 months after the first buyers had purchased their phones, the first batch of 25,000 finished smartphones was shipped to customers in Europe.

2.2 The Fairphone Idea

Fairphone wants to address the increasing human and environmental costs of technology and the growing need to know what actually goes into the making of a product. It wants to view the economy from the perspective of "fair." By producing a smartphone, Fairphone sets out to make the entire value chain transparent and raise awareness for the numerous components, minerals, and production steps that are currently unclear to buyers, and often also to companies. Its smartphones are designed, created, and produced to uncover the story behind sourcing, production, distribution, and recycling processes and to make interventions that create positive impact.

Fairphone's operations are based on drivers that differ greatly from "normal" mobile phone producers. It does not merely want its practices to be "less bad" while securing more market share, but it wants to create positive impact with its activities – promoting and pioneering a framework which it believes will lead to business models that support an economy based on fair principles. Fairphone wants to motivate the entire technology industry to scale up best practices, act responsibly, and truly create positive impact. Social change, not financial growth or technological innovations, are therefore at the forefront of its actions. This can also be seen in Fairphone's business model (Fig. 1).

Fairphone decided to take a commercial approach because it believes that only by being part of the industry and designing, creating and producing a technological product can its value chains be laid completely open and changed for the better. Fairphone therefore understands the need to connect the development of a product to its value chain when making choices for sustainable growth.

Nevertheless, "fair" is a subjective human value and its perception is often in the eye of the beholder. Therefore, "fairness" is taken as a starting point for conversations, hoping to lead a debate for change, it centers on the human and environmental benefits and takes a holistic approach to the activities that make up our economy. In fact, Fairphone emphasizes that the phone is not currently "fair." Actually, the name "Fairphone" was not chosen to indicate that the product already

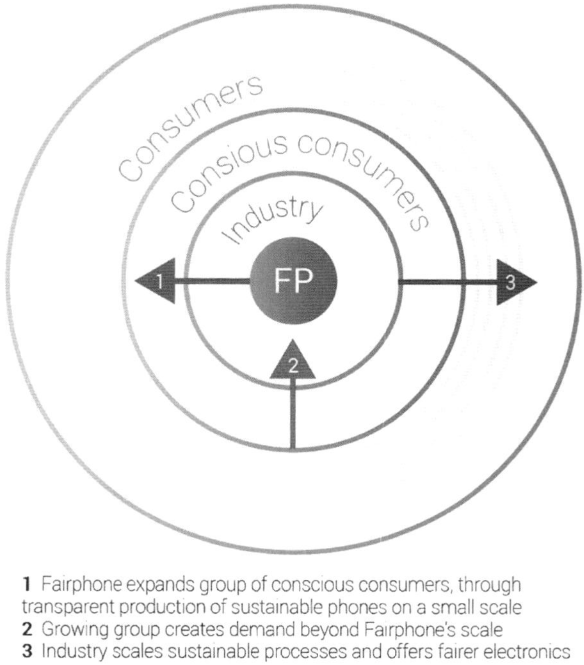

Fig. 1 Fairphone's business model

is fair, but to state its ambition and stimulate conversations about what "fair" is. Fair denotes the strategy and not the phone itself.

Fairphone sees itself as being on a journey towards a fair economy and has so far been able to translate its vision into becoming a social enterprise, a product, and a growing movement. Smartphones were selected as a focal point because they are ubiquitous products which are used globally in developed and developing countries alike. A phone is a perfect symbol of our interconnected world, and simultaneously shows a paradox in the consumer's lost connection to the source; we no longer understand how a phone is made nor the social consequences of its production. Being much more complex than a toaster, a smartphone consists of thousands of different parts (Guardian 2014a) each with their own function and purpose, and accompanying production processes and value chains. It is therefore a great storytelling device to uncover these processes and to start redesigning our economy with different values in mind.

3 Fairphone: Designed for Sustainability

The design of an electronic product is absolutely crucial. Design determines its purpose, content, and use as well as options for recycling and disposal. Design can be used to determine the product's features and extended to involve the processes that are involved in its conception. A designer can pinpoint problems in the value chain and take responsibility for its impact by developing sustainable solutions. Owing to the connected nature of the design process, it becomes visible that all supply chains are connected and cannot be considered separately.

From the beginning, Fairphone's ambition was to address the entire lifespan of a mobile phone – all steps and elements that are connected through its initial design. Responsible, open design is the guiding principle throughout the process, considering sustainability in every step before the first product is actually produced, and therefore a driver for positive change. While design determines the components for all electronic devices, Fairphone wants to select components according to sustainable principles as well as functionality. Therefore, the decision to encourage social innovation drives product design.

Designing a phone from scratch is enormously expensive and was not feasible, nor the priority for the first generation Fairphone with its small production run and ambitions of independence. With this in mind, Fairphone decided to use a licensed exterior hardware design with additional technology upgrades for the first Fairphone.

Although the licensed model offered certain restrictions, Fairphone was able to select many features that extended its lifespan and improved reparability and recyclability. In doing so, Fairphone strived to reduce the environmental and social impact of e-waste and mining. The underlying belief is based on the maxim from the Maker's Bill of Rights that states "*If you can't open it, you don't own it.*" Currently, many consumer electronics are impossible to open and repair, leading to a constant demand for new products. This leads consumers to become dependent on manufacturers. If they cannot fix or tinker with the things they bought, they never truly own (or control) the products they buy. Another contributing factor is the frequently discussed matter of planned obsolescence, which "*…is a business strategy in which the obsolescence (the process of becoming obsolete – that is, unfashionable or no longer usable) of a product is planned and built into it from its conception. This is done so that in the future the consumer feels a need to purchase new products and services that the manufacturer brings out as replacements for the old ones…*" (Economist 2009).

To increase reparability, Fairphone decided that the phone's batteries should be removable and replaceable using normal screws. While the electronics industry generally only sells spare parts to other businesses, Fairphone was the first to offer spare parts like batteries or camera modules directly to customers through their online shop. To facilitate repair, Fairphone partnered with iFixit, an organization renowned for its repair guides for electronics, to draft repair manuals that customers can download.

To reduce its environmental impact, the Fairphone was shipped with minimal packaging. The phone did not include a charger. Because it can be charged with a USB cable, a common adapter for other electronic products, Fairphone asked buyers to check their existing cables before ordering one. Making people more aware of the things they own, and not shipping accessories by default, was a first step in taking responsibility for the potential e-waste Fairphone is producing.

Likewise, including dual SIM capability allowed consumers to have only one mobile phone where previously two phones were needed – for instance for private and for professional use.

Another way Fairphone started to take responsibility for adding more products to the electronics supply chain was to add a premium of EUR 3 on each phone it made in the first batch. These funds are used to set up projects in countries where safe electronic waste recycling does not yet exist. With the funds from the sales of the first phone, Fairphone is collaborating with Closing the Loop, a foundation which buys scrap from Ghana and ships it to Europe for proper recycling. The sales of the first batch of phones allowed Fairphone to contribute to the recycling of 75,000 phones (around 4 tones) of scrap phones at a recycling plant in Belgium. The ultimate ambition is to use the minerals "mined" from these phones for the production of future Fairphones, effectively "closing the loop" in the supply chain and embracing the ideas of a circular economy. Initially, this is likely to be a "mass-balance" system. The recycling partner will sell the recycled material on the world market and Fairphone reaches out to suppliers using recycled materials.

The second generation Fairphones, scheduled for production in 2015, will be designed from scratch rather than using a licensed model. This move will provide even greater transparency in Fairphone's supply chain and the organization will be able to extend its social and environmental impact. A major design focus for the next generation phone is longevity and reparability. The design phase intends to apply open design thinking, inviting input from a group of users and community members (NGOs, buyers, partners, etc.). An example of this is the inclusion of iFixit at the beginning of the design trajectory. Partnering with an organization that understands repair will make it easier to incorporate internal modularity in the final design and influence the choice of components.

4 Influencing the Supply Chain

The complexity of smartphones means every single mineral and component has its own unique origins and supply chains, making tracing "one" smartphone supply chain nearly impossible. In addition, supply chains are not static, companies can easily change partners and suppliers making it harder to trace materials all the way to the source. Yesterday's supply chain may not be today's supply chain. One of the most challenging issues in commodity supply chains is that there are no established tracking and tracing systems that enable minerals to be followed to their source. It is therefore extremely difficult to tell where materials actually originate from, under what conditions they were mined and who benefited from mining processes. Through its website, Fairphone is taking the first steps, to expose its entire value chain to uncover all the processes that lay underneath. For instance, Fairphone lays open the costs of producing a Fairphone (Fig. 2). Throughout its supply chain, Fairphone aims to select partners that share its values, allow room for improvement, and are open to collaborative approaches. It wants to work with suppliers that offer transparency into their business operations, first and second tier suppliers, and by doing so help trace

Fig. 2 Cost breakdown of the first Fairphone (Fairphone, 2013a)

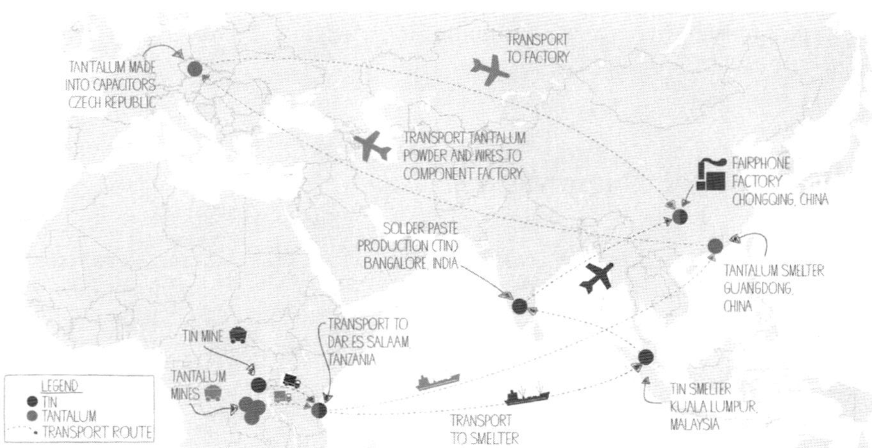

Fig. 3 The Tin and Tantalum Route. (Fairphone 2013b)

minerals and materials back to the source. Transparency creates positive change in an opaque industry and allows end users to better understand where their products come from, who is involved, and what the underlying conditions are. This can be illustrated by Fig. 3, which shows the global tin and tantalum route.

4.1 Fairphone's Approach to Mining

As the result of the Dodd–Frank Act and subsequent reporting requirements, many companies have started to completely avoid mineral purchases from the DRC, which is why one can even talk about a "de facto embargo" here (Jeffrey 2012). Countries like the DRC that rely heavily on mining revenues will have a multitude of residents with serious economic and social problems if all mining companies left or stopped working. In fact, as a result of regulations in line with the Dodd–Frank Act, smuggling has caused up to two million artisanal Congolese miners to be out of work (Seay 2012). For example, the Malaysia Smelting Corporation (MSC), which used to buy up to 80 % of its tin from the eastern DRC, left the country, causing huge effects on miners and their families, who depend on this income for their subsistence (Seay 2012). While the Dodd–Frank Act is certainly a step in the right direction, Fairphone believes that the conflict itself should remain part of the debate. Conflict-free should not mean that businesses boycott conflict areas as a potential source for their minerals, as pulling out of these areas can have adverse effects, such as decreased employment in key economic activity and pushing mineworkers into the informal sector. That often gives them the choice to either be a mineworker that is exploited by (armed) superiors, or start exploiting others as well. To fully address the problems of conflict materials, Fairphone therefore believes that there should be better global tracking, legislation, and investments made in high-risk areas to provide safe and fairly paid mining.

Fairphone's aim is to only use conflict-free and fairly mined minerals in its products. To find a solution, it has taken the first step in conflict-free sourcing from the DRC by joining the Conflict-Free Tin Initiative (Solutions Network 2014a) and the Solutions for Hope Project (Solutions Network 2014b). These two initiatives were brokered and funded by the Dutch Foreign Ministry and developed further through a multi-stakeholder initiative of Tata Steel, Blackberry, Fairphone, Royal Philips, Motorola Solutions, Apple, and others (Guardian 2013a). They create demand for tin and tantalum sourced from conflict-free certified mines in the DRC.

Through these initiatives, Fairphone has the potential to support small-scale miners and contribute to economic development, improved labor conditions, and regional stability. For example, local people have the opportunity to work for the Conflict-Free Tin Initiative rather than the local militia, enhancing their safety and ensuring a steady income. Fairphone hopes that by participating in these initiatives and encouraging them to expand the scope of improvement (focusing also on fair wages and child labor), it will pave the way for increased demand for responsible mining initiatives, transforming the mining sector in regions that face ongoing or high risk of conflict.

As a result of its cooperation with the Conflict-Free Tin Initiative and the Solutions for Hope Project, Fairphone can confirm the origins of two minerals – tin and tantalum – used in the soldering paste and capacitor respectively of its first generation smartphone. However, it cannot trace any of the other 30 to 38 minerals that were used. What is more, while two minerals are conflict-free, Fairphone cannot guarantee that no child labor was used in its sourcing. When it comes to labor rights and health and safety conditions in the mines, there is much room for improvement.

While already the first generation Fairphones laid open the value chain, the second generation will go even further in this process. The aim is to increase traceability of minerals as well as source minerals for components from fair trade mines and conflict-free mines in the Congo. Fairphone is currently researching how to use fair trade gold in its phones, which, if successful, will make it the first technology company to do so.

4.2 *Manufacturing and Assembly*

Fairphone wants to establish reliable, transparent relationships with its suppliers to ensure safe conditions, fair wages, and collective bargaining power. Fairphone promotes the conventions of the Ethical Trading Initiative (ETI) and International Labor Organization (ILO) which cover a broad range of topics such as labor rights, preventing forced as well as child labor, equal remuneration, fair working hours, and the right for collective bargaining. Though many countries have signed these conventions, the actual situation is often very different; workers are not being paid a living wage, they work for long hours and do not have the right to be part of labor unions or otherwise raise their concerns.

Rather than taking a compliance approach, Fairphone wants to work collaboratively at uncovering the complexity behind these systemic issues and making a change, step by step. For the production of the first generation phone, Fairphone researched the possibility of manufacturing in Europe. In the end, this idea was abandoned as the supply chains for the production of mobile phones are mainly in Asia. Producing in Europe would require a large initial investment – something Fairphone did not have, or want to attract, given its aim to stay independent. In addition, the relatively low product quantities made it impossible for Fairphone to be produced in Europe. More importantly, even if the phone could be produced in Europe, its components would still come from Asia. Producing in Asia was the best choice to support Fairphone's mission of creating positive impact where change is needed urgently.

In selecting a manufacturer, Fairphone wanted to ensure that this partner understood Fairphone's values and ideals. The right partner needed to be interested in a long-term relationship and should be willing to open up their own supply chain to Fairphone. After visiting several production sites in China, Fairphone selected Guohong, a small factory in central China that produced phones for the larger, better known electronics brand, Changhong, but did not have experience producing phones for the European market. Its workforce consists of less than one thousand people, and its location in central China means it employs a local workforce and fewer migrant and temporary workers. Despite the low quantities Fairphone expected to make (20,000 at the time), Guohong was keen to work with a start-up, even though Fairphone had never made a phone before and had no substantial funds to show. After receiving the first down payment, Guohong opened up its factory to a social assessment program and agreed to publish the results because it maintained that it was interested in enhancing the situation for its employees.

Although Fairphone has put in a significant amount of work into deciding on a manufacturer, the production circumstances cannot currently be described as being satisfactory or fair. Seeing it as work in progress, Fairphone decided that the initial focus should be on better representation channels and improving working conditions through training. Together with Guohong, Fairphone set up a Worker Welfare Fund. The factory's workforce elects worker representatives who then design projects for the fund (e.g., skills development training, bonus pay-out, entertainment/leisure, improved conditions on the work floor) on their behalf. This can motivate workers to identify and express their needs and increase the dialogue with peers and superiors. The worker representatives draw up and present proposals to a board that consists of a Fairphone, a Guohong management, and a worker representative. The workforce is trained and guided before, during, and after the election of the worker representatives, as to most this is unchartered territory.

5 Engaging with Stakeholders

Engaging positively with its multiple stakeholders is a guiding principle for Fairphone. It does so very proactively and takes input, comments, and requirements into account in its actions and decision-making processes.

5.1 The Power of Consumers

Fairphone believes that consumers can have immense power when they gather together as a group. Consumer demand, caused by purchasing decisions, can influence companies in a multi-billion dollar industry by collectively demanding a review of products and supply chains in terms of social and environmental values. In this process, Fairphone sees itself as a campaign; creating more awareness and working with already conscious consumers to increase the group of people who can make informed purchasing decisions by understanding the story behind the production of their products.

Fairphone's website informs visitors about the numerous procedures that are part of producing a smartphone. The gradual unraveling of the complexity behind production shows the intricacies in sourcing, manufacturing, design, transportation, and recycling. It demonstrates that changing systems is a process that takes time, investment, and bringing together multiple stakeholders.

Storytelling is a dialogue, with Fairphone providing several channels to tell the narrative and listen to informed customers and followers who make comments and suggestions in a public online debate. It is a platform for community dialogue, including everyone from consumers and designers to government institutions, NGOs, and telecommunication providers. Consumers become drivers of change and are given tools to talk about the situation within social media platforms and amongst their friends, exponentially increasing the reach of the debate in online spaces that are open for the world to see.

Customers and followers actively shape the Fairphone brand as they share their stories online with #WeAreFairphone. By joining the waiting list for a new phone, people become part of Fairphone even before owning a phone. Besides raising awareness and providing information through social media, Fairphone encourages consumers to use their mobile phones for a longer period of time, agreeing that the most ethical and environmental phone there is, is the one you already own.

5.2 Dialogue with Stakeholders

A concrete example of multi-stakeholder dialogue that directly influences Fairphone's decisions is the working groups that it has created for the Worker Welfare Fund and Life Cycle Assessment (LCA). People with backgrounds in sociology, ecology, sustainability, ethics, business, design, and engineering are actively involved and meet regularly to give Fairphone informed, multi-perspective advice, and guidance to develop activities and plans for the future.

The model for the Worker Welfare Fund was codesigned with a working group of individuals from the labor rights movement and research institutes, including SOMO (Stichting Onderzoek Multinationale Ondernemingen), a strong and critical advocate for improved working conditions in the electronics sector. With this group, Fairphone defined key focus points and challenges. Currently, a big part of the workforce feels it has no means to express grievances or suggest improvements. By improving worker representation channels, the fund aims to motivate and empower workers to have a voice and provide financial means to invest in the things they value. Members of the working group have visited the factory to interview workers and guide the discussion with management. The fund is allocated for training and skills development, improved working conditions, increased wages, and a better work/life balance. Elections for the fund were held in summer 2014. This is the first step in a long line of improvements that must be made to work toward fairer electronics manufacturing.

The working group centered on Fairphone's LCA aims to offer insights into the ecological footprint of the phones' entire life cycle – from production and distribution to use, reuse, and recycling. By undertaking this analysis, Fairphone should be able to identify areas in the supply chain – so called "hot spots" – it can focus on to decrease its environmental impact. The first assessment focused on CO_2 emissions. The results of the assessment, for example that changing transportation methods can lead to a decrease in emissions, will be taken into account in the design, production, and business processes of the current and next generation of

Fairphones. Among others, Fairphone cooperates with Karsten Schischke from the Fraunhofer Institute in Berlin, Germany, to select focus areas such as packaging or transportation.

For Fairphone, its community is a great source of knowledge and expertise and consists of a collection of people who work in a variety of industries and are willing to positively contribute to the project. In that sense, the Fairphone brand and company are formed through active engagement from the outside-in. Actively encouraging people to criticize and debate its progress and inner workings allows its future growth to be shaped by the outside. Simultaneously, Fairphone lives sustainability "inside-out." Its ambition to establish social change and a fair economy motivates Fairphone to act according to its own fair principles.

6 Outlook

Fairphone has been able to create a market for a smartphone that lays open the story of its sourcing, production, and recycling and inspires people to consider what is behind their purchases. It hopes to drive a movement that grows the demand for products made fairly. This should, in turn, inspire the industry as a whole to internalize this thinking in its practices. Fairphone hopes to move the discussion of sustainability to the very core of business in the belief that by allowing the true impact of our activities to drive decision making, systemic change will happen. Major mobile phone company investors have started to approach Fairphone to discuss sustainability and share knowledge. This indicates that, despite being a relatively small enterprise, Fairphone has started to have an effect on society and the economy and punches well above its weight.

Being "fair" is a debate that will never have a one-sided answer, but Fairphone hopes that its buyers understand that their phone is not an end product with a "fair" certification, but rather an investment in change and a statement for fair electronics. It is a company that sees consideration for people and the environment as a natural part of doing business and it is a movement that connects people who want to see positive change in our economy. Fairphone's mission is a journey and everyone can join the process and contribute to creating positive social impact and systemic change.

In the future, Fairphone also needs to deal with the challenges and opportunities of its own growth. Being something in between a corporation and an NGO can be a balancing act. Nevertheless, Fairphone learns as it grows and will acquire new expertise and knowledge in the process.

During the production of its second generation of smartphones, Fairphone hopes to increase its phone's fairness, specifically aiming to incorporate more conflict-free and fairly mined minerals. It moves towards improving its value chain, fostering better working conditions, and reducing environmental destruction caused by mining and e-waste. Fairphone also aims to introduce social assessments and other impact measurement tools such as key performance indicators to better track and trace its production steps and the progress which is being made towards enhanced fairness.

People and their conscious or unconscious decisions are and will be what shapes Fairphone's identity; its brand. The slogan "We are Fairphone" illustrates that it is all of us who are in control of shaping our economy and the way it works.

Bibliography

Akormedi, M., Asampong, E., & Fobil, J. N. (2013). Working conditions and environmental exposure among electronic waste workers in Ghana. *International Journal of Occupational and Environmental Health, 19*(4), 278–286.

BBC. (2013). *Mobiles 'to outnumber people next year', says UN agency*. http://www.bbc.com/news/technology-22464368. Accessed May 14, 2014.

Bryceson, D., & Jonsson, J. B. (2014). Mineralizing Africa and artisanal mining's democratizing influence. In D. F. Bryceson, E. Fisher, J. B. Jonsson, & R. Mwaipopo (Eds.), *Mining and social transformation in Africa: Mineralizing and democratizing trends in artisanal production* (pp. 1–22). Abingdon: Routledge.

CMF. (2012). *Crowdfunding in a Canadian context*. http://www.cmf-fmc.ca/documents/files/about/publications/CMF-Crowdfunding-Study.pdf. Accessed April 10, 2014.

CNN. (2013). *Bangladesh building collapse kills at least 123, injures more than 1,000*. http://edition.cnn.com/2013/04/24/world/asia/bangladesh-building-collapse/index.html?hpt=hp_t3. Accessed April 28, 2014.

Economist. (2009). *Planned obsolescence* http://www.economist.com/node/13354332. Accessed May 13, 2014.

EUROPA. (2010). *Waste electric and electronic equipment*. europa.eu/legislation_summaries/environment/waste_management/l21210_en.htm. Accessed April 11, 2014.

EUROPA. (2014). *EU proposes responsible trading strategy for minerals from conflict zones*. European Commission, Press releases database, 5 March 2014. http://europa.eu/rapid/press-release_IP-14-218_en.htm. Accessed May 23, 2014

Fairphone. (2013a). *Cost breakdown of the first fairphone*. http://www.fairphone.com/wp-content/uploads/2013/09/Fairphone_Cost_Breakdown_and_Key_Sept2013.pdf. Accessed May 14, 2014.

Fairphone. (2013b) *Tin and Tantalum Road Trip*. http://www.fairphone.com/2013/11/08/tin-and-tantalum-road-trip/. Accessed November 23, 2014.

Financial Times. (2014). *EU plans voluntary rules on conflict mineral imports* http://www.ft.com/cms/s/0/b4823608-a47e-11e3-9cb0-00144feab7de.html#axzz32YCBWAgc. Accessed May 22, 2014.

Gibb, H. & O'Leary, K. G. (2014). *Mercury exposure and health impacts among individuals in the artisanal and small-scale gold mining community: A comprehensive review*. Environmental Health Perspectives, National Institute of Environmental Health Sciences. http://www.colombiapuntomedio.com/Portals/0/Archivos2014/DepositoDocumen2014/Informe-sobre-mercurio-en-la-miner%C3%ADa_OMS-1.pdf. Accessed May 23, 2014.

Guardian. (2013a). *"Conflict free" minerals from the DRC will only be possible if companies stay*. http://www.theguardian.com/sustainable-business/conflict-free-minerals-drc-companies-stay. Accessed May 6, 2014.

Guardian. (2013b). *Toxic "e-waste" dumped in poor nations, says United Nations* http://www.theguardian.com/global-development/2013/dec/14/toxic-ewaste-illegal-dumping-developing-countries. Accessed April 8, 2014.

Guardian. (2014a). *How sustainable is your smartphone* http://www.theguardian.com/sustainable-business/ng-interactive/how-ethical-is-your-smartphone. Accessed April 8, 2014.

Guardian. (2014b). *Social enterprises must not prioritise social aims over viability* http://www.theguardian.com/social-enterprise-network/2014/feb/17/social-enterprises-must-prioritise-viability?CMP=new_1194. Accessed May 6, 2014.

Jeffrey, J. C. (2012). Tungsten is forever: Conflict minerals, Dodd-Frank, and the need for a European response. *New England Journal of International and Comparative Law, 18*, 503–514.

Le Billon, P. (2003). Getting it done: Instruments of enforcement. In I. Bannon & P. Collier (Eds.), *Natural resources and violent conflict: Options and actions* (pp. 215–286). Washington, DC: World Bank.

Lepawsky, J. (2014). The changing geography of global trade in electronic discards: Time to rethink the e-waste problem. *The Geographical Journal*. http://onlinelibrary.wiley.com/doi/10.1111/geoj.12077/abstract

Melcher, F., Sitnikova, M. A., Graupner, T. (2008). Fingerprinting of conflict minerals: columbite-tantalite ("coltan") ores. *SGA News Number* 23. http://e-sga.org/fileadmin/sga/newsletter/news23/SGANews23.pdf.

MIT. (2014). *Environmental risks of mining*. http://web.mit.edu/12.000/www/m2016/finalwebsite/problems/mining.html. Accessed April 10, 2014.

Narine, M. (2013). From Kansas to the Congo: Why naming and shaming corporations through the Dodd-Frank Act's corporate governance disclosure won't Solve a human rights crisis. *Regent University Law Review, 25*(2), 351–401.

OECD. (2014). *OECD due diligence guidance for responsible supply chains of minerals from conflict-affected and high-risk areas*. http://www.oecd.org/corporate/mne/mining.htm. Accessed May 27, 2014.

Schluep, M. et al. (2013). Insights from a decade of development cooperation in e-waste management. In L. M. Hilty, B. Aebischer, G. Anderson, W. Lohmann (eds.) *Proceedings of the First International Conference on Information and Communication Technologies for Sustainability*. ETH Zurich, 14–16 February 2013, pp. 45–51.

Schmidt, C. (2012). Mercury pollution from artisanal and small-scale gold mining. *Environmental Health Perspectives, 120*(11), A424–A429.

Seay, L. E. (2012, January) *What's wrong with Dodd-Frank 1502? Center for Global Development* (Working Paper 284). http://www.cgdev.org/files/1425843_file_Seay_Dodd_Frank_FINAL.pdf.

Solutions Network. (2014a). *Conflict free tin initiative*. http://solutions-network.org/site-cfti/. Accessed April 10, 2014.

Solutions Network. (2014b). *Solutions for hope: A platform to support responsible sourcing, peacebuilding, and community development*. http://solutions-network.org/site-solutionsforhope/. Accessed April 10, 2014.

StEP. (2014). *What is e-waste* http://www.step-initiative.org/index.php/Initiative_WhatIsEwaste.html. Accessed April 25, 2014.

TED. (2011). *How i built a toaster–From scratch*. http://www.ted.com/talks/thomas_thwaites_how_i_built_a_toaster_from_scratch.html. Accessed April 10, 2014.

TIME. (2013). *More people have cell phones than toilets, U.N. study shows*. http://newsfeed.time.com/2013/03/25/more-people-have-cell-phones-than-toilets-u-n-study-shows/. Accessed May 14, 2014.

UNEP. (2011). *Basel convention on the control of transboundary movements of hazardous wastes and their disposal*. http://www.basel.int/Portals/4/Basel%20Convention/docs/text/BaselConventionText-e.pdf. Accessed April 11, 2014.

US Government Printing Office. (2010). *Dodd–Frank Wall Street Reform and Consumer Protection Act*. http://www.gpo.gov/fdsys/pkg/PLAW-111publ203/html/PLAW-111publ203.htm. Accessed May 14, 2014.

Waag Society. (2014). *Organisation*. http://waag.org/en/organisation. Accessed April 8, 2014.

WHO. (2013). *Mercury and health*. Fact sheet No. 361. http://www.who.int/mediacentre/factsheets/fs361/en/. Accessed April 10, 2014.

Yang, H., Huo, X., Yekeen, T. A., Zheng, Q., Zheng, M., & Xu, X. (2012). Effects of lead and cadium exposure from electronic waste on child physical growth. *Environmental Science and Pollution Research, 20*(7), 4441–4447.

SAP AG & StarShea Limited (Ghana): Sustainable Value Creation Through Collaboration with Companies, NGOs, and Intermediaries

Heino Kantimm

1 Introduction

SAP has been working with the nonprofit organization PlaNet Finance in terms of a commitment to the shea value chain in Ghana since 2009. What started as a project back in 2009 is now a company in its own right in Ghana, set up as a social business in accordance with the thinking of Prof. Muhammad Yunus. Why? What does SAP, the world's biggest business software provider, have to do with the processing and trading of shea nuts and shea butter? This was and still is the predominant question when people mention this particular project or the social business.

The aim of the article is to answer this question. Specifically, it will demonstrate how the core competencies of various interest groups can be brought together in an intelligent way for delivering shared value creation. It should be noted that this is only one example of the wide range of work done by SAP in the field of sustainability. The aim is therefore not to provide a comprehensive picture of SAP's sustainability strategy and social commitment, but to describe by means of a specific example how a strategic requirement led to an idea for a project that has now become an IT-powered social business.

There will be a special focus on the various roles played by the groups and partners involved. A tangible model will be used to discuss the issues and decision-making processes involved in a model encompassing various stages, starting from the preparatory stage, followed by the design stage, the completion stage, and ultimately, the stage in which long-term success is ensured. In essence, this model can be applied to other companies, even if the roles played will certainly vary from sector to sector and from company to company.

H. Kantimm (✉)
SAP AG, Dietmar-Hopp-Allee 16, 69190 Walldorf, Germany
e-mail: heino.kantimm@sap.com

M. D'heur (ed.), *Sustainable Value Chain Management*, CSR, Sustainability, Ethics & Governance, DOI 10.1007/978-3-319-12142-0_4

2 SAP and Sustainability

2.1 The SAP Company

SAP is the world's number one business software provider and its headquarters are located in Walldorf, Germany. Since its incorporation in 1972, innovation and growth have made SAP the world's leading business software provider. Using SAP applications and services, more than 232,000 customers of every size and from every business sector around the world are able to run profitable operations, adapt continuously to new requirements and achieve sustainable growth (SAP 2013).

Close collaboration with customers and partners has been part of the SAP business model since it was founded, and is a major reason behind the company's success. SAP's partners can be companies that develop additional solutions for SAP products, partners that help customers to implement the software and use it productively but even partners that act as SAP software resellers located close to customers across all industries, especially small to medium-size enterprises. Shared value creation through collaboration with partners is almost part of SAP's DNA.

SAP has set itself the objective of reaching one billion people with SAP software by 2015. The strategy for achieving this includes massive use of mobile applications (SAP 2013).

2.2 The SAP Corporate Sustainability Strategy

Combining philanthropy and business development is part of SAP's corporate sustainability strategy. SAP started to address sustainability in a focused way in 2009, introducing both a sustainability strategy and the appropriate organizational structure (SAP 2009a). Special attention was paid to the fact that SAP is able to operate sustainably in a dual way. As the world's biggest provider of business processing software, SAP is able to help other companies with their sustainability strategies, generating considerable leverage. On the other hand, SAP also has a duty to set a good example in terms of exemplary sustainable behavior.

This dual strategy was the result of an initial materiality analysis. According to the Global Reporting Initiative (GRI), issues are material, i.e., essential for companies, "that have a direct or indirect impact on an organization's ability to create, preserve, or erode economic, environmental, and social value for itself, its stakeholders, and society at large" (GRI 2013).

In answering the question of whether SAP could set a good example to others, the company set itself an ambitious goal to return to the level of greenhouse gas emissions from 2000 by the year 2020. This happened as it became clear that, of all possible environmental issues to be addressed, energy consumption is the most important to SAP in environmental terms. However, it was difficult to answer this question in social terms, especially the question of whether SAP could become a model for sustainable CSR. Porter and Kramer provided the basic academic framework for addressing this issue in their article published in the Harvard Business Review in 2006 (Porter and Kramer 2006). It was primarily a matter of how to get past a purely philanthropic approach, instead achieving the combination of a positive impact on society and successful business development.

3 Preparatory Stage

3.1 Materiality Analysis as the Basis for a Successful Project

In order to answer the question of what "material" means in social terms, it is necessary to begin with an analysis of corporate strategy and internal core competencies. It is also essential to isolate the specific positive social effect desired; for example, even the United Nations Millennium Development Goals (UNDP 2000) define a scope for action that is far too wide for any company to be able to commit to all of them.

SAP is a growth-oriented company, where growth is driven by its employees' powers of innovation. Future growth depends not only on having well-trained company employees, but equally on customers having well-trained employees who are well-versed in working with information technology. SAP's economic success is based on a functioning economic system composed of a large number of companies. Training and support for entrepreneurship are therefore also a focus of SAP's social commitment.

An excellent knowledge of business processes and value chains across all industry sectors is one of SAP's core competencies. Structuring these processes using standard software is central to its business model. New technology and trends, such as using mobile devices even for business applications, are just as important as the internet as a platform for business processes and the use of cloud storage for the requisite data.

For SAP, the criteria for credible projects (CSR or social business projects) that take into account SAP's core competencies and objectives on the one hand and address current social issues on the other, are therefore located at the intersection of new technology, such as mobile applications, and activities that nurture businesses and provide IT and software training. The biggest positive impact on society is therefore felt in market segments where the activities within this overlap are not already catered for by other initiatives.

3.2 Connecting with the Base of the Pyramid as a Solution

"We believe that small and midsize businesses in emerging markets will generate long-term global economic growth. In this context, digital inclusion is a corporate responsibility for SAP as well as an opportunity for growth and development (SAP 2009b)." This statement in SAP's Sustainability Report for 2008 forms the basis for identifying the key factor. The billions of people at the lowest income level, known as the Base of the Pyramid (BOP), not only constitute an important market for sales (Prahalad 2004), but also the basis for a large number of agricultural supply chains. The overwhelming majority of people forming the BOP live in developing and emerging countries (Heeks 2002).

Most business activity occurring at the BOP is informal (Gradl et al. 2012). Millions of agricultural smallholders have little more than a tiny plot of land, where they grow or harvest cocoa, coffee, cotton, shea nuts, cashew nuts, or a whole host of other raw materials. Only a fraction of smallholders are organized into cooperatives or similar groups (ADM 2013). Products are usually sold via a multilayered and often non-transparent network of intermediaries over long distances between farms and seaports. Products then very often undergo industrial processing overseas, ultimately finding their way to consumers in industrial nations via major food corporations and retailers (Lovett 2004).

Smallholders are therefore the lowest and weakest link in the end-to-end value chain. Profit margins are on the edge of survival, even in terms of local conditions in developing countries (Hammond et al. 2007). There are many different reasons why smallholders are in such a weak position. They often lack the capital for making the small-scale investments that could raise productivity considerably, such as buying an oil press, for example. Even the purchase of seeds can represent an enormous hurdle in many cases. In many instances, micro-financing institutions (MFI) can form part of the solution thanks to the products that they offer (Vodafone 2011).

The absence of knowledge and training constitutes a further challenge. It is a major problem for smallholders if they have no knowledge of efficient cultivation and processing methods for improving yields or basic understanding of business processes. Illiteracy is very widespread in a number of regions, making it harder to establish fair commercial relationships with intermediaries.

The lack of access to information is a third major reason for the disadvantaged position that smallholders find themselves in. The lack of pricing transparency also works against smallholders. The dramatic rise in mobile phone usage in developing countries is an opportunity to combat this informational disadvantage, e.g., accessing market prices via a mobile phone.

The above requirements and considerations point to the necessity of optimizing the existing value chain by means of a project that includes smallholders, enabling them to benefit from value creation. In order to achieve this aim, the existing barriers already referred to need to be eliminated through micro-financing, better education, and the use of information technology.

Selecting a suitable target country is also important, as many different regions and countries of the world need their value chains to be optimized. The decisive factor behind choosing Ghana as a target country was political and economic stability and easy communication, as the national language is English.

3.3 *The Right Implementation Partners*

SAP has a very good understanding of the top of the pyramid, but does not have the necessary knowledge or resources to understand the specific requirements at the BOP. It was therefore necessary to find a suitable partner with a similar objective and complementary skills and resources. SAP decided to enter a global partnership with the international nonprofit organization PlaNet Finance in 2009 (SAP 2009a). The aim of PlaNet Finance is to help people affected by poverty to develop income-generating activities to improve their long-term livelihoods (PlaNet Finance 2013). Currently running projects in 50 different countries, PlaNet Finance has access to a large network of micro-financing institutions. PlaNet Finance has local offices in a number of rural areas in developing countries, giving it is a very precise knowledge of the processes and requirements on the ground.

The global partnership facilitated the next step, which involved turning the rough idea of optimizing existing value chains between the BOP and the official business sector, turning them into practical reality and developing proposals for what practical implementation might look like.

4 Design Stage

4.1 *A Brief Introduction to the Shea Tree*

Shea nuts are the fruit of the shea tree, which only grows in the Shea Belt in West Africa (Lovett 2004). The tree grows wild and it has not yet been possible to cultivate it. Women have the centuries-old task of gathering the shea fruit from around their villages when it ripens in spring. The nuts are then removed from the fruit by hand and dried, before the kernel is removed (Lovett 2004).

The kernel has a high fat content, which the women extract via a labor-intensive manual process and then make into shea butter. There is a long tradition of shea butter in West Africa. It is used for cooking and also has excellent qualities to use as a skin cream.

Only about a third of all the nuts gathered are exported via a network of intermediaries – mainly as nut kernels, which are then often processed outside the country. There is very little export of handmade shea butter. Around 90 % of exported kernels go into the food industry, as shea butter is a cocoa butter equivalent (CBE), making it an important ingredient of chocolate production. Only around 10 % of international demand comes from the cosmetic industry, where shea butter is used as an additive in cosmetic products (Lovett 2004).

It is of central cultural and economic significance that only women have control of the proceeds from selling shea nuts, whereas farm income (if there is any) is traditionally a male preserve.

4.2 Practical Project Ideas

PlaNet Finance was able to use its local experience and knowledge to put forward a proposal to conduct a more detailed analysis of the shea value chain in Ghana. But PlaNet Finance still lacked any in-depth understanding of the exact processes that would be required to make a decision at that point. PlaNet Finance carried out a focused feasibility study to ascertain the current situation as accurately as possible and talk to all parties involved, which would then enable it to recommend a practical project. The open-ended feasibility study was extremely important, as too many projects in developing countries impose prefab solutions, making them doomed to failure right from the start (Dodson et al. 2012).

The feasibility study resulted in proposals for initiatives that would specifically improve the situation of women in Ghana as they saw it. This basically delivered the following essential findings:

Training

- Given that they had mainly worked alone in the past, it was proposed that women should receive support through self-organization. Groups could be trained within villages, then regional groups, and finally interregional groups – a kind of virtual cooperative. Working in groups was not a new concept for the women, but the idea had never been applied to the processing of shea nuts before. An important factor was not to dictate the organization of each group in advance, but to allow the women to organize themselves through a democratic process. The formation of virtual groups would provide the network with greater bargaining power, putting it in direct contact with major international customers and giving it access to better prices.
- In turn, grouping would be a necessary precondition for carrying out product and business training sessions more efficiently. For example, the feasibility study confirmed that even the simplest rules for processing shea nuts can help improve product quality considerably and that international customers are also willing to pay a premium over the regular market price for higher quality.

Micro-finance

- Another finding of the feasibility study was that women were not in a good bargaining position when selling their shea kernels because the spring shea harvest comes at precisely the most difficult time of the year in terms of finance, making any form of income vital for survival. Intermediaries use this pressure to their advantage, pushing prices down even further. If women were not forced to sell at almost any price, they could benefit from the usual massive price rise for kernels in the summer, partly caused by the natural drying of the nuts. Micro-financing would be a helpful way of tiding the women over in their time of need until such time as they had access to attractive prices for shea kernels on local markets.

Information Technology

- Mobile phones are now widely available even in the remotest and poorest areas of Ghana and Africa. This provides the opportunity of sending women brief messages with information on prices, putting them immediately in a better bargaining position with regard to potential buyers. Setting up a professional web site (StarShea 2013) would raise the profile of the entire women's network with a wide range of different international buyers. Although the development of an information system for the simple recording of transactions and management of orders and logistical information would have no direct impact on how the women work, it would be very useful for buyers, giving the women's network a competitive advantage that would be reflected in higher prices.

4.3 A Clear Understanding of Roles and Objectives

In addition to defining work packages in detail, it was equally important to set measurable project objectives at an early stage once the preliminary study had been completed. To avoid any conflict of interest, all parties involved needed to work towards achieving the same objectives. This was not to be underestimated where partners with totally different business backgrounds were involved in the same project.

Apart from having a shared understanding of objectives, it was also vital to have a shared understanding of roles. In addition to the question of "what" you would like to achieve is the question of "how" you are going to achieve it. For example, one option for SAP would have been to use its expertise to develop a concept for a solution that would then be implemented by a local IT provider; alternatively, with even less direct engagement, SAP could have restricted itself to the role of advisor, supporting the project through finance and expertise alone. With this scenario, the IT side of things could easily have taken care of and probably in a much more cost-effective way for SAP. However, this would have meant a missed opportunity for SAP in terms of learning about software development and the direct relationship that this learning curve had with the company's innovation strategy.

5 Completion Stage

5.1 The Benefits of a Wide Eco-System of Project Partners

By focusing on the core competencies of both PlaNet Finance and SAP, both parties were able to create the basis for effectively incorporating their individual ecosystems into the project. PlaNet Finance has an extensive network of partners, especially on the ground. This meant that existing communications channels via two Ghanaian micro-financing institutions (MFIs) could be used for effective access to villages and therefore the women. This enabled the MFIs to strengthen their core business through their connection with the project and be of benefit to the women at the same time.

SAP has a wide-ranging network of customers across all industry sectors. Many of the corporations at the upper end of agricultural value chains, such as the food industry as well as the agricultural industry, are also SAP customers. Good business relationships in one area can therefore make it easier to open doors in other areas. For example, Olam International, which is not only a SAP customer but also an important player in the shea value chain, was acquired at an early stage as the first international customer for the products made by the women's groups. This meant that a network of partners was able to cover the entire value chain effectively, ranging from the work carried out with the women through to the international end customer.

5.2 Development and Validation

The remainder of this article mainly addresses the solutions developed by SAP in its role as project partner:

The results of the feasibility study were fed into the design of a software solution that could potentially support the process. Software solutions can be used in many different ways: one of the most difficult issues was to ascertain where true value-add could be created for the women through the use of IT, rather than implementing it merely for its own sake. The situation was also aggravated because of the continuous influx of new knowledge on a daily basis and changing insights into processes as well as circumstances. This meant that the software design had to be flexible enough to adapt continuously to changing requirements.

An SMS-based price information system had to be part of the overall solution, because the women wanted it that way. From the viewpoint of the MFIs, it was important to ensure efficient process mapping along the whole trajectory, starting from the receipt of an order, splitting a large order amongst different groups of women, the logistics, and finally, payment. The women's network did not have the necessary training for entering into business relationships with international buyers in the short- and medium-term. The MFIs therefore provided support at the interface between international buyers and the women's network and the MFIs could only carry out such complex tasks with the support of IT.

The resulting software solution was designed as a prototype. A major objective was for the technology to create added value for the women. Another objective was for SAP to ascertain the factors involved in developing software in an environment unfamiliar to SAP, i.e., with target groups not used to IT, connectivity problems, narrow bandwidths, etc. The intention was to identify important findings that could be useful for core development and addressing future markets.

All the project partners believed that, in addition to each party conducting its own assessment of interim results, it was also important to obtain an independent evaluation of the impact of the activities carried out. Led by Prof. Hau Lee, a team from the Stanford Graduate School of Business reviewed the project and produced a catalogue of recommendations. It was possible to demonstrate many positive effects, such as increases in income from shea products ranging from 58 to 82 % (Rammohan 2010). The wide geographical distribution of the women, meaning potentially higher administration costs and the possible risk of state intervention such as market regulation, came under close scrutiny.

Alongside this external assessment, an interim internal assessment of the software prototype confirmed that the expected costs would outweigh any potential benefits provided by the solution and that an even simpler solution had to be found. In finding a simpler solution, the software application focussed on increasing the use of mobile phones for data capture compared to the past and prioritizing the traceability of the value chain end-to-end over the optimization of order processing in terms of increased economic value-add. A second prototype was developed. Inexpensive barcodes were stapled to the sacks that the women used to store the kernels in their villages. When the sacks were collected, the barcode was simply read with a mobile phone, recording additional data such as weight or simple quality parameters. This solution meant that each individual sack could be traced back to an individual woman. This was particularly important to international buyers. This created the basis for demonstrating not only compliance with social standards but also fair payments too, for example. The women saw this added value in the form of competitive advantage and higher prices as a result.

5.3 *Communication and Employee Commitment*

The question of when and how information is to be released for such a project needs to be tackled very early, as does the question of how external and internal communications are to be differentiated. If information is released externally too soon, this might make people think that the main focus is on improving a company's reputation, leaving it open to accusations of greenwashing (Oxford 2013).

However, improving a company's reputation internally should be seen as a much more effective approach than any improvement that might be obtained via external communications. Detailed internal communications should therefore be prioritized within the organization at an early stage. Projects with a social focus boost the emotional connection between a company and its employees, who feel proud and communicate a positive image of the company to those closest to them.

However, the desire of many employees to work actively on this type of project, either as part of their job or voluntarily, is far more powerful than the emotional connection created through simply being made aware of a project. SAP's direct commitment not only as a provider of capital but also as a driver of innovation opened up lots of different opportunities to involve employees actively in the project. This is why www.starshea.com was created solely by volunteers from the SAP Marketing and Corporate Communications teams, for example. This enabled them not only to apply their professional skills, but also to acquire valuable new experience important to their professional development within the company and the company as a whole. Other volunteers worked hard on the design and development of the IT solution.

6 Ensuring Long-Term Success

6.1 What Happens When a Project Ends?

By definition, projects have a beginning and an end, and this was also true for the global partnership between SAP and PlaNet Finance, which was designed to run over a period of 3 years. However, people soon began to wonder what should happen at the end of the 3 years. Too many grant-based projects come to an end once the sponsorship period is over, because investments and infrastructure are no longer maintained or because buildings and equipment slowly fall into disrepair or are repurposed. Extending the sponsorship period is one short-term solution, but is by no means a long-term option.

SAP started thinking about the next stage very early on in the project partnership. Inspired by the activities of other major companies such as BASF and Danone, SAP considered taking the project forward as a social business. The social business concept is strongly influenced by the work of Prof. Muhammad Yunus and refers to a profit-oriented company with the sole corporate purpose of solving a social problem. Profits are not distributed to investors, but remain within the company to support the work it does for society. Foundation capital can be paid back, but there are no dividend pay-outs (Yunus 2009). Both BASF and Danone founded social businesses like this in the form of joint ventures with Grameen (Yunus 2009) in 2008.

Two basic questions need to be answered to turn a grant-based project into a social business:

1. Is there a business model that makes it possible to work in aid of society whilst still covering costs? This obviously means that costs need to be calculated with precision and realistic income potential ascertained compared to the project.

2. What would the ideal social business model look like and what would be the role of the two partners already working together and any other possible partners?

The Yunus Social Business – a spin-off of the Grameen Creative Lab in Frankfurt, Germany – set itself up as a consultancy company specializing in answering this kind of question (Yunus Social Business 2013). Grameen Social Business, SAP and PlaNet Finance joined forces to design just such a business model, running a number of local workshops in Ghana, conducting a detailed analysis of existing value chains and carrying out countless interviews with local experts and stakeholders. The core of the model consisted of creating a marketplace as an intermediary between the women and international export markets. The marketplace also enabled women to participate in the activities of the marketplace and provide the appropriate product quality.

This meant that the solutions developed under the project provided an important competitive advantage compared to traditional intermediaries, which focus on maximizing costs at the expense of these women. The marketplace was financed by the transaction fee, which was high enough to cover costs but low enough to ensure that the women involved still had a clear rise in income. If enough women participated with the associated number of transactions, then a profitable model could be generated.

6.2 *Turning the Project into a Social Business*

In addition to the question of the social business model, the issue of the operator model also needed to be clarified. The project team proposed a model to the SAP/PlaNet Finance committee, whereby PlaNet Finance would have 100 % ownership of the social business. SAP would continue developing the software and also cover the operational costs through an interest-free loan. The social business would pay back the loan once it had become profitable and its business operations were stable.

The social business called StarShea Ltd was registered in Ghana in 2012. PlaNet Finance and SAP selected and appointed a CEO together. This laid the foundations, enabling StarShea Ltd to operate in the market long-term and independently of grant money.

6.3 *Social Business: Present and Future*

The aim of the project was to bring 3,000 women into a single network over a period of 3 years. In the form of a social business, the network numbered 15,000 women by the beginning of the harvest season 2014. This number could double in just a few years. Successful scaling is vital for achieving profitability.

The product mix mainly consisted of nut kernels and various processed forms of shea butter. Selling shea butter was a more attractive option for the women because it is a higher form of value creation, therefore leading to higher income expectations and a greater sense of personal pride. The focus throughout the project remained on selling kernels, as demand was continuously high. The market for shea butter in terms of the cosmetics industry was much smaller, but the social business still managed to become a leading global supplier of hand-crafted shea butter within just 1 year.

The next stage involved organic and Fair Trade certification for their products in order to increase margins for the women further. Both of these labels were obtained in 2013. Today, StarShea is an established brand in the supply chain of the cosmetics industry and counts many of the key players in that industry as its customers.

The biggest challenge for the social business in the mid-term will be to ensure that the marketplace remained attractive to the women, so that they remained willing to sell their products via the social business and pay the transaction charge. However, the women always have to have the choice of using other sales channels too.

A long-term option would be to make them owners of the social business – partially at least. Prof. Yunus also refers to this as a type II social business (Yunus 2009).

If, when and how such a step is to be taken is up for debate according to the current situation.

7 Lessons Learned and Next Steps for Scaling

7.1 Creating a Win–Win–Win Situation

From the present perspective, all major stakeholders have seen positive results so far – even though the continuing effects are naturally still to be seen over an entire year.

SAP so far sees three reasons for the positive outcome:

1. First, the project and social business provide an excellent platform for developing innovative ideas and solutions for future markets and testing them in a real environment. As demonstrated by the project, developments in technology can be validated at an early stage and adapted as required. At the same time, these solutions help the world to run better and improve people's lives. This is at the core of the SAP corporate vision (SAP 2013).
2. Second, following from the first reason, there is an increase in employee commitment levels, whether through passive pride or active involvement. Active involvement also contributes to personal development, additionally benefiting the company at the same time.

3. Third, positive public reporting can have a positive impact on relationships with partners, customers and potential customers.

A social business can greatly improve the social and economic situation of women in Ghana, without changing traditional working methods or behaviors significantly. Nor are there any dependencies that might even make their situation worse if the social business failed.

The social business provides PlaNet Finance with the opportunity to continue project activity with a social aim, meaning that it is not always exposed to the risk of losing future financing.

7.2 Successful Partner Collaboration

There are both risks and opportunities with any collaboration between an industrial corporation and an NGO on a grant-based project where each sees the other as an equal partner. The most varied core competencies can complement each other very well if they are used appropriately, but this requires both sides to contribute considerable resources, especially in terms of personnel. Intense project work is part of an NGO's business model, but more of an exception for an industrial corporation, as the Corporate Social Responsibility departments that would normally be responsible are generally rather streamlined.

Differing corporate cultures also play a major role in any collaborative undertaking. Different ways of thinking and approaches can ultimately be very enriching and generate a greater number of innovative solutions, but this also requires openness and tolerance from both partners. Frictional losses need to be taken into account, especially at the start of the process.

With a portfolio approach, where the industrial corporation distributes grants to various projects and also to NGOs, objectives are often determined jointly. Implementation and decision-making for how the objectives are to be achieved are mainly left to the NGO, which then reports regularly on project progress. By way of a contrast, a project involving equal partners requires decisions to be made and cooperatively carried forward. In addition to increased input in terms of coordination, this also principally involves a detailed discussion of the objectives on all sides and a willingness to make compromises.

7.3 Future Development

Two different aspects need to be considered with regard to the issue of how SAP is to carry the social business forward.

Aspect one is technology: SAP has been a partner in the African Cashew Initiative since 2009 (SAP 2011). Important function modules were developed during this research project, which could then be applied to the shea project. The research project was extended by another 3 years in 2012, with the intention of expanding the piloted solution in terms of the functions and regions covered, as well as in terms of reaching additional agricultural value chains and increasing user numbers on a massive scale. The knowledge acquired would be used as a basis for the desired aim of "productization," i.e., turning a pilot solution into a commercially available product. If the solution succeeded, it could then be added to the SAP product portfolio. Traditional customers in the agricultural industry, but also in the public sector – governments, for example, could use the solution to help millions of smallholders throughout the world acquire a competitive advantage.

Aspect two is the question of social business approach: StarShea Ltd is now well on its way to becoming a successful and financially independent social business. The focus of input of both PlaNet Finance and SAP therefore remains on ensuring that this continues. As has been demonstrated, social business is generally a long-term solution for a grant-based project. The resources required for making a social business successful cannot be underestimated, however. This means that although a social business is always an option, it will not necessarily always be the solution which is chosen.

Bibliography

ADM. (2013). *ADM cocoa confirms its commitment to improving livelihoods of farmers at World Cocoa Conference*. ADM press release from Jan 8, 2013. http://www.adm.com/en-US/products/Cocoa/news_trends/_layouts/StoryDetail.aspx?ID=33&l=/en-US/products/Cocoa/. Accessed April 2, 2013.

Dodson, L., Sterling, S. R., & Bennett, J. K. (2012). Considering failure. In *Proceedings of the fifth international conference on information and communication technologies and development – ICTD'12, 56*. New York, NY: ACM Press. doi:10.1145/2160673.2160681.

Gradl, C., Kükenshöner, C., Schmidt, J., & Ströh de Martinez, S. (2012). *Growing business with smallholders*. Bonn: Deutsche Gesellschaft für Internaltionale Zusammenarbeit (GIZ) GmbH.

Global Reporting Initiative (GRI). (2013). *Materiality in the context of the GRI reporting framework*. https://www.globalreporting.org/reporting/guidelinesonline/TechnicalProtocol/Pages/MaterialityInTheContextOfTheGRIReportingFramework.aspx. Accessed April 2, 2013.

Hammond, A., Kramer, W. J., Tran, J., Katz, R., & Walker, C. (2007) *The next 4 billion: Market size and business strategy at the base of the pyramid*. Washington, DC: World Resources Institute, International Finance Corporation.

Heeks, R. (2002). *Failure, success and improvisation of information systems projects in developing countries* (Paper No. 11). Manchester: Institute for Development Policy and Management.

IFAD. (2011). *Rural poverty report 2011 – New realities, new challenges: New opportunities for tomorrow's generation*. Rome: Quintly.

Lovett, P. (2004, November). *The shea butter value chain. Production, transformation, and marketing in West Africa*. WATH technical report no. 2. Daka: WATH.

Oxford. (2013). Greenwashing. http://oxforddictionaries.com/definition/english/greenwash. Accessed April 2, 2013.

Planet Finance. (2013). *L'expert de la microfinance*. www.planetfinance.org. Zugegriffen April 2, 2013.

Porter, M., & Kramer, M. (2006). Strategy and society: The link between competitive advantage and corporate social responsibility. *Harvard Business Review, 84*(12), 78–92.
Prahalad, C. K. (2004). *The fortune at the bottom of the pyramid*. Upper Saddle River, NJ: Wharton School Publishing.
Rammohan, S. (2010). *The shea value chain reinforcement initiative in Ghana*, by SAP, PlaNet Finance, Grameen Ghana and Maata-N-Tudu, CASE study, Stanford Global Supply Chain Management Forum. Stanford Graduate school of Business.
SAP. (2009a). *SAP increases focus on sustainable business*. SAP press release from 2 March 2009. http://www.sap.com/corporate-en/press.epx?pressid=10988. Accessed April 2, 2013.
SAP. (2009b). *Digital inclusion. SAP sustainability report 2008*. http://archive.sapsustainabilityreport.com/2008/priorities/digitalInclusion.html?sct=view. Accessed April 2, 2013.
SAP. (2011). African Cashew Initiative helps small-scale farmers in Ghana through IT. In *SAP Community Network*. Published on 29 July 2011. http://scn.sap.com/community/research/blog/2011/07/29/african-cashew-initiative-helps-small-scale-farmers-in-ghana-through-it. Accessed April 2013.
SAP. (2013). *Integrierter Bericht 2012. Vision, Mission, und Strategie*. http://www.sapintegratedreport.com/2012/unternehmensleistung/vision-mission-und-strategie.html. Accessed April 2, 2013.
StarShea. (2013). www.starshea.com. Accessed April 2, 2013.
UNDP. (2000). *The millennium development goals – 8 goals for 2015*. http://www.undp.org/content/undp/en/home/mdgoverview.html. Accessed April 2, 2013.
Vodafone. (2011). *Connected agriculture – The role of mobile in driving efficiency and sustainability in the food and agriculture value chain*. Policy.
Yunus, M. (2009). *Creating a world without poverty: Social business and the future of capitalism*. New York: PublicAffairs.
Yunus Social Business. (2013). http://www.yunussb.com/. Accessed April 2, 2013.

Nanogate AG: Sustainable Value Creation in Technology Companies

Ralf Zastrau

1 Trust as the Basis of Corporate Success

Nanogate is a leading international integrated systems provider for high-performance surfaces. Nanogate's expertise and technology give surfaces new functions, make production processes more efficient, equip plastics with additional properties, reduce the amount of energy used by heating and filters and improve textile performance.

Surviving sustainably as a company is impossible unless reliable relationships based on trust are created and nurtured with cooperating partners. Therefore Nanogate's success as a technology company must depend on its relationships based on trust with customers, investors, employees, technology partners, suppliers, government authorities, trade associations as well as with the public in general. From the day our company started out with four employees in 1999 we have taken sustainability very seriously. In a complex environment it was important to create shared value for all our partners and, at the same time, to survive economically in a new, complex nanotechnological environment, at a time when no business models were known.

However, the achievement of Nanogate AG's corporate goal – to be perceived as the best provider of high-performance surfaces – was and is, in our view, dependent not merely on whether we are classified as one of the market's most competent and reliable players and whether we offer our partners attractive value creation.

Rather, we are, and always have been, convinced that there is a very special business growth opportunity and distinctive feature in going the extra mile and assuming overall responsibility. This understanding is also reflected in our growth strategy (see Fig. 1). Therefore from the earliest days of Nanogate AG, we wanted

R. Zastrau (✉)
Nanogate AG, Zum Schacht 3 66287, Göttelborn, Germany
e-mail: ralf.zastrau@nanogate.com

M. D'heur (ed.), *Sustainable Value Chain Management*, CSR, Sustainability, Ethics & Governance, DOI 10.1007/978-3-319-12142-0_5

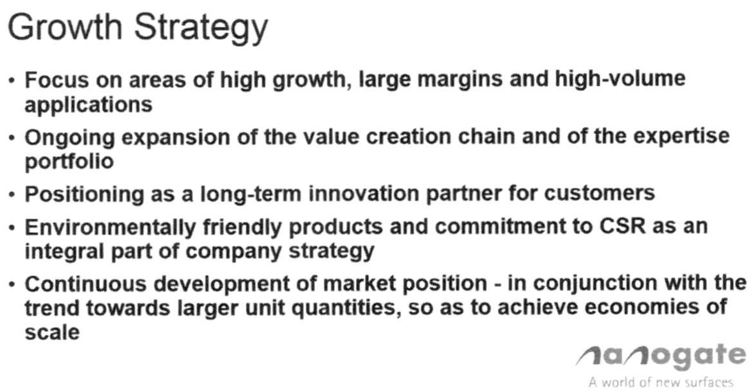

Fig. 1 Nanogate growth strategy. *Source*: Nanogate (2013a, b)

to be seen as a company that addresses social issues in a credible manner, puts respect and integrity at center stage, and is prepared to take extra care concerning value chain creation. This means that from the outset it has been important to make forward-looking contributions to the anticipated discussion about the impact of technology and the sustainability aspects of nanotechnology.

2 Early Embedding in the Corporate DNA

Since the beginnings of our company to today we have seen an average annual growth rate in double digits and an increase in our staff to over 400 people. Throughout this time period, we have aimed to embed our understanding of our own role in creating trust into our central corporate values and to implement it in our day-to-day business.

However, like every other company, we find ourselves in an environment in which it is not easy to establish a broad measure of trust in the core processes and value creation of a company and in its wider societal activities. Dynamic changes in competitive conditions, constant pressure on earnings and costs and the concomitant decision-making conflicts make the implementation of discerning value systems in corporate practice an ongoing challenge. As it is typical for a midsized company in Germany, Nanogate AG must meet these challenges with a manageable budget but also with a high degree of credibility.

However, concrete possibilities for taking responsibility with few resources but a high level of commitment can always be identified in practice. Accordingly, right back during its start-up phase, Nanogate AG was able to exploit a very specific opportunity. As early as in our first financial year (1999) we became joint initiators of Nanosafe, an EU-wide technology assessment programme, then the largest of its kind, and were subsequently intensively involved in this over a period of several years. The main aim of the programme was to achieve safe production and use of nanomaterials throughout Europe (Nanosafe 2013). This was a major commitment on the part of Nanogate AG, which was not an easy undertaking because of the difficult economic situation at the time and the fact that the company was at that stage still incurring significant losses.

In addition to Nanosafe's important contribution to all aspects of the nanotechnology discussion taking place then, the successes and benefits achieved as a result have substantially increased Nanogate AG's credibility, boosted its reputation with almost all its various cooperating partners, established important international networks for the company, and, last but not least, given it valuable knowledge for future product development and value creation strategies. However, there was now also an appreciably better in-house understanding that we were not just paying lip service to the wider company commitment to which we aspired, but that it is being realized, even under difficult conditions. In fact, according to our experience, it is particularly in challenging situations that a high level of credibility can be developed in the company, excellent opportunities occur for practising the skills needed to cooperate with different partners, and internal management processes can be improved.

Because of its positive experiences early on, in the following years Nanogate AG was able to further develop its understanding of how to link social responsibility directly with the company's core business, and to allow this to have a systematic influence on the different areas of business management. Specifically, the keynote ideas of our value-based management have become evident in our product and partnership strategies, also combined with the claim that we enable the creation of shared value with all our partners.

3 From Company Vision to Measurable Successes

Nanogate staff give constant attention to one core question:

> *How can we become and continue to be an exceptionally successful company which is trusted by our partners?*

In order to respond to this question, we must consistently implement our contributions to value creation in such a way that building trust is linked to concrete operational and strategic corporate goals and to the requisite wider commitment.

It is Nanogate AG's understanding that this happens on three key levels:

- In the core operational and strategic processes
- In networked projects between companies, civil society, politics, and administration
- In wider activities which shape the general social and political sphere of operations

3.1 First Level: Core Operational and Strategic Processes

We are convinced that the basis for a company's ideas of responsible value creation and partnership strategies is first and foremost the embedding of sustainable value creation in its core processes. This first level is of fundamental importance for all further considerations and is always the natural starting point for all further endeavors.

To do justice to this level, Nanogate AG, at that time still a very small company – the first nanotechnology company in Europe in fact – introduced a comprehensive quality management system early on (in 2000). This was in line with international standards, underwent an external audit, and was then developed further. During the subsequent years the next step was the introduction of an integrated, certified environmental management system – likewise as a "Nano-pioneer" within Europe. This provided us with successful tools which even today make important contributions to the optimization of our value creation and our relationships with external partners and suppliers (Fig. 2).

Active responsibility becomes evident, however, especially in the formulation and implementation of sustainable product strategies. Starting with the use of particularly sustainable raw materials, Nanogate AG clearly focused its development on enabling the creation of products which cause lower levels of environmental pollution than traditional solutions. This has included introducing better and more environmentally friendly production processes and concepts that increase durability or contribute to a more efficient use of resources. Within this context, consideration of the whole life cycle of the products that use Nanogate's technologies plays a major role. Nanotechnologies are especially helpful in significantly reducing impact on the environment at the stage when the consumer uses the product. In our business environment these general possibilities are also currently dealt with in public discussion under the heading "Green Nano." Over the years Nanogate has been able to put such keynote ideas into practice in many ways,

Fig. 2 Environmentally friendly products as sales drivers. *Source*: Nanogate (2013a, b)

setting measurable targets and linking them with commercial success. One example is the surface systems for printers made by reknown manufacturers, which make the industrial cleaning processes significantly easier, while saving costs and dispensing with aggressive cleaning agents and methods. Innovative building surfaces have also been created (for example, in the Stockholm urban tunnel, the longest in Europe). These minimize aggressive dirt adhesion from vehicles and make cleaning much simpler without the need for agents that pollute the environment. Maintenance intervals are extended considerably, saving costs and contributing to the protection of the environment. Further examples of sustainable product developments promote for instance improved diesel engine combustion processes and the reduction of emissions in exhaust systems.

Our latest innovation is our successful development in the field of energy efficiency coatings last year. These improve the effectiveness and energy efficiency of traditional and forthcoming heating systems at the same time as minimizing maintenance costs. This is a special issue, not just within the framework of the current CO_2 debate: it also offers great commercial potential for a technology company such as Nanogate AG. In the meantime the innovative concept of an energy efficiency coating has been brought into large-scale use with leading industrial partners, and a new, exceptionally resource efficient generation of heating systems has recently been successfully positioned on the market. Current demand, with the results already achieved in practice in industry, serves as a particularly solid example of the measurable contributions made by shared and sustainable value creation in the core areas of our company. Our product innovations in the area of glazing are particularly newsworthy. Our objective was to create lightweight, multifunctional plastic materials with glass-like properties. In the automotive sector, for instance, innovative plastics with glass-like properties and high optical quality are high in demand. They offer new design options and interesting applications, and because they are so light, fuel consumption and hence CO_2 emissions are reduced.

Overall, the development of environmentally friendly products is an essential part of Nanogate's company strategy and has always been viewed as a central criterion within the context of our innovation process. One of our operational targets is that in future Cleantech applications and systems with clear environmental contributions will account for substantially more than 50 % of our turnover (Nanogate 2013a).

Since sustainability and social commitment form part of our corporate philosophy, we are constantly seeking opportunities to apply our knowledge and our products to solving social problems. Examples of this include our collaboration with Professor Yunus to use nanotechnology to improve living conditions in Bangladesh, projects with German Doctors e.V. to provide medical care in developing countries, and our cooperation with Adlens, a manufacturer of adaptive eyewear technology. The cooperation with Adlens in particular shows the opportunities for employing products to solve social problems. Quality of life and sight are seriously reduced for over a billion people because they do not have access to the appropriate vision aids. The British manufacturer Adlens, in cooperation with its affiliated charitable organization, Vision for a Nation, aims to make its products available to remedy this situation and to improve the sight of as many people as possible in a simple and afforable manner. Adlens' glasses use a unique fluid injection technology. It allows the user, with the aid of ophthalmic opticians or trained health personnel, to adjust the visual strength of glasses using a small adjustment wheel on the spectacles frame. Adlens is a global company focusing on sustainable technologies. Nanogate enhances the glasses with a high-performance surface. The transparent coating protects the lenses from scratches, increasing their durability and allows for flexible use over a long period. The inkjet technology used by Nanogate subsidiary GfO ensures the technological precision with which the complex plastic components can be coated exactly and without requiring a mask, while remaining commercially feasible and producing perfect optical results (Nanogate 2013b).

3.2 Second Level: Networked Projects Between Companies, Civil Society, Politics and Administration

From the earliest years of our company's development we have been involved in the second area as well – particularly in regional projects like the Saar Environmental Pact, in voluntary work with numerous business organizations (for example with Nanobionet e.V. www.nanobionet.de) or in terms of our wider cooperation with educational establishments promoting an understanding of technology.

Today Nanogate AG is still particularly involved in the field of education. This becomes evident due to our high percentage of trainees, our wide-ranging practical placements, the support we provide for those engaged in writing theses, and our participation in research projects and other exciting assignments at a regional level and beyond, actively involving many of the company's employees. Further examples are the company's establishment of Nanocamps for school pupils, the introduction of taster days, including some for very young people, the active help given to student companies, assistance with setting up a Summer of Science, introduction of a regional Learning Festival, involvement in establishing nanotechnology as a seminar subject for sixth form students, and the promotion of further education events, exhibitions and conferences on all aspects of the key topic of nanotechnology. Under this heading we would also include the Saarland Empowering Nano initiative.

Concerning this second area of activity, our company sees it of central importance to embed nanotechnology permanently as a strategic competence in our local area, the Saarland, a federal state in Germany, and to give sustainable support to regional structural change. Nanogate AG therefore views the active contributions we make to this in various networked projects as a particularly important aspect of the key concept of sustainable value creation. It is in this context that we see our initiation of and involvement in the network "Verantwortungspartner Saarland," which brings together entrepreneurs, social institutions and the public sector to pool their efforts in site development. Since the network was established, well over 100 collaborators have been enlisted and some superb individual projects have been implemented. A further benefit is that extensive, resilient networks and partnerships of trust have come into being.

We regard our commitment as a direct investment in Nanogate AG's social capital. Many other contributing companies are, just like us, convinced that the work achieved has a direct and positive effect on the formation of a resilient company value culture, on leadership training, and on staff loyalty and employee development.

3.3 *Third Level: Wider Activities Which Shape the General Sphere of Social and Political Activities*

Nanogate is continuing to expand its involvement at the third level, that is, in shaping wider conditions. Current examples of such projects at regional level and beyond are our active involvement in the formulation of the innovative CSR strategy for the Saarland on the initiative of the regional government, our commitment to the nationwide Bertelsmann Foundation initiative, Companies for the Region, and our membership of the Nano Commission set up by the Federal Government.

A further example of our perception of value creation on this third level is our support of the Oeko-Institut [Institute for Applied Ecology] in the uniform evaluation of nanotechnology's potential for sustainability. We have been participating in a project, run by the Oeko-Institut and sponsored by the German Federal Environment Agency and the German Federal Ministry for the Environment, to develop a new Nano-Sustainability check. This new analysis matrix aims to identify environmental benefits or drawbacks as well as risks and challenges for the market introduction of products and nanomaterials at a very early stage.

We believe that, in addition to the importance of joint and sustainable value creation for the core business activity of a company, it is at this third level that interesting opportunities arise for companies to participate in making key decisions. Important contributions to knowledge and learning are constantly being generated, which can and must be actively used within the context of in-house opportunity and risk management.

4 Those Who Come Too Late...

Nanogate AG is convinced that where companies pay genuine attention to social responsibility this will be a key factor for overall corporate success and business development in the future.

The boundaries of responsibility are currently being newly defined and called into question across society as a whole: traditional ways of seeing things are changing as a result. Prevailing megatrends are inevitably leading to an acceleration of this debate and lay the foundations for a new definition of the responsibilities required of companies:

(a) Global value creation processes with unforeseen degrees of freedom for companies to optimize their individual earnings potential;
(b) A new dimension of transnational and (system-oriented) business relationships that allow only limited political and social mechanisms of intervention;
(c) Fundamental revolutionary social challenges such as climate change, demographic development, or irresponsible public debt.

The debate about the future role the state can and should play in these issues is ongoing and universal, as are changed social expectations. This means that the progressive interpretation of the concept of Corporate Social Responsibility, together with the way in which a company and its partners understand sustainable value creation, will have a lasting influence on the "correct" way of handling responsibility. It can already be predicted that companies are going to be required to broaden their concept of management to include the subject of "responsibility management."

Companies which grasp these emerging changes and market requirements faster and better than others will be able to take advantage of the opportunities already there and gain a competitive edge. Nanogate AG therefore believes that the credible implementation of a well-defined strategy, based on the creation of trust between all cooperating partners, is directly linked to future corporate success.

However, sustainable value creation only works when it produces specific, measurable benefits for society and for the company itself, which will be the case when its implementation is understood properly. Thus a company's specific involvement in its region can lead to a directly improved reputation, greater employee satisfaction and loyalty, and better local conditions. Furthermore, stronger commitment to direct employee needs, such as health, work–life balance, and continuing professional development, consistently leads to reduced absenteeism, higher corporate identification, and greater success in obtaining new specialist and management personnel. Nanogate AG can give clear evidence of the contribution CSR has made to measurable results, such as those gained from the use of anonymous employee surveys. What energy consumption, active climate protection, or renewable energies are concerned, CSR contributes in a directly measurable way to a company's reduced use of resources, cost cutting, product innovation, and brand building. Not least, sustainable value creation in a company's active sales and procurement produces measurable effects in terms of higher customer satisfaction, more efficient production processes, and more successful development of new market segments. Issues worth mentioning here are the special attention and care paid to product quality and responsibility, active and responsible involvement of partners during value creation and sensitive customer management.

Overall we advocate a new, broadly-based culture of responsibility, which finds genuine reinforcement in day-to-day business. Cost effectiveness and sustainable value creation are not contradictory, but interdependent.

Given this basic understanding, small and midsized companies in particular have always provided a clear model for credible corporate social responsibility: they think in generations, are fully involved in their region and as such constitute the backbone of our society. Nanogate also sees itself as a "responsible midsized company."

5 Outlook

Since it was established, Nanogate AG has understood the acceptance of responsibility as a tool for the sustainable development of the company and has, as a result, undergone intensive learning processes. Today we are more than ever convinced that sustainable value creation and partnership strategies provide important answers to the core question, "How can we become and remain an exceptionally successful company?" In addition, we see the current debate about corporate social responsibility as a genuine opportunity for an urgently needed, improved collaboration between politics, business, and civil society. This means that for networked midsized companies there is unique potential in pooled regional projects – a potential that can be directly converted into individual competitive advantages.

Nanogate AG's history of success in all areas of the business has encouraged us to continue our route of sustainable value creation and responsibility and to view it as an integrated core area of our company's activity.

Bibliography

Nanogate. (2013a). *Das Wertschöpfungsprinzip*. http://www.nanogate.de/de/. Accessed June 15, 2013.

Nanogate. (2013b). *Press release: Nanogate will supply technology for the global leader in variable focus eyewear*. http://www.nanogate.de/de/. Accessed June 15, 2013.

Nanosafe. (2013). *Nanosafe*. http://nanosafe.org. Accessed June 15, 2013.

Audi: Raw Materials, Road, Recycling – How Life Cycle Analysis Influences Product Development

Peter F. Tropschuh and Martina Biendl

1 Corporate Responsibility in the Audi Group

Futurologists often refer to sustainability as a megatrend. For Audi, however, corporate responsibility is about using long-term thinking and action to safeguard the fundamentals of the company – and along with them, the basis for an automotive company's business. It is in the inherent interest of the Audi Group that driving an automobile remains both socially acceptable and practical for everyday use. The company will therefore design individual mobility so that it meets the requirements for sustainability. In addition to its social responsibility, Audi views corporate responsibility[1] (CR) as also involving a more general societal responsibility that includes the economic responsibility a company has for its environment (cf. Aachener Stiftung Kathy Beys 2013). This is why corporate responsibility is embedded as one of four areas of action in the Strategy 2020 "Audi – the premium brand" – under the heading of "We live responsibility." This area of activity relates to the three pillars of sustainability: society, the environment, and economy (see Fig. 1). In addition, the corporate goal of "Sustainability in products and processes" forms the basis of the Audi Strategy House. This includes the aim to make all products and processes sustainable across the entire value chain.

[1] Definition: "The responsibility of enterprises for their impacts on society" (European Commission 2011, p. 6).

P.F. Tropschuh • M. Biendl (✉)
Corporate Responsibility, AUDI AG, 85045 Ingolstadt, Germany
e-mail: cr@audi.de

M. D'heur (ed.), *Sustainable Value Chain Management*, CSR, Sustainability, Ethics & Governance, DOI 10.1007/978-3-319-12142-0_6

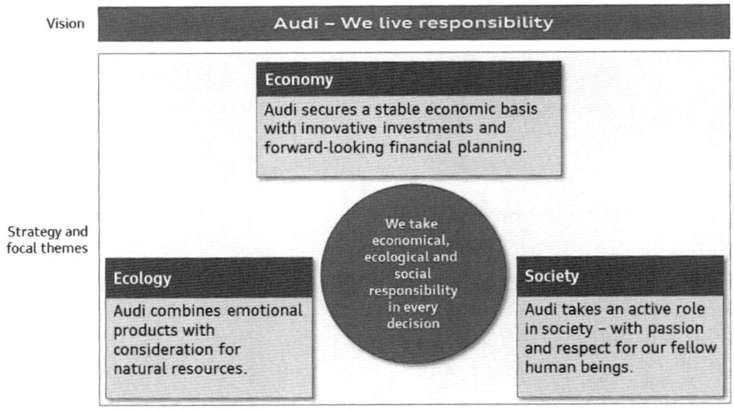

Fig. 1 Audi corporate responsibility strategy

1.1 "We Live Responsibility": The CR Strategy of Audi

For the brand with the four rings, corporate responsibility means keeping the consequences in mind when making any decision. Audi wants to be a good employer, remain competitive, inspire its customers, and protect the environment long term. The goals of the CR strategy include the long-term aim of enabling carbon-neutral mobility with Audi products, conserving resources throughout the entire value chain, and strengthening responsibility toward employees and society. The Corporate Responsibility department consolidates all activities throughout the Group that relate to sustainability. It is the assigned responsibility of the Chairman of the Board of Management.

In its CR strategy, Audi has defined five key issues for operational implementation of the corporate responsibility mission:

- Product responsibility
- Environmental protection
- Responsibility to employees
- Responsibility to society
- Responsible operations

The company has based its CR strategy on a stakeholder-management approach. The German Federal Ministry of Labor (BMAS) defines a stakeholder as "an individual or group that has an interest in a decision or activity of an organization" (BMAS 2011, p. 28). Audi has developed a multiphase process that provides for a systematic examination of the assessments and recommendations of stakeholder groups regarding all relevant issues. This is, after all, "precisely where the actual challenge lies. Companies face the question of how to achieve balance (of the key issues) in their case, and how the individual areas are to be weighted" (Schulz 2012, p. 280).

Fig. 2 Materiality matrix of AUDI AG: product as most relevant aspect (cf. AUDI AG 2013a, p. 13)

The results of the stakeholder surveys form the basis of Audi's sustainability efforts. The Stakeholder Management System introduced in 2012 is oriented on the Stakeholder Engagement Standard AccountAbility 1000, or AA1000SES (cf. AUDI AG 2013a, p. 72). This set of standards is applicable worldwide and enables an assessment of sustainability management and sustainability reporting practices. An initial analysis identifies the stakeholders that are relevant for the company. For instance, stakeholder groups within the Audi Group include customers, employees, investors, public policymakers, the scientific community, members of nongovernment organizations, and residents (cf. Tropschuh et al. 2013, p. 147).

Audi employs various forms of dialogue to communicate with groups of stakeholders. Involving these reference groups serves to reveal critical areas that require action, and to develop new guidelines, codes of conduct, or specific measures. The results and consequences of the stakeholder dialogue are incorporated into the sustainability strategy and form an important foundation for future decisions made by the company. The assessment of the relevance of significant sustainability issues are compiled in a materiality matrix, which is regularly reviewed and supplemented with new findings. The current materiality matrix shows that product-related aspects possess the highest relevance for Audi (see Fig. 2).

1.2 The Vision and Goals of the Corporate Responsibility Strategy

Audi has set carbon-neutral mobility as a long-term goal. To make this vision a reality, Audi has set specific goals for sustainability activities: the focus is on the development of alternative drive systems and synthetic fuels, on solutions for mobility for the future, on carbon-neutrality in production and on responsible organization of the value chain. Audi considers the entire life cycle of the car here, from its production to its use and, ultimately, its recycling (see Sect. 2). In all of these phases, the environmental compatibility of the products and production facilities and the environmentally sound use of natural resources play a decisive role. AUDI AG makes advanced technologies available worldwide for this purpose, and enables their use along the entire process chain. The company partners with the communities and with public policymakers at all of its sites, which allows the company to make a sustainable contribution to positive social and environmental development. Environmental management at AUDI AG – in cooperation with vendors, service providers, trade partners and recycling companies – ensures that the environmental compatibility of the cars and production sites is continuously improved.

Overarching all sustainability goals is the mindset that corporate decisions should be made with consideration of economical, environmental, and social aspects. Guidance is provided by various internal guidelines and codes, including the Audi Code of Conduct, the compliance guidelines for the Audi Management Board and numerous works agreements. The company is committed to and acts in accordance with the Universal Declaration of Human Rights, the principles of the International Labor Organization (ILO) and of the OECD, the principles of the Declaration of Rio on the Environment and Development, and the UN Convention against Corruption. In February 2012, AUDI AG joined the United Nations Global Compact; in 2014, Audi submitted its declaration of conformity with the German Sustainability Code from the German Council on Sustainable Development (cf. Rat für nachhaltige Entwicklung 2014).

2 Corporate Responsibility in the Life Cycle of a Vehicle

Mobility is always intertwined with environmental impacts. Global warming, scarcity of raw materials, mandates regarding emissions, and rising fuel prices all increase the importance of environmentally compliant product design. The automotive industry has primarily reacted by reducing fuel consumption with increasingly lightweight construction and more efficient engines, thus reducing the carbon emissions in the utilization phase of their cars. With its cars, Audi not only considers fuel efficiency, but also analyzes the environmental impacts of its products throughout their entire life cycle – from the production of raw materials and of

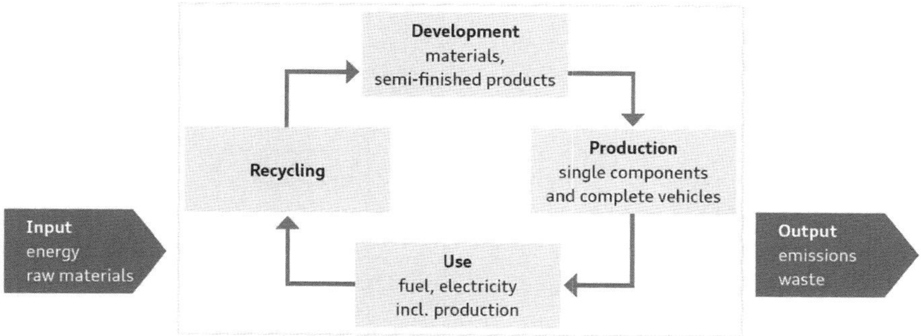

Fig. 3 Phases in the motor vehicle life cycle assessment. *Source*: AUDI AG (2014a, p. 7)

the cars to their operation and recycling.[2] To achieve the greatest transparency with respect to the company's environmental impacts, the Audi Group has determined its Corporate Carbon Footprint and is the first premium carmaker to be certified in accordance with the worldwide ISO 14064 standard. For this, the company analyzed in detail the emissions in all of the phases of an automobile's life cycle. In the certified Corporate Carbon Footprint, the emissions of Scopes 1 and 2 as well as all of the Scope 3 emissions of the Audi Group were reported. The technical experts at DEKRA Certification GmbH and DEKRA Certification Inc. reviewed and confirmed the data sources and calculations that served as the basis of the reporting. With the Corporate Carbon Footprint, Audi has made all of the company's greenhouse gas emissions along the entire value chain transparent. As a result, the company will be able to analyze these values more systematically in the future and to reduce them more effectively (cf. AUDI AG 2014e).

In order to evaluate specific environmental impacts over the entire life cycle, Audi compiles a certified Life Cycle Assessment (LCA) for each new model line in accordance with ISO 14040 ff. LCAs provide important indicators, such as the amount of greenhouse emissions in the production phase. This is helpful in the preliminary concept specifications for the product design phase, when irreversible decisions are made that have an impact on the selection of materials, the production, and the entire supply chain. The goal is to reduce the environmental impacts of each model in comparison with its predecessor. Good examples of this are the current Audi A6 and Audi A3. Both models feature improvements in all relevant environmental categories. The new Audi A3 features a better environmental footprint than its predecessor starting with the first mile driven (cf. AUDI AG 2013b). The results of the assessment consequently form the basis for product decisions and development in the interest of sustainability, and help reduce resource and energy consumption.

The phases of the life cycle (see Fig. 3) have varying degrees of impact on the environmental compatibility of a car. While production of raw materials and

[2] In the scientific discussion, cf. Schulz 2012, p. 271, among others.

production of a car with a conventional engine make up around 20 % of its assessment, the utilization phase makes up about 80 %. The input for recycling of the car, at 1 % of greenhouse emissions, is also considered. The influence of each of the phases can vary over the course of time and depending on the type of drive system, of production, and of the materials used. The following sections describe how the company influences the phases of the life cycles of its cars, the concepts that are currently being developed, and the visions Audi is pursuing to make carbon-neutral mobility possible.

2.1 Phase 1: The Extraction of Raw Materials and the Car Production

The first phase of a car's life cycle involves the production of raw materials and of the car itself. The upstream chain prior to the processing of a material or raw material in the company has a major impact on the ecological footprint of a product. This includes supplier management and logistical processes. Audi has a direct influence on its own production, primarily through site-related environmental protection measures such as careful use of resources, energy production and consumption, and through using the most efficient processes and concepts possible in the production process.

2.1.1 Supplier Management and Logistics

Supplier Management

In addition to demanding a high level of performance and competitiveness from its suppliers, the Audi Group expects them to adhere strictly to the company's sustainability standards. These apply to the suppliers for the entire Volkswagen Group, since all Group brands rely on a shared pool of suppliers in order to optimize potential synergies. All suppliers have agreed to adhere to the "Volkswagen Group requirements regarding sustainability in its relationships with business partners."[3] The concept of sustainability is aligned with principles laid down in the UN Global Compact, the OECD Guidelines for Multinational Enterprises and the relevant conventions of the International Labor Organization (ILO).

Requirements placed on suppliers include the use of an environmental management system, avoiding harm to the environment and to health during the production process, and the prohibition of child and forced labor. In addition, Audi requires that working hours and compensation fulfill national legal requirements and minimum standards. And naturally, Audi expects its suppliers to ensure sustainability

[3] Available at http://www.audi.com/cr

on the part of their own suppliers. In this way, the Volkswagen Group has firmly established the importance of environmental and social standards in its business relationships with suppliers. The concept has been continuously and systematically developed further since its introduction in 2006.

Prior to submitting offers, potential suppliers must actively confirm the sustainability requirements of the Group on the supplier platform. To lend the sustainability requirements even more weight, in 2014 they became an integral part of the contract with suppliers, and elements of a code of conduct for business partners were added. In addition to the environmental and social standards applicable up to now, they include a right to extraordinary notice of cancellation in case of violations and greater consideration of efforts to fight corruption, money laundering, and import and export controls as well as statements regarding free competition.

For Audi, the focal point of all measures is the positive development of business relationships and dialogue with our suppliers. If monitoring activities reveal a need for improvements in environmental, corporate governance, or social standards, Audi offers its support – including the involvement of Audi experts from areas such as work safety or environmental protection, if needed. Moreover, in early 2012 a mandatory eLearning qualification module on the topic of sustainability was made available in nine languages on the corporate business platform of the Volkswagen Group.

Audi is taking yet another step in its commitment to sustainable supply chains, and is actively participating as part of the lightweight construction strategy in the creation of sustainability standards for aluminum, one of the most important materials for Audi. Coordinated by the environmental organization International Union for Conservation of Nature (IUCN), Audi is a member – along with prominent companies in the aluminum development and processing industry – of the Aluminum Stewardship Initiative (ASI), which was founded in 2012. The goal of the initiative is to develop a worldwide standard for sustainable aluminum by the end of 2014. This will specify the environmental, social welfare and governance standards that apply to all stages in the mining of the raw material, its production, and its processing. With this, Audi is seizing the opportunity to influence the entire value chain for a highly critical material – one of the goals of the Audi corporate responsibility strategy.

Logistics

Audi places great importance on sustainable logistics, and continuously improves processes in order to conserve resources during the transport of structural components and cars, as well as during warehousing. In 2013, AUDI AG delivered 1,575,480 cars worldwide. Group-wide, Audi has more than 60 % of these cars transported by freight train. Among models from Ingolstadt, the figure is more than 70 %, 46 % of which are transported on trains powered by electricity generated from renewable sources. Since August 2010, Audi has been using so-called "green" trains that provide completely carbon-neutral transport between Ingolstadt and the loading port in Emden (cf. AUDI AG 2013a, p. 44). Here, Audi was able to lower emissions of CO_2 in 2013 by 7,107 tons. Since October 2012, "green" trains have also been operating on the Neckarsulm-Emden line – the most heavily frequented transport route between Neckarsulm and a loading port – saving 3,979 tons of CO_2 per year.

Audi was awarded the 2012 Logistics Sustainability Prize for its efforts to consistently pursue sustainability in its logistics. Presented by the Austrian Logistics Association and the German Logistics Association, this international accolade recognizes those transport strategies featuring resource-efficient approaches that, in some cases, already boast a zero-carbon footprint.

2.1.2 Site-Related Environmental Protection

The increased concentration of greenhouse gases in the atmosphere is leading to a warming of the Earth and, consequently, to climate change. In addition to natural processes such as the rotting of biomasses, forest fires or volcano eruptions, energy produced and consumed by humans is the main cause of the increase in CO_2. This is why efficiency has become the top priority for Audi when it comes to using resources and producing and consuming energy. As part of its environmental policy, therefore, Audi is continuously working to develop environmentally efficient processes and concepts for production, to proactively avoid damaging impacts on the environment with a focus on the efficient use of resources and energy, and to continuously improve the environmental compatibility of products and sites. The goals resulting from this environmental policy include:

- Reducing energy consumption and thus greenhouse gas emissions
- Efficient use of valuable raw materials
- Reducing water consumption
- Reducing waste

The company is therefore working hard to improve the carbon footprints of its manufacturing sites. The premium manufacturer is aiming for a 25 % reduction in its specific CO_2 emissions by 2018, based on its emission levels in 2010. The plant in Ingolstadt is already up to 70 % carbon-neutral and is thus setting new benchmarks in terms of conserving energy and resources in the production process. The site at Brussels, for instance, has been using renewable hydroelectric power since April 2012; in Neckarsulm, the engine testing center was designed in consideration of energy and environmental aspects and is equipped with state-of-the-art ventilation and air-conditioning systems, including integrated heat recovery. During test runs of the combustion engines, generators can reclaim up to 86 % of the kinetic energy and convert it into electricity. With respect to supplying energy to Audi's German sites, there are plans to reduce carbon-dioxide emissions by 40 % by 2020.

Environmental management systems at the sites ensure that environmental aspects are considered early on in all corporate decisions. The Eco-Management and Audit Scheme (EMAS) of the European Union is consistently observed at all sites of the Audi Group. Furthermore, the Ingolstadt, Neckarsulm, Győr, and Sant'Agata Bolognese sites already conform to the new DIN EN ISO 50001 standard, which stipulates especially demanding requirements for systematic and ongoing reductions in energy consumption. The sites where the Audi Group produces as part of the VW Group's network also satisfy the requirements of an environmental management system. The production sites at Bratislava (Slovakia), Martorell (Spain), Aurangabad (India), and Changchun (China) are certified according to the global standard DIN EN ISO 14001.

An annually published environmental declaration on all EMAS sites includes the specific environmental program for the site along with specific goals for improving the site-related environmental protection, as well as facts and figures on environmental aspects.[4]

The Group-wide environmental protection measures also include environmentally friendly data centers ("green IT"). The IT center completed in 2012 in Ingolstadt functions here as an example of best practices: thanks to the indirect free cooling of the servers, energy consumption in the IT center was reduced by one third and carbon emissions were reduced by 9,000 tons each year. Low-loss transformers enhance efficiency. Moreover, flywheel accumulators – which boast a considerably longer service life than conventional batteries – can supply electricity during a power outage. TÜV Rheinland has confirmed the effectiveness of the various measures and has awarded the data center the Premium Certificate for the highest level of energy efficiency. The goal of wastewater-free production is being pursued just as persistently. Where possible, Audi uses process water instead of potable water. Rainwater is collected on rooftops and parking lot surfaces, for instance, routed to the plants' own storage reservoirs and storage sewers and, after use, is cleaned in a treatment plant. As a result, 216,013 cubic meters of rainwater was collected and processed into process water in 2013.

In 2015, a membrane bioreactor (MBR) will be employed, which cleans wastewater first biologically and then with ultrafiltration technology. MBR technology will reduce the annual requirements for freshwater for production at the Ingolstadt site by up to 40 %, and the amount of wastewater is expected to drop by up to 50 %.

2.2 *Phase 2: Vehicle Utilization*

Once the production of the vehicle is finished and it is delivered to the customer, the second phase of its life cycle begins: utilization. With around 80 % of the total resulting emissions according to the current state of the technology, this phase has the greatest effect on the environmental footprint of an automobile.

[4] Download the declaration at http://www.audi.com/content/dam/com/EN/corporate-responsibility/enviroment/audi-environmental_declaration-2013.pdf

For this reason, in this phase the premium carmaker pays attention to increasing the efficiency of conventional drive systems while also searching for alternative types of drive systems and synthetic fuels. "For intelligent mobility that conserves resources, the Volkswagen Group deliberately relies not on a single technology but rather on multiple technologies – and is also setting itself apart from its competitors with its broad approach: from hybrid and natural gas to an electric drive system" (VW Group 2014, p. 19). All developments and measures are considered in an overall context of enabling carbon-neutral mobility and are aligned with the Audi Strategy 2020.

2.2.1 Boosting Efficiency and Alternative Drive Systems

Optimizing Conventional Drive Systems

With its TDI and TFSI engines, Audi has achieved important milestones in boosting the efficiency of combustion engines. The goal is to develop smaller, more economical engines that are meanwhile more efficient and feature high torque – so-called "right-sizing." High torque allows for a longer transmission ratio and therefore lower revs, without impairing the car's dynamic capabilities. The individual technologies for even better use of the fuel are consolidated in the modular efficiency platform, which incorporates technical fields that go beyond the drive system and that further increase the efficiency of the car as a whole. The elements of the efficiency platform are being implemented incrementally in new Audi cars during model changes or in product upgrades. As a result, almost all of today's Audi models come with a start-stop system as standard equipment. When the Audi Q5 was revamped in 2012, efficiency measures reduced the consumption of the TDI and TFSI engines by 15 %. In the current Audi A3 Sportback, these and other technologies have helped to reduce consumption by an average of 9 % compared to the previous model.

Reducing a car's weight by 100 kg can lower fuel consumption by 0.3 l per 100 km. Lightweight construction therefore constitutes an essential basis for boosting the efficiency of the entire vehicle system. Audi has set the goal for itself to make each new Audi model lighter than its predecessor.

Alternative Drive Systems

The combustion engine has completely dominated as the drive system for cars for more than 100 years. The supply of fossil fuels once appeared to be endless. Today we have a better understanding of the global contexts. For this reason, Audi is developing alternative drive systems that will continue to make mobility a responsible and attractive option in the future. New car models normally have a lead time of 4–5 years and are sold to the market for 7 years. Audi models have an average product lifespan of almost 20 years. With these long spans of time, it is necessary to anticipate or prepare for future developments. Gasoline and diesel continue to be important fuels. Now, electricity and natural gas are increasingly being looked at as energy sources for the automotive drive systems of the future.

The current hybrid models from Audi offer a combination of electric and combustion engines in the vehicle's drivetrain: the A6 hybrid, the A8 hybrid, and the Q5 hybrid quattro. The electric motors are powered by lithium-ion batteries.

In keeping with its electrical mobility strategy, Audi introduced the first plug-in hybrid model to the market: the Audi A3 Sportback e-tron. Its average fuel consumption is just 1.5 l of gasoline per 100 km (156.81 US mpg), which equates to CO_2 emissions of 35 g/km (56.33 g/mile). It offers a 50 km (31.07 mile) driving range in electric mode; when combined with the additional 890 km (553.02 miles) of range supplied by the gasoline engine, its total driving range is an impressive 940 km (584.09 miles). The battery of the A3 Sportback e-tron has an energy capacity of 8.8 kWh and can be charged in less than 2.5 h using an industrial electrical outlet (cf. AUDI AG 2013c). Along with the market introduction of the model, Audi offered clean electricity – known as "Audi energy" – to its customers in Germany. The cooperating partner is the Hamburg energy provider LichtBlick SE. "With Audi energy, the A3 e-tron is completely emissions-free when operated electrically" (AUDI AG 2014c). Audi is successively and systematically transferring the technology from the A3 Sportback e-tron to its midsize and luxury class model lines.

As a complement to the developments in the area of electric mobility, the A3 g-tron is a part of Audi's new, integrated, and sustainable mobility concepts. Its engine can be operated with natural gas, with the e-gas produced by Audi, or – if a CNG station is not available – with gasoline. Audi e-gas is synthetic methane that is produced at the Audi e-gas plant in Werlte, Germany – the world's first industrial power-to-gas plant (see Sect. 2.2.2). An A3 Sportback g-tron fueled by Audi e-gas is currently the most environmentally friendly form of long-distance mobility. This is because the production of Audi e-gas involves the binding of precisely as much CO_2 as will be later released by the A3 g-tron in gas-powered operation. Under these conditions, the vehicle offers carbon-neutral mobility. With normal consumption of between 3.2 and 3.3 kg per 100 km (5.1–5.3 kg per 100 miles, depending on transmission) and corresponding CO_2 emissions of 88–92 g/km (141.6–148.1 g/mile), the Audi A3 g-tron operates in consumption ranges that were previously possible only with city cars. In pure gas operation, ranges of over 400 km are possible.

The buying principle for Audi e-gas is straightforward and corresponds to how green electricity is distributed: Audi records the quantities of gas that the customer pays for using their Audi e-gas fuel card and ensures that exactly this amount of Audi e-gas is fed into the German natural gas network. The card can be used to purchase Audi e-gas at over 650 fuel stations in Germany.

2.2.2 Renewable Energies and Synthetic Fuels

When it comes to the limits of improving efficiency, Audi has decided to also make renewable energies and synthetic fuels available. In the total life cycle assessment for a vehicle, the primary energy used plays a decisive role, since the utilization phase exhibits the greatest leverage effect. For this reason, Audi places a focus on the development of renewable forms of energy – called Audi e-fuels – that can help achieve carbon neutrality in the entire mobility system, from the source to its use in the vehicle. For this, Audi is playing a part in the development and production of third-generation renewable fuels, which do not compete with the production of food or animal feed. The strategic goal of these projects is to use CO_2 as a raw material for fuels and thus to significantly improve the overall footprint. Audi is specifically considering the synthetic fuels e-gas, e-gasoline, e-diesel, and e-ethanol. In January 2014, Audi e-ethanol underwent its first-ever test cycle in the pressure chamber and in the glass engine. The results showed that the fuel displays some properties that perform even better than fossil fuels. Thanks to its chemical properties, fewer pollutants are produced in the combustion of e-ethanol than is the case with bio-ethanol. It does not contain any olefins or aromatic hydrocarbons. As a result, synthetic Audi e-ethanol provides a more effective mixture formulation, cleaner combustion and lower emissions (AUDI AG 2014b).

Audi e-gas project

The Audi e-gas project represents another important component and a pillar of the AUDI AG product sustainability strategy, and extends far beyond the automotive industry. In June 2013, Audi became the first carmaker in the world to open its own power-to-gas plant with the startup of the plant in Werlte, Germany (see Fig. 4). This project shows a way that electricity can be stored in large amounts, over long periods and independent of location – one of the biggest challenges in the German energy transition. The plant first splits water into oxygen and hydrogen. In a second step, the hydrogen is reacted with CO_2 to produce synthetic methane, or Audi e-gas. It is virtually identical to fossil natural gas and will be distributed through the German natural gas network to the CNG filling stations. The plant's energy conversion efficiency level in converting electricity to methane is around 54 %. The waste heat is consistently used in the heat sinks of the biomethane plant, which as a result no longer needs to burn raw biogas during operation of the power-to-gas plant in order to treat the biogas, to disinfect the waste, or to heat the fermenter. In all, this is expected to result in a 65 % energy conversion efficiency level, and a target value of 70 % has been set for further plants with improved electrolysis technology. Audi is leading a project funded by the German federal government to explore ways to optimize plant technology and the power-to-gas system as a whole.

Audi's plants are expected to generate about 1,000 tons of e-gas each year, which will involve the bonding of 2,800 tons of CO_2. This roughly corresponds to

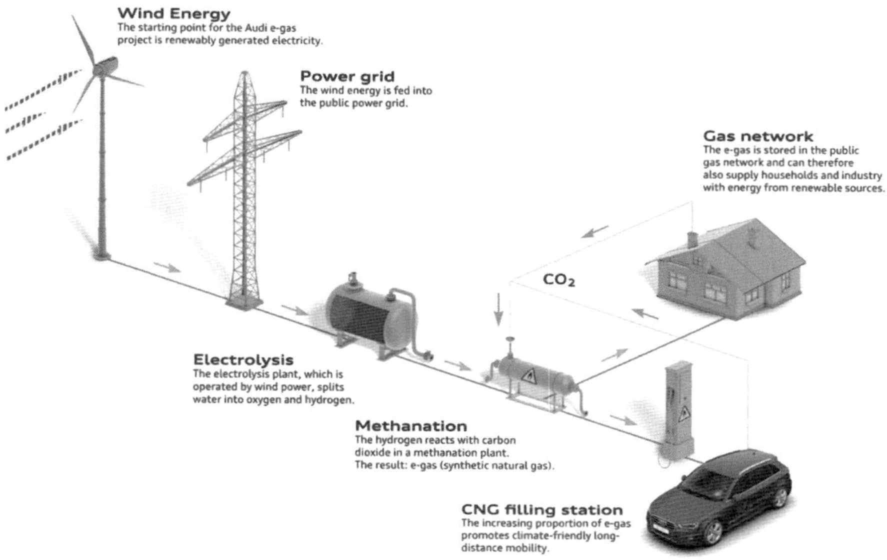

Fig. 4 Audi e-gas plant in Werlte. *Source*: Audi MediaServices (2013)

the amount that a forest of over 220,000 beech trees absorbs in 1 year. The Werlte facility will generate enough CO_2-neutral e-gas to power 1,500 new Audi A3 Sportback g-tron vehicles for 15,000 km (9,320 miles) every year (cf. VW Group 2014, p. 101). With the e-gas project, Audi is participating in and driving the energy transition. Major German energy utilities have since taken up the idea of power-to-gas cogeneration and are following Audi with initial projects of their own.

Audi e-ethanol, e-diesel, and e-gasoline

In parallel with the e-gas plant in Werlte, Audi also operates a research facility in Hobbs, New Mexico, USA, for the production of e-ethanol and e-diesel, in partnership with Joule. At this facility, microorganisms use water (brackish, salt, or wastewater), sunlight, and carbon dioxide to produce high-purity fuels for test purposes. The so-called oxygenic photosynthesis process has been modified so that the organisms convert the carbon dioxide directly into ethanol or long-chain alkanes. The latter are a key component of diesel fuel.

Audi e-ethanol has the same chemical properties as bioethanol, which is already established in the market, but is manufactured from plant-based biomass. Audi e-ethanol can be used in any admixture with conventional gasoline, including in today's established engine fuels E5 and E10, as well as in the fuel E85 that is widespread in Scandinavia and North America and that contains 85 % ethanol. Synthetic e-diesel is distinguished by its high degree of purity. It burns more cleanly than diesel made from petroleum because it contains no sulfur, nitrogen, or aromatic hydrocarbons. Using the yield of e-ethanol or e-diesel per unit area for comparison, these synthetic fuels outperform current biofuels made of rapeseed, corn, or sugar beets many times over. In addition, land that is too barren to be used for food production, such as desert regions, can be used to produce these fuels.

In early 2014, Audi also began researching the nonfossil fuel gasoline. For this purpose Audi entered into a strategic partnership with the French biotech company Global Bioenergies. This innovative technology will in the future allow renewable fuels to be produced without creating competition with food production or farmland.

2.2.3 Driving Behavior and Driving Safety

Aside from technical conditions, there is one more decisive factor that influences fuel efficiency: the person behind the wheel. The driver determines up to 30 % of fuel consumption through his driving behavior. Audi therefore strives to help the driver drive in environmentally friendly, fuel-efficient and safe manner – for example by developing and offering various driver assistance systems. These include Audi active lane assist, which warns drivers if they leave their lane, adaptive cruise control (ACC) with Stop & Go, which automatically maintains the distance from other cars, and the night-vision assistant that detects people and animals in the dark. In addition, online services show the way to free parking spaces.

Audi has set itself the goal of achieving piloted driving in cars during this decade. Electronic systems are expected to take over the steering of the car in certain situations and thus to make driving more convenient and, above all, safer. Drivers, meanwhile, will always be able to decide for themselves whether they want to drive or employ the driving pilots. One of the upcoming features is piloted parking, which was presented last year at the International Consumer Electronics Show in Las Vegas, Nevada (US). With the help of this system, an Audi can park itself independently – the driver doesn't even need to be in the car. Another aid will be the traffic jam assistant, which offers drivers relief in slow-moving traffic. At speeds between 0 and 60 km/h (37.28 mph), the system takes over the steering. It also accelerates and brakes autonomously. Technically, Audi is working to achieve a central driver assistance domain architecture. This will merge all of the available sensor information together in a central driver assistance control unit (zFAS).

While networking between the car and its environment already plays a major role in the automotive industry, its importance will continue to increase steadily, providing the driver with optimal support in the future and enabling the upcoming driver assistance systems to become reality. Car-2-X communication will make it possible for cars to communicate with each other and with the infrastructure. This will make driving a car safer and more efficient. Audi is currently acquiring new knowledge in this area with the Audi Urban Intelligent Assist (AUIA) project. The Audi Urban Future Initiative focuses its attention even further into the future, examining the interrelationship of individual transport and urban infrastructure (see Sect. 3).

2.3 Phase 3: Recycling

With an influence on roughly 1 % of the life cycle assessment of a car, recycling – that is, the return of used materials to the material cycle – is of little importance in terms of numbers. As mentioned in Sect. 2, the influence of each of the product life cycle phases can vary over the course of time and depending on the type of drive system, of production, and of the materials used. Audi is also convinced that responsible use of resources and the actual use of secondary materials in the production of structural components are essential for the future. Many highly coveted materials are found in cars. The company therefore strongly supports research into new ways to make materials reusable.

Currently, Audi can already prove the recyclability of 95 % of a vehicle's weight. The metals in particular can almost entirely be recycled for reuse. Recycled plastics are used for bumper covers, wheel arch liners, and lining of the spare wheel well. It is essential here that the recycled plastics comply with the same quality standards as primary raw materials. The reconditioning of old assemblies such as starters and alternators from used cars guarantees additional savings. The goal of recyclability also applies for supposed waste. This has almost been achieved, for instance, at the Ingolstadt plant. More than 95 % of all waste is recycled, and individual materials like steel scrap are almost completely reintroduced into the recycling loop. The recycling rate at the plants in Neckarsulm, Győr and Brussels is 90 %.

In addition, Audi has made it a goal to conserve natural resources through new recycling concepts for closed-loop material cycles and is participating in the development of a recycling concept for carbon fiber-reinforced plastics. In contrast to established materials like paper, glass, or metal, researchers are still in the early stages when it comes to the ability to recycle carbon. As an expensive material, carbon fiber-reinforced polymer (CFRP) is in direct competition with aluminum in the automotive industry. Audi is working together with research institutes and other industrial partners in the "MAI recycling" research project to find new concepts and techniques for reusing carbon fibers. The main issue for the research initiative is to make carbon practical for use in volume production and to "develop the Munich–Augsburg–Ingolstadt region into a European center of excellence for CFRP light-weight construction that covers the entire value chain of CFRP technology and helps the represented partners in the key technology of CFRP to achieve a leading position in the world market" (Carbon Composites e.V. 2013).

A similar project is concerned with accepting and recycling returned vehicles with high-voltage batteries. The high-voltage components in electric vehicles are likewise expensive products and contain scarce and valuable raw materials. In addition to developing suitable recycling concepts, these components also carry another, completely new aspect for the automotive industry. It is already clear that traction batteries that are no longer powerful enough to meet the mobility needs in cars still have sufficient capacity to be used in stationary applications. Audi is working on "second-life" concepts in which the batteries can be used in energy stores, for instance, until the actual end of their working life. The valuable substances they contain should be reintroduced into the manufacturing process of new traction batteries in order to conserve these scarce resources. As part of the LithoRec II research project, Audi is participating in the development of a recycling method that aims to ensure environmentally sound recycling of lithium-ion batteries. The goal is to use various processing steps to reclaim as much as possible of the valuable raw materials contained in the battery cells, including lithium and electrolytes.

3 Conclusion

Audi was able to further reduce the fuel consumption of its cars in 2013. A total of 150 models achieved CO_2 maximum emissions of 140 g/km (225 g/mile); of these, 63 drive system variants emitted up to 120 g/km (193 g/mile) (cf. AUDI AG 2014d). Consumption and emissions were further reduced in order to achieve the European Union's climate goal of 95 g of CO_2 by 2020 (cf. Amt für Veröffentlichungen der Europäischen Union 2014, p. 18). This goal will be difficult to achieve, however, through further increases in efficiency or downsizing of combustion engines alone. Audi is pursuing the vision of carbon-neutral mobility with all of its strategic decisions for new vehicle projects, new types of drive systems and synthetic fuels. In doing so, Audi has set a high goal that the company is seeking to achieve with an integrated approach: assessing environmental influences throughout the entire vehicle life cycle in order to recognize problematic influences as well as promising parameters, which can be adjusted to achieve sustainable mobility. Only in this way can the company develop new, progressive ideas for the future. Along the way, the power-to-gas plant in Werlte and the Audi A3 g-tron represent milestones in the energy transition. For the first time, a successful bidirectional connection between the electricity and gas grids has been achieved. Surplus electricity and operating reserve can be stored in the natural gas grid and made available for various uses. Now it is up to public policymakers to reach decisions about the future that will allow this storage technology for renewable energies to be developed to maturity in the actual grid environment.

In considering the mobility of the future, one must also investigate what the world of the future will look like. The world is changing at a rapid pace, and population is growing ever more quickly. Audi launched the Audi Urban Future Initiative 4 years ago to discuss the issues resulting from this development and to find new solutions. This forum brings together specialists from various disciplines, cultures, and perspectives. Their joint discussions target an analysis of the challenges of tomorrow's mobility in the megacities of the world and seek to find possible solutions. Besides covering technical aspects, the findings and ideas resulting from this process also incorporate social, environmental and aesthetic aspects. For Audi, making individual mobility available in the future as well means more than just examining the right technologies. It means placing individuals and their needs, as well as responsible use of resources and conserving the environment, at the center of all corporate activities.

Bibliography

Aachener Stiftung Kathy Beys. (2013). *Corporate Social Responsibility*. http://www.nachhaltigkeit.info/artikel/corporate_social_responsibility_unternehmerische_1499.htm. Accessed 24 Mar 2014.

Amt für Veröffentlichungen der Europäischen Union (2014) *Amtsblatt der Europäischen Union*. L 103. Luxembourg, 4 May 2014.

AUDI AG (2013a) *Audi Corporate Responsibility Report 2012*. Ingolstadt.

AUDI AG (2013b) *Audi involved in standard for sustainable aluminum*. Press Release, Ingolstadt, 28 February 2013.

AUDI AG (2013c) *Audi A3 Sportback e-tron at the starting line*. Press Release, Ingolstadt, 10 September 2013.

AUDI AG (2014a) *Audi e-gas project. Life Cycle Assessment*. Ingolstadt.

AUDI AG (2014b) *Audi e-fuels pass the test in glass engine*. Press Release, Ingolstadt, 31 January 2014.

AUDI AG (2014c) *Audi energy: Audi and LichtBlick offer green electricity*. Press Release, Ingolstadt, 7 March 2014.

AUDI AG (2014d) *Audi 2013 Annual Report*. Ingolstadt.

AUDI AG (2014e) *Maximum transparency: Audi receives DEKRA certificate for carbon footprint*, Ingolstadt, 15 April 2014.

Audi MediaServices (2013) *Heavy goods transport Audi e-gas project*. Image A3g130027. Ingolstadt, 25 January 2013. https://www.audi-mediaservices.com/publish/ms/content/en/public/fotos/2013/01/25/A3g130027.html. Accessed 11 Apr 2014.

BMAS. (2011). *Die DIN ISO 26000. Leitfaden zur gesellschaftlichen Verantwortung von Organisationen – Ein Überblick*. Bonn: Bundesministerium für Arbeit und Soziales.

Carbon Composites e.V. (2013) V*om Verschnitt zur Wiederverwertung: MAI Recycling nimmt Fahrt auf*. Press Release, Augsburg, 5 December 2013.

European Commission (2011) *Communication from the commission to the European parliament, the council, the European economic and social committee and the committee of the regions. A renewed EU strategy 2011 – 14 for Corporate Social Responsibility*. Brussels.

Rat für nachhaltige Entwicklung (2014) *Deutscher Nachhaltigkeitskodex, AUDI AG*. Berlin. http://datenbank.deutscher-nachhaltigkeitskodex.de/DNKProfil/DNKHome.aspx?CompanyID=5373&lang=de&year=2012. Accessed 7 Mar 2014.

Schulz, O. (2012). Nachhaltige ganzheitliche Wertschöpfungsketten. In A. Schneider & R. Schmidpeter (Eds.), *Corporate social responsibility: Unternehmensführung in Theorie und Praxis*. Berlin/Heidelberg: Springer Gabler.

Tropschuh, P. F. et al. (2013) CSR-Kommunikation in der Automobilindustrie. In P. Heinrich (ed.) *CSR und Kommunikation. Unternehmerische Verantwortung überzeugend vermitteln*. Management-Reihe Corporate Social Responsibility. Berlin/Heidelberg: Springer Gabler.

VW Group (2014) *Group Sustainability Report 2013*. Abridged. Wolfsburg: Volkswagen Group.

Symrise and Vanilla: Tradition, Strategy, and Total Commitment

Stephan Sielaff, Christina Witter, and Clemens Tenge

1 Proven Sustainability: By Tradition

Symrise is one of the most sustainable companies in Germany, as demonstrated by its win of the German Sustainability Award 2012.[1] Symrise convinced the jury in the category "Germany's Most Sustainable Initiative," with its well-established commitment to sustainable vanilla cultivation and vanilla farmers in Madagascar. One of Symrise's predecessor companies – Haarmann & Reimer – laid the foundations for this success when it synthesized vanillin successfully for the first time in 1874.

The founding fathers of Symrise developed the very first approaches to sustainable business activity that many years ago. They chose the site for their headquarters in Holzminden, Germany, with great forethought – large volumes of raw materials grew in the immediate vicinity. This meant that lavender, Reseda, and lupin plantations were located near the company right from its early days, from which natural plant extracts could be obtained. This kept transport distances between the fields and the production site short and manufacturing costs low. The company also had access to European transport routes via the river Weser.

[1] This award is an initiative of the Deutsche Nachhaltigkeitspreis e. V. foundation in cooperation with the German Federal Government, trade associations, civic organizations and scientific institutes. Since 2008, the award has been given to leading companies that prove business success is compatible with social responsibility and environmental stewardship.

S. Sielaff (✉)
Archroma - Archroma Management GmbH, Neuhofstrasse 11 (AB 01/09), 4153 Reinach, Switzerland
e-mail: stephan.sielaff@archroma.com

C. Witter • C. Tenge
Symrise AG, Mühlenfeldstrasse 1, 37603 Holzminden, Germany
e-mail: christina.witter@symrise.com; clemens.tenge@symrise.com

M. D'heur (ed.), *Sustainable Value Chain Management*, CSR, Sustainability, Ethics & Governance, DOI 10.1007/978-3-319-12142-0_7

Today, Symrise produces thousands of products for the fragrance and flavoring sectors, including cosmetic active ingredients, raw materials, and functional ingredients. A balance between customer proximity and availability of raw materials, personnel, and technology is vital when choosing site locations, and Symrise has purposely applied these criteria to its international production sites where important core competencies are concerned. Similarly, the backward integration within the citrus value chain in Brazil and vanilla value chain in Madagascar provide further proof of this approach, alongside the availability of plant material in the immediate vicinity of the site in Holzminden, Germany, where Symrise processes the raw materials locally. Short transportation distances are still part of the Symrise culture, as they have been for over a 100 years. This is because supplier–producer proximity provides more than just logistical benefits. All of Symrise's partners in the supply chain are aware of the factors influencing value creation and how they can help to control and optimize them (German Sustainability Award 2012).

2 Towards an Integrated Corporate Strategy

2.1 *Strategy and Objectives*

Symrise sees economic success as being inextricably linked with sustainable and socially responsible corporate management. Symrise has determined a set of values reflecting its corporate culture as a whole, forming a stable basis for the way it thinks and acts as a company. "Because we care" is the guiding principle for its voluntary commitment to a holistic interpretation of entrepreneurial activity.

Companies must consider sustainability factors at all levels if they want to continue increasing corporate value and differentiate themselves on the market. Symrise's corporate strategy is based on the three pillars of growth, efficiency, and portfolio. Sustainability provides the framework for implementing this strategy and is a fixed component of the Symrise business model, enabling it to acquire a distinct competitive advantage.

Symrise has the long-term aim of positioning itself as the most successful company in the fragrance, flavoring, and functional ingredients sector, strengthening its market position and consolidating its independence. This means that sales are intended to grow faster than average market volumes over the long term, which will allow Symrise to win market share and outperform its competitors.

Symrise is well on the way to combining economic, ecological, and social objectives, but still has a long way to go. The Group has therefore set clear goals and guidelines for business development and sustainability (Fig. 1).

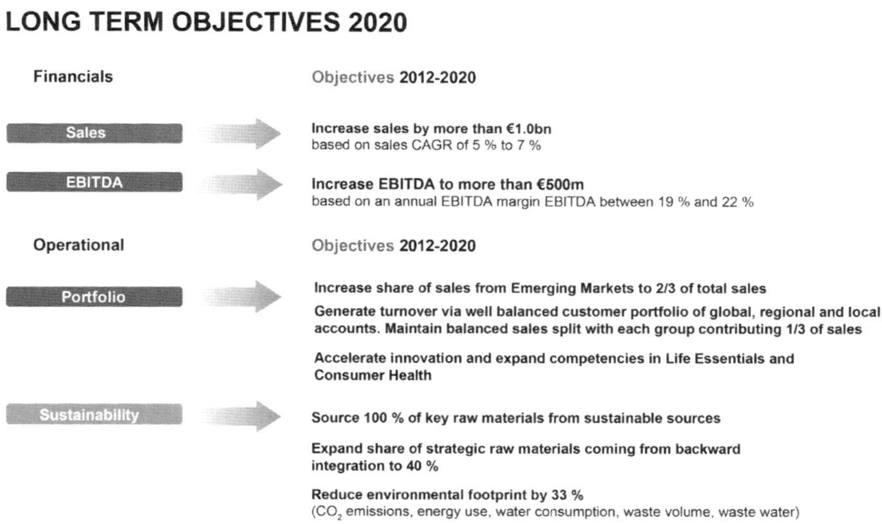

Fig. 1 Long-term Symrise goals for 2020. *Source*: Symrise AG (2013)

2.2 Corporate and Financial Reporting

Integrated strategies need to resonate in all corporate publications and reports. Symrise changed its reporting procedure for the 2012 financial year radically compared to the reporting system in place in 2011 in order to achieve this and demonstrate its approach more clearly. The corporate report 2012 provides an integrated depiction of Symrise's performance in 2012 – from both the economic and the sustainability viewpoint. Entitled "Sharing Values," the report shows how Symrise considers the requirements of various stakeholder groups at all stages of value creation and generates value for all concerned.

Symrise has developed its reporting strategy continuously since publishing its first corporate social responsibility report entitled "We have the Future in Mind" in 2006. Symrise combined its financial and sustainability reporting for its 2012 corporate and financial report, introducing integrated reporting for the first time. The Symrise Group is focused on the continuous consolidation of sustainability within the business model. The objective is to provide information on how Symrise activity generates sustainable added value for all relevant stakeholder groups. Symrise has used the report to address customers, employees, investors, and other stakeholders, thus creating a platform for dialogue with the company.

Symrise has aligned itself with the currently applicable and internationally recognized G3 guidelines of Application Level A of the Global Reporting Initiative (GRI) (GRI 2013) to provide transparency and a wider context for the data and information in its corporate report. The Global Reporting Initiative (GRI) has developed a comprehensive framework for sustainability reporting that can be applied worldwide. Symrise decided to adopt this particular reporting framework because it affords customers, analysts, and investors the highest degree of comparability with other reports. The GRI standard also provides Symrise with the optimum platform for illustrating its commitment to sustainability in a transparent, structured, and measurable way.

Along with the reporting guidelines, this reporting framework provides principles and indicators that organizations can use for measuring their economic, ecological, and social performance. Symrise's reporting conformed to the requirements of Application Level A – the highest level – of the GRI [see page 118 of the Corporate Report (Symrise 2013)]. The 2012 report also constitutes the company's first progress report (Communication on Progress – COP) (UN 2013) with reference to the United Nations Global Compact. Symrise is also the only manufacturer of fragrance and flavoring ingredients issuing this level of comprehensive, transparent reporting on its business activity – as confirmed by the GRI.

Symrise became a member of the United Nations Global Compact, the world's biggest voluntary CSR initiative, in 2012. In doing so, Symrise AG has made a commitment to the Compact's objective of harmonizing business activity with social and environmental concerns.

3 Sustainable Business Activity

Symrise sees responsible corporate management as an important component of its integrated strategy. It is the prerequisite and essential basis for both corporate success and sustainable value creation. Success is based first and foremost on the trust that customers, employees, and investors have in companies. The aim is therefore to acknowledge and strengthen that trust.

3.1 Stakeholder Dialogue

The success of Symrise as a company is essentially dependent on the fact that corporate processes are aligned as accurately as possible with the requirements of those with a valid stake in the company. Customers, employees, shareholders, external lenders, neighbors, politicians, NGOs, and business partners count amongst the most important stakeholder groups.

Active exchange with these stakeholders helps Symrise to take expectations on board and develop convincing solutions. Symrise uses a materiality analysis as a central tool for assessing issues in terms of their strategic relevance for the company and its stakeholder groups. In recent years, Symrise has examined the global management of all stakeholder groups in all of its segments. Workshops have been conducted to establish the actual or potential influence of corporate behavior on various stakeholders and vice versa.

Symrise uses many different media and communications channels to hold dialogue with its stakeholders – both internally and externally. The approach taken towards this dialogue is intensive. Thus, 2012 saw another employee survey conducted at the largest and most important company locations around the world, giving over 80 % of employees the chance to evaluate the strengths and weaknesses of their company as well as an opportunity to be involved in shaping it.

Symrise also uses the Supplier Ethical Data Exchange (SEDEX 2013) platform as an important tool for assessing suppliers as well as making internal information available to customers. Like many of its suppliers, competitors, and customers, Symrise provides information on working conditions, employee rights, health and safety, environmental issues, and ethical business practices via the platform provided by the nonprofit organization SEDEX.

In the 2012 financial year, Symrise conducted a comprehensive international survey of its major stakeholder groups, including over 600 customers, employees, investors, suppliers, neighbors, media representatives, NGOs, rating agencies, and industry associations, asking them to share their views on sustainability issues and performance within the individual areas of the value chain with Symrise.

Symrise conducted a critical review of its materiality matrix based on the results of the survey, adapting it where necessary. The areas where stakeholders have particularly high expectations of the company were discussed and evaluated by the Sustainability Board (Fig. 2).

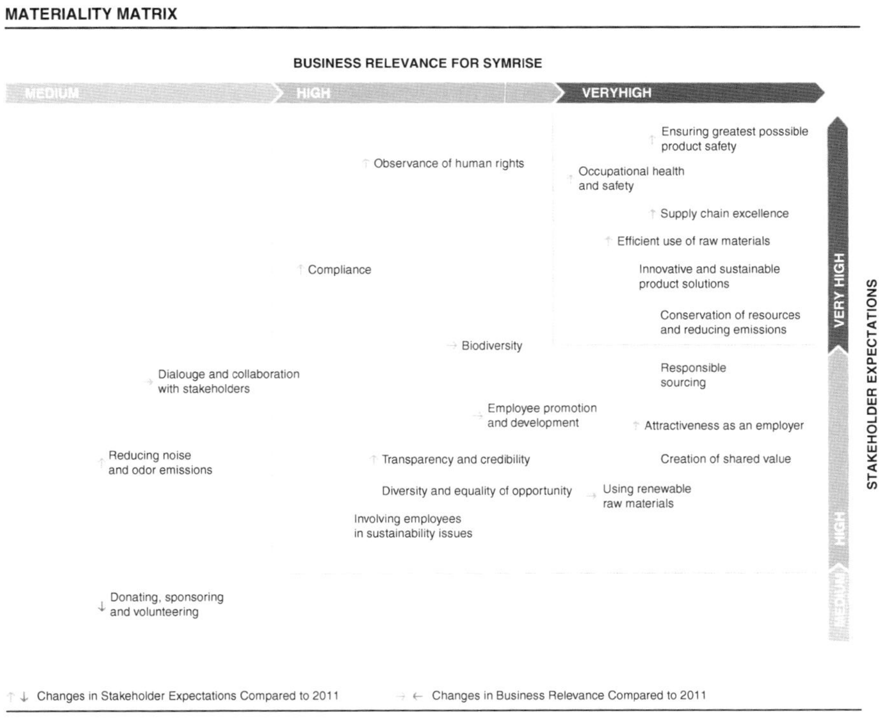

Fig. 2 Symrise materiality matrix. *Source*: Symrise AG (2013)

The results of the evaluation of Symrise's sustainability performance reveal that about three quarters of the respondents are satisfied or very satisfied with the strategic approach to sustainability; furthermore, they rate Symrise's performance as good or excellent for this aspect compared to that of competitors.

There is generally a strong correlation between those issues considered as relevant by the most important stakeholder groups and a positive evaluation of Symrise's performance in these areas. Symrise was rated with good to very good performance across all areas on average.

3.2 Sustainability Within Core Business

3.2.1 From Strategy to Implementation

Cross-functional control is required to ensure that the company incorporates all of the issues identified as essential in stakeholder dialogue in its strategies and processes. With this in mind, Symrise created its interdivisional Symrise Sustainability Board in 2009. It is made up of senior employees and sees to both the development and implementation of sustainability-related issues along the whole extended value chain and makes sure that the concerns of major stakeholders are taken into account. As a committee ensuring the control and implementation of corporate sustainability strategy, the Sustainability Board is made up of Symrise managers only. Thanks to constructive exchange with external experts across the entire sustainability spectrum, the Sustainability Board receives both feedback and inspiration on an ongoing basis.

Sustainability objectives set by the Sustainability Board are implemented directly within the business divisions. To facilitate this, the Executive Board and Sustainability Board appointed sustainability managers for the Flavor & Nutrition and Scent & Care divisions in 2011. Joined by representatives from Corporate Compliance and Corporate Communications, they are known as the Sustainability Core Team at Symrise. Led by Corporate Compliance, the Sustainability Core Team manages the company-wide sustainability strategy. The CEO of Symrise AG is directly responsible for the strategy. The Executive Board receives quarterly reports on the progress of all sustainability activities. Corporate sustainability objectives also feed into individual management targets and salaries.

A global network of volunteer sustainability ambassadors raises awareness of our sustainability strategy locally and within the organization as a whole. "Care Teams" at all regional company headquarters and numerous other sites throughout the world also use targeted initiatives to help embed sustainability even deeper within our business model in a range of different ways.

3.2.2 Risks and Opportunities

For Symrise, sustainable value creation is just as much about eliminating potential dangers to the environment, society, and the company itself as about identifying and exploiting opportunities created by global megatrends at an early stage.

The Symrise risk management system incorporates sustainability issues as a subcategory of all risk categories – from environmental and sector risk to sourcing and product risk all the way to operational risk. Sustainability issues specifically cover the risk arising from advancing climate change and the progressive depletion of finite raw materials. Consequences such as fluctuations in the availability of raw materials, rising energy prices, and stricter government regulations are a potential risk to sustainable corporate success and are therefore specifically considered as part of the sourcing strategy, product development, and end-to-end value chain processes.

Having said that, megatrends such as climate change and resources depletion are also leading to a series of potential opportunities – such as increased sales for certain sustainability-related product families due to changing consumer behavior. Symrise is also convinced that its policy of implementing demanding global integrated environmental and social standards to adapt to regulatory frameworks in emerging and developing countries offers it a competitive advantage.

3.2.3 Responsible Sourcing

Sourcing raw materials is naturally the first aspect that needs to be addressed in terms of sustainable value creation. Divergent stakeholder expectations and many practical challenges must be faced when implementing sustainable sourcing principles. Symrise applies forward-looking business practices to address these challenges in a responsible and future-oriented way (Fig. 3).

Symrise uses some 10,000 natural and synthetic raw materials from over 100 countries in the manufacture of its products. They form the basis for the production of 30,000 or so different products for the fragrance and flavoring sectors, including cosmetic active ingredients, raw materials, and functional ingredients. The vast majority of these products are creative compositions made from many different raw materials. Apart from carriers and solvents, Symrise's products

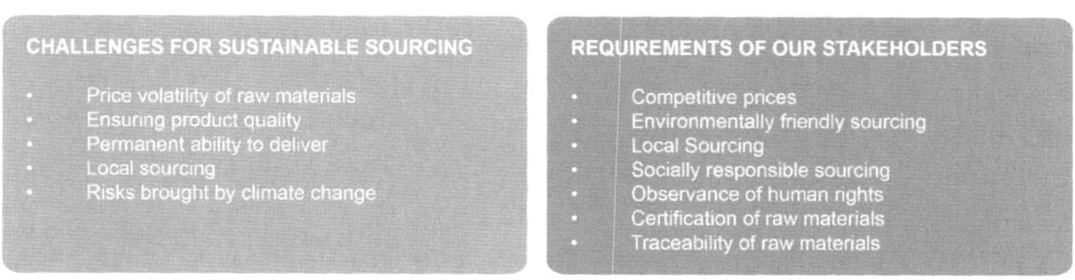

Fig. 3 Sustainable sourcing – the trade-off. *Source*: Symrise AG (2013)

mainly consist of value-adding preliminary products, i.e., aromatic components and active components, and these form the main focus of Symrise's sourcing strategy.

Long-term growth is only possible if companies are able to ensure the long-term availability and quality of raw materials and maintain competitive pricing. Given that some of the critical raw materials for Symrise are only used in the fragrance and flavoring industry and certain primary products are only required in small quantities, options for acquiring market power and influence are extremely limited in some supply chains. This is a major challenge – in terms of both sourcing materials and supplier management.

On the one hand, these raw materials are competing for space with foodstuffs and biological fuel production. On the other hand, most of the primary products come from countries with problematic social conditions. Given that this context affects the quality, availability, and price of the raw materials used, Symrise needs to consider long-term economic viability as well as responsible sourcing practices if Symrise is looking to establish and roll out the highest ecological and social standards across all of its purchasing processes. Symrise has redesigned its purchasing department's organizational structure in response to this complex and dynamic set of factors.

Symrise has introduced Balanced Scorecard Systems for the optimum control of purchasing and production processes. This enables corporate processes to be monitored via KPIs. The Production Scorecard is aligned with the Balanced Scorecards currently in use and applies the five basic analysis and control criteria: Occupational Safety, Service, Quality, Cost, and People/Motivation. The individual Scorecards also have 20 Performance Indicators (PIs) (Fig. 4).

Symrise has continued to develop and optimize the definition and measurement of these KPIs internally over the past few years. These efforts are designed to provide division managers with the most powerful and targeted KPIs for the continuous improvement of internal operational processes (OpEx – operational excellence).

In 2012, Symrise adapted its Production Scorecard System for use in raw materials purchasing. The Purchasing Scorecard analyzes individual purchasing groups, namely similar and related materials, such as juices and spices. The Purchasing Scorecard, as opposed to the Production Scorecard, has only four core areas: KPIs for the Purchasing Group, Financial KPIs, Service, and Quality. There are a number of PIs in addition to the KPIs, such as single sourcing percentage, for example (Fig. 5).

Long-term strategic thinking is prioritized over short-term cost-savings. The responsible sourcing model also follows this approach: it is based on long-term

Fig. 4 Balanced Scorecard: Production. *Source*: Symrise AG (2013)

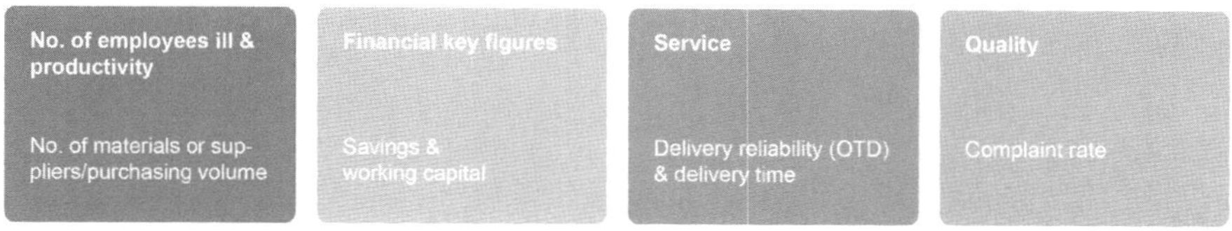

Fig. 5 Balanced Scorecard: Purchasing. *Source*: Symrise AG (2013)

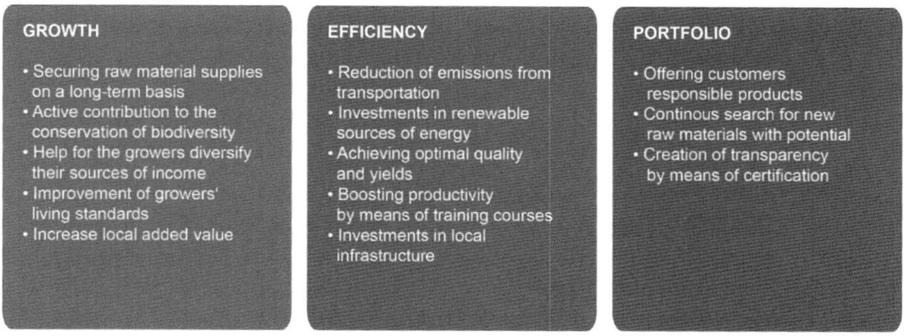

Fig. 6 Symrise AG strategic pillars. *Source*: Symrise AG (2013)

supplier contracts and cooperations, local social and ecological commitment – and strategic partnerships with key customers.

Responsible sourcing draws on the three strategic Symrise pillars (Fig. 6).

Symrise purchased around 200,000 tonnes of raw materials for product processing in 2012 alone. Process losses are relatively small, occurring mainly in extraction and distillation processes. Wherever possible and advisable, Symrise tries to reduce losses through the introduction of new procedures, continuous improvement of existing procedures and recycling.

3.2.4 Sustainable Raw Material Base

End customers in the fragrance and flavoring industry are increasingly looking for natural and renewable raw materials; however, it is impossible or nearly impossible to replace many of the synthetic raw materials with renewable ones, given the current situation. In fact, already more than half of the raw materials in the Symrise portfolio are natural in origin and this percentage is on the rise. However, refraining from the use of synthetic raw materials entirely is not only economically unfeasible in many cases, but would also entail complex ecological and social consequences. The increased use of renewable raw materials involves not only more intensive land use, but also entails competition for land used to grow food and the preservation of biodiversity, which is in turn an important prerequisite for cosmetic and medicinal product innovation.

The production of synthetic raw materials often consumes less energy than harvesting and transporting the natural version. The amount of value-adding substances found in plants is also highly dependent upon climatic and geological conditions – a risk that does not apply to synthetic production. The use of fossil resources in the manufacture of chemical products represents only a minute portion of global consumption. Symrise is therefore aiming to further reduce its dependency on finite raw materials. The company has therefore focused one aspect of its research on identifying alternative ways of obtaining synthetic substances.

The decision regarding whether to use a natural substance or its synthetic equivalent depends on the individual product. Regardless of the source of the raw material, Symrise prioritizes a sustainable approach to resources and tries to maximize yields. The company also requires its suppliers to adopt this strategy. Only this kind of far-reaching, integrated approach can guarantee economic success into the future.

3.2.5 Supplier Management

A systematic approach to supplier management is required to ensure that suppliers adhere to corporate guidelines with regard to quality management, product safety policy, and environmental management. The predecessors to Symrise, Haarmann & Reimer, and Dragoco, had already introduced a supplier management system. An increasing focus is being placed on sustainability factors. Symrise has therefore updated and greatly extended processes with regard to these factors in recent years. Joint implementation with key suppliers is the top priority.

In addition to providing information in the areas of quality, product safety, and the environment, new suppliers are required as part of a detailed approval process to submit information on their sustainability programmes, the adoption of CSR in their supply chains and the consideration of ethical standards. An honest and open response to the supplier acceptance questionnaire is a precondition for entering into a business relationship. Symrise supports the protection of universal human rights, the refusal of any form of child or forced labor and adherence to legal occupational safety and health protection requirements. Symrise also requires its suppliers to allow their employees the right of assembly and collective bargaining, to adhere to at least statutory requirements for working hours and employee pay and to adopt a clear policy against discrimination and ill treatment.

These standards are part of the company-wide risk management system. Symrise conducts systematic risk and performance evaluations for all its suppliers based on economic, ecological, and social factors. See Fig. 7 for risk analysis criteria.

This analysis produces a weighted global risk matrix, serving as a basis for supplier audit planning. Carriers and packaging suppliers are taken into consideration as well as raw materials suppliers. Where suppliers have a significant risk factor, where raw materials are of critical importance or where problems have been revealed during the evaluation process, Symrise conducts audits in line with a globally defined procedure. These audits focus on product safety, quality

Fig. 7 Supplier base risk analysis criteria. *Source*: Symrise AG (2013)

- Purchasing volume with the supplier in question
- Strategic importance of the raw material
- Number of alternative sources of supply
- Result of the supplier qualification
- Supplier's country of origin

management, occupational safety, health, the environment, and CSR. If a supplier fails to meet one or more of the standards set by Symrise, they are informed in detail and joint measures are devised and implemented to support the supplier. If the supplier fails to meet the required standards, a sanction will be imposed in the form of reduced order volumes placed with that supplier. As a last resort, the business relationship with the supplier is ended.

In the medium term, Symrise plans to ensure that essential ethical and ecological standards are adhered to for all raw materials used. Symrise uses international platforms such as the Supplier Ethical Data Exchange (SEDEX) to achieve this aim. Symrise renewed its membership with the SEDEX nongovernmental organization in 2011 with the aim of shoring up the supply of sustainable raw materials and systematically identifying risk. Along with many of its suppliers, competitors, and customers, Symrise posts information on working conditions, employee rights, health and safety, environmental issues, and ethical business practices on the SEDEX platform. Symrise has made its internal ethical data available to many of its customers via SEDEX since 2006. All Symrise production sites are now registered with SEDEX. Symrise has been an "AB Member" since 2011 and has started to evaluate its own suppliers in this way. The Group selected its first suppliers based on its own risk analysis. Symrise first checks whether suppliers are already SEDEX members and requests access to their SEDEX data. If they are not yet on SEDEX, Symrise advises them to join. The company will be evaluating its 250 top suppliers according to SEDEX criteria by 2015, covering over 80 % of its purchasing volumes.

3.2.6 Standards and Certifications

National and international legislation for the protection of natural resources needs to be part of the purchasing policy of any sustainability-oriented company, and this has been par for the course for Symrise for a long time now. The Group only permits the use of plant- and animal-based raw materials listed under CITES (Convention on International Trade in Endangered Species of Wild Fauna and Flora) (CITES 2013) if a guarantee is given that the supplier can demonstrate that the raw materials are from a sustainable source. Symrise makes considerable effort to source raw materials with certified origins whenever natural raw materials are used as an essential ingredient.

Some 1,500 Symrise raw materials are available worldwide for use in organic and fair trade products. This represents around 15 % of the range of raw materials available in the EAME region (Europa, Africa, and the Middle East). Given that Symrise products often only constitute a small percentage of the final product volume-wise, customer demand for certified products has so far mainly been directed at particularly iconic ingredients for superior quality end products, such as vanilla, for example. The growing consumer demand for certified responsible products is providing clear orientation for product development and raw material selection. The percentage of certified products will most probably continue to increase. In 2012, Symrise helped over 1,000 vanilla farmers with whom the company works directly to acquire Rainforest Alliance certification in Madagascar.

3.2.7 Local Commitment

When companies like Symrise purchase their raw materials locally and simultaneously integrate themselves within the local economy by producing locally and creating value, they can exert influence on factors such as cultivation methods, fruit selection, and harvest times, generating benefits such as high yields, secure supplies, top quality, and the highest levels of traceability. Symrise also supports local socio-economic structures and contributes to local value creation. This helps Symrise to reduce its costs, lower its transport miles, and reduce its climate-damaging emissions considerably.

The most obvious example of this is vanilla cultivation in Madagascar. The island country off the coast of East Africa is known for its exquisite spices, including vanilla, pepper, and cloves. It has a unique level of biodiversity: around 80 % of the indigenous plants grow nowhere else in the world. As the only company in its industry, Symrise has a site in Madagascar and is committed to preserving the biodiversity there.

4 Vanilla Cultivation in Madagascar

4.1 From Orchids to Ice Cream

Vanilla is the most popular sweet taste direction in the world. Symrise customers use over 500 different vanilla flavorings. What would ice cream be like without the vanilla flavor? Unthinkable. That is why Symrise has very good reasons for committing to sustainable vanilla cultivation in Madagascar.

Vanilla planifolia, to give it its botanical name, is an evergreen orchid. It can grow up to three metres high. Food manufacturers mainly use Bourbon vanilla, which grows on the Bourbon islands of Madagascar, Réunion, Mauritius and the Comoro Islands. The farmers in Madagascar pollinate the vanilla flowers with little wooden sticks, as there are no bees or hummingbirds on the island to do this job.

Around 80 % of the vanilla processed around the world comes from Madagascar. But the route is long and arduous before the orchid fruits can begin their metamorphosis into the brown gold prized so highly by countless consumers. Vanilla farmers have to pollinate 500 flowers to produce a single kilo of vanilla beans.

Vanilla also grows deep in the primeval forest. Plantation cultivation is impossible or at least generates negligible profits. So thousands of smallholders earn their livings from tiny plots of land in the heart of Madagascar's rainforest. They nourish and care for the orchids for an entire year, finally harvesting the ripe beans and treating them in a traditional fermentation process.

As fascinating as the world of vanilla might be, Madagascar is also a country facing many challenges. Over half of the ever-growing population lives on less than one dollar a day. Although 80 % of the people are farmers, productivity levels simply are not high enough. Many farmers are forced to cut down more of the forest just to survive. The unique biodiversity is gradually disappearing as a result.

Barely 10 % of the population has access to electricity or clean drinking water. There are just three hospital beds for every 10,000 people and hardly anyone can afford to see a doctor. The quality of schooling is also poor and there are not enough teachers. Only half of the students leave with any qualifications after 5 years. The international and national transport infrastructure is also inadequate. The roads are eroded and very few are asphalted. Political uncertainty and climate threats such as cyclones add to the problem.

All of these factors make purchasing through intermediaries difficult, to the detriment of the farmers and the quality of their products. Symrise has taken up the challenge of making this market a sustainable one – for itself and particularly for the local farmers.

There is a tradition behind such commitment. Symrise and vanilla have gone hand-in-hand since the company was founded and the first vanillin was synthesized in 1874. Back then, two young scientists discovered a process that was to lead the way for an entire industry. Wilhelm Haarmann and Ferdinand Tiemann discovered that they could obtain vanillin from coniferin, the liquid extracted from the cambium of coniferous trees in two simple chemical steps. In 1875, Haarmann founded a vanillin plant in his home town of Holzminden, Germany, to exploit this synthetic route on a large scale. This was the first factory in the world to manufacture a synthetic flavoring.

Working with Karl Ludwig Reimer, Haarmann managed to produce vanillin in a much cheaper process using eugenol, a component of clove oil, a few years later. Reimer joined the company, which went on to trade under the name Haarmann & Reimer. Vanilla was as valuable as gold at the time and only a few people could afford the natural flavor. Today there is a return to the natural source of vanillin. This is because consumers around the world want the natural taste. Symrise therefore sources a large part of its raw materials for the most popular sweet flavor from the vanilla bean.

Symrise is committed to Madagascar because this is the only place where the world's favorite Bourbon vanilla thrives. The availability of natural vanilla forms the basis of competition. Symrise therefore decided 7 years ago to gradually extend the vertical integration of its vanilla value chain, in order to ensure a long-term supply of this limited commodity. Symrise acquired a majority holding in Aromatics Sarl, a vanilla producer in Madagascar, in 2006. This was followed by total acquisition in 2008 and the vertical integration of this French company.

Symrise has always kept its position as a pioneer in mind, and was the first company to invest in Madagascar – years before the first competitors came along.

4.2 The Traditional Supply Chain

If one looks at a typical vanilla value chain, one thing becomes clear: companies in the fragrance and flavoring industry and their customers usually have no direct access to the farmers, who sell their vanilla to intermediaries. These intermediaries then sell larger volumes of vanilla on to processors and refineries. It is not until this point that vanilla finds its way via exporters to flavoring manufacturers that use the vanilla to make vanilla extracts and flavorings for the global market. The flavoring manufacturers supply these extracts and flavorings to the food industry for use in a wide variety of products: ice cream, yoghurt, chocolate, and many more. This system makes it almost impossible to trace the product back to the vanilla farmers or exert any influence on social or ecological factors (Fig. 8).

4.3 The Sustainable Symrise Supply Chain

The Symrise vanilla supply chain naturally starts with the smallholders – with one major difference. As the first flavor producer, Symrise has made a comprehensive investment in a sustainable working relationship with the vanilla farmers based on

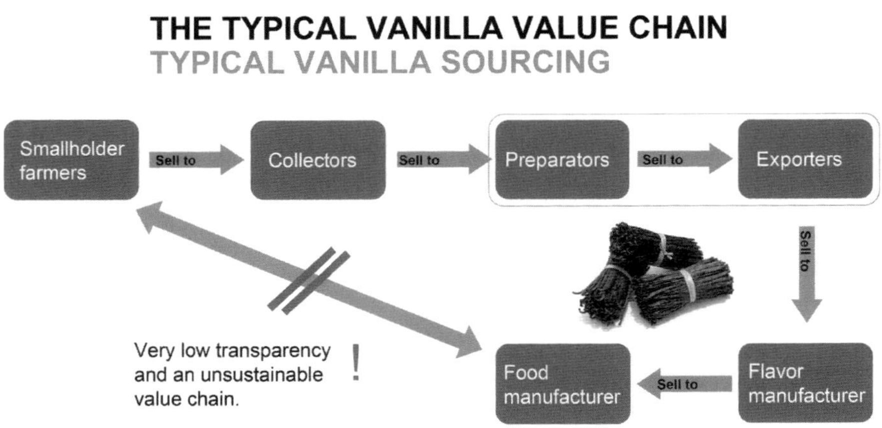

Fig. 8 Traditional value chain for black vanilla beans in Madagascar. *Source*: Symrise AG (2013)

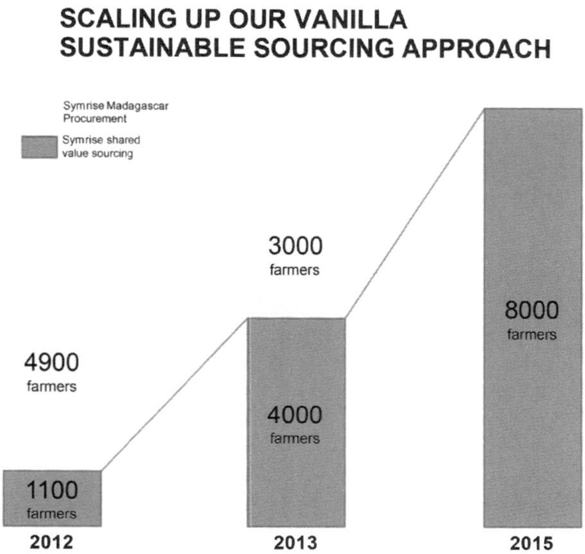

Fig. 9 Development of the sustainable approach to sourcing. *Source*: Symrise AG (2013)

trust. The company worked DIRECTLY with over 1,000 farmers in 2012. Only a year later – 2013 – this figure increased fourfold and is set to be eight times as high by 2015 (Fig. 9).

Symrise employees maintain a close relationship with the farmers to ensure that their collaboration is as effective and profitable as possible in terms of value creation for both sides. Symrise pays a premium price above the global market rate to the farmers for them to have a better standard of living. This is designed to help them achieve greater independence. Symrise has succeeded in significantly improving the quality of the vanilla beans through its close working relationship with the farmers. This has given rise to a totally sustainable initiative that also has a firm economic basis (Fig. 10).

The key benefit of this model is that it has eliminated the intermediary stage involving collectors and processors. After harvesting, farmers deliver their green vanilla beans directly to Symrise or partners working with Symrise. It is at this stage that the much-prized black gold is obtained through a fermentation or extraction process from the vanilla beans. With this model, only those participants of the supply chain who make a direct contribution to value creation work together.

Symrise also ensures that as many manufacturing stages as possible are carried out in Madagascar itself, thus making a comprehensive contribution to local value creation. Well over a hundred Symrise employees live and work on the island, processing over 10 % of the national vanilla harvest of around 2,000 tonnes. Symrise has set up its own purchasing organization on Madagascar as well as a warehouse complex for storing and processing vanilla beans. It also has the only vanilla extraction plant on the island.

The Symrise approach to sustainability also ensures that consumers can enjoy vanilla products that are not only of premium quality but that are also of proven

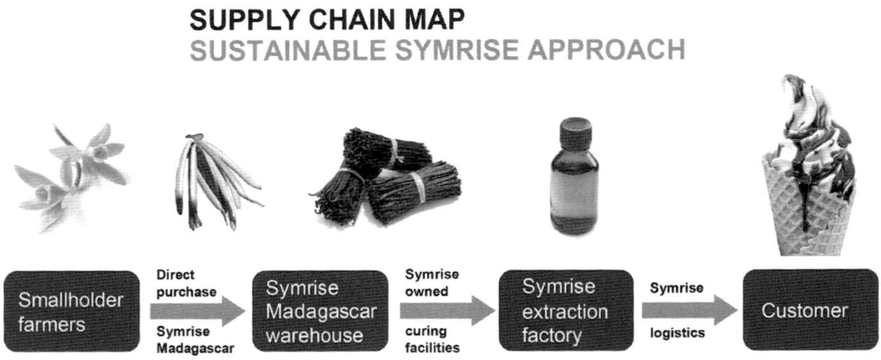

Fig. 10 Symrise supply chain for black vanilla beans in Madagascar. *Source*: Symrise AG (2013)

sustainable origin. Local investment was therefore an important and logical next step within the sustainability strategy.

4.4 *Sustainability as a Way of Life*

Symrise shows that economic success can be combined successfully with sustainability as a way of life in Madagascar. There are various local initiatives of direct benefit to the vanilla farmers. Experts from NGOs are also on board, providing valuable information and support. They bring their expertize to the table, working with Symrise to analyse the socio-economic situation of the farmers on the ground. They also make a contribution to the extensive knowledge that Symrise has acquired in terms of the vanilla business. The company then uses this additional knowledge to develop effective sustainability strategies for each market. There is also a benefit for the NGOs, as they acquire new expertize through their collaboration with the industry.

Symrise has also been working with the German Society for International Cooperation (GIZ) since 2010, in order to gain an insight into and help improve the living conditions of vanilla farmers in Madagascar. GIZ helps Symrise to acquire a better understanding of sustainability issues, providing tools and experience in the field of rural development. The partnership with GIZ is just the beginning, and the collaboration is intended to continue for a long period of time. This is the only way Symrise can ensure the high quality and availability of vanilla over the long term.

Special loans for buying rice, the staple food on Madagascar, bridge the gap until the next vanilla harvest. The farmers would otherwise have to sell their beans while they were still unripe in order to feed their families. This would affect vanilla quality and yields considerably. Symrise employees also provide information on alternative crops that Malagassi people can grow in order to extend their earning capacity. Grown as a catch crop, vetiver is a valuable source of oil for perfume production, for example.

Symrise is also committed to the village communities of the vanilla farmers. The Group pays half of the families' health insurance and helps finance the numerous little village schools. This enables many people to see a doctor for the first time in their lives or send their children to school.

Madagascar also has one of the most biodiverse ecosystems in the world. Symrise has therefore set up a programme that makes a considerable contribution to species protection on the island. Symrise keeps a systematic record of the various plant species on the 1,000 or so farms in the initiative. The farmers also can attend meetings held in their villages, where they learn to appreciate the biodiversity of their homeland and how to avoid slash-and-burn techniques.

These activities have improved farmers' lives for the long term. The Symrise vanilla business is also seeing the benefit, with two-digit percentage growth. Symrise has therefore set the tone for the global vanilla market. Through this approach, the Group has been able to improve vanilla quality to such an extent that it can finance the initiatives described above. Symrise employees use sustainably sourced raw materials to produce the premium quality flavor compositions that make numerous well-known brand products so distinctive and popular.

Symrise makes an important contribution to local value creation in this way. Symrise customers value and support this approach. Long-term engagement and further investment are therefore on the future Symrise agenda.

4.5 Ensuring Traceability

Seamless control of the supply chain is a key element of the Symrise model. It enables employees to trace the journey of every single vanilla bean from plant to flavoring. This involves meeting five challenges:

4.5.1 Challenge 1: From Manual Labor to Electronic Data Collection

Vanilla farming relies on traditional methods in Madagascar. There is no electricity in the primeval forest – just pen and paper. This means that Symrise employees have to record all source data by hand. Employees use GPS to locate and survey individual farms. Employees then go on site to record comprehensive data ranging from agricultural business performance to family relationships by hand. Information is then transferred to a database at a local office in Madagascar. Once the deliveries are made to the plants, the information is transferred into Symrise's ERP system. Symrise has also worked with a software house in recent years to develop its own SMRS subsystem. One of the benefits of this system is its ability to guarantee improved traceability along the value chain (Fig. 11).

By taking manually collected data from the upstream Madagascar databases at the beginning of the supply chain, important data can be transferred from an undeveloped data environment for use in a highly complex IT application. This

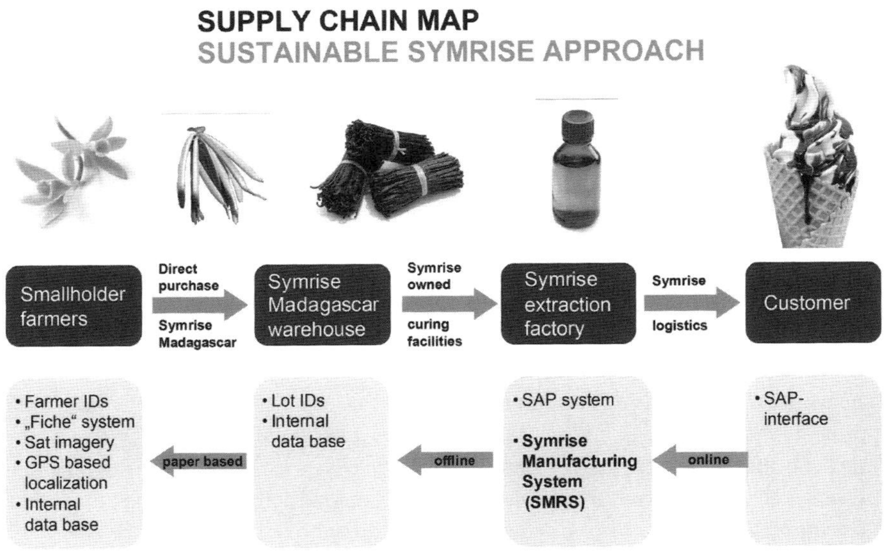

Fig. 11 The Symrise supply chain: integrated processes and systems. *Source*: Symrise AG (2013)

constitutes a huge advantage for Symrise customers. They obtain vanilla that is 100 % traceably certified, thus safeguarding their biggest asset – their brands.

4.5.2 Challenge 2: 6,000 Smallholders

Unlike the orange plantations in Brazil, where three or four major growers dominate the market, vanilla is typically a product of small farms in Madagascar. The farmers hardly ever form cooperatives beyond their village communities. In total, Symrise sources its vanilla from over 6,000 independent farmers – partly via middlemen but increasingly directly from farmers. The Group manages this complex supply chain entirely on its own – from purchasing to processing. Symrise is the only flavoring manufacturer with this unique concept. There are no modern communication channels like the Internet in the rural areas of Madagascar. This means that the only approach is through the tireless personal dedication of employees out in the field.

4.5.3 Challenge 3: Seasonality

Vanilla is an agricultural product that can only be harvested once a year. This makes it vulnerable to the usual qualitative and quantitative fluctuations of biological raw materials. The vanilla market is particularly volatile at the moment – a situation requiring long-term strategic sourcing decisions to be made. This makes accurate sales forecasts vital. Symrise has a modern approach based on Supplier Managed

Inventories (SMI) as part of its Synchronize™ initiative. This model involves Symrise carrying out inventory management on behalf of its customers, meaning that specific demand can be assessed at an early stage. The result: accurate forecasts and strategic sourcing plans.

4.5.4 Challenge 4: The Complex Supply Chain

Given that vanilla harvesting only takes place once a year between June and August, Symrise would need an area the size of several hundred football fields to process its entire vanilla requirement on its own. The company therefore processes some of its vanilla on its company premises and buys the rest in an already fermented state. Symrise has established detailed documentation procedures so that it can trace the route taken by the vanilla from farmer to processor.

4.5.5 Challenge 5: Cost Pressure

A glance at the shelves of local supermarkets and delicatessens makes it clear that vanilla beans are relatively expensive in most developed countries. At the same time, manufacturers like Symrise are subject to cost pressures from all sides within the supply chain. It is for this reason that Symrise has developed and patented a procedure for the efficient procurement of vanilla extracts. The company applies this procedure at its extraction plants in Madagascar. The technology makes it possible to extract highly prized vanillin from vanilla beans in an inexpensive way while reducing CO_2 emissions. Symrise has obtained a strategic advantage in the supply chain through its local presence and patented procedure (Figs. 12 and 13).

Forming direct partnerships with farmers is a major objective. Symrise is the only multinational company that has located its entire value chain – starting with obtaining the beans from farmers and ending with extraction – in Madagascar. Symrise is also investing in the future. It is bringing a new building and plant complex on line in 2014. By taking this step, the company will further improve its vanilla supply chain and extraction procedure, generating greater value locally at the source of the raw material.

4.6 *Shared Values: Sustainable Development Objectives in Madagascar*

Symrise plans to make further strategic investments over the next few years. New buildings and facilities will be built in the SAVA[2] region. New measures will be

[2] Sava is one of the 22 regions of Madagascar. It belongs to the (old) province of Antsiranana in the north-east of the island. The Sava region is divided into four districts: Sambava, Antalaha, Vohémar, and Andapa. The name of the region was created from the first letters of the four districts.

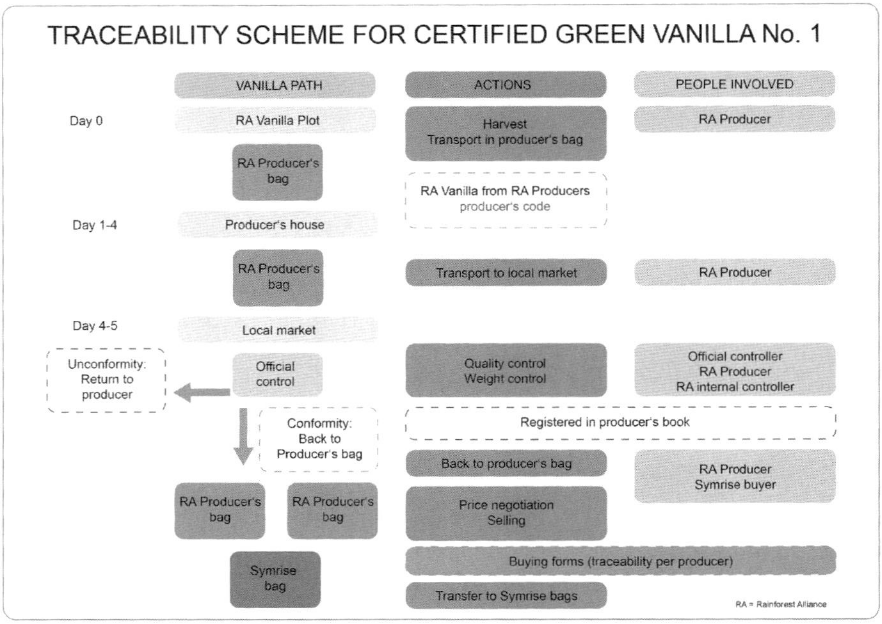

Fig. 12 Vanilla traceability I. *Source*: Symrise AG (2013)

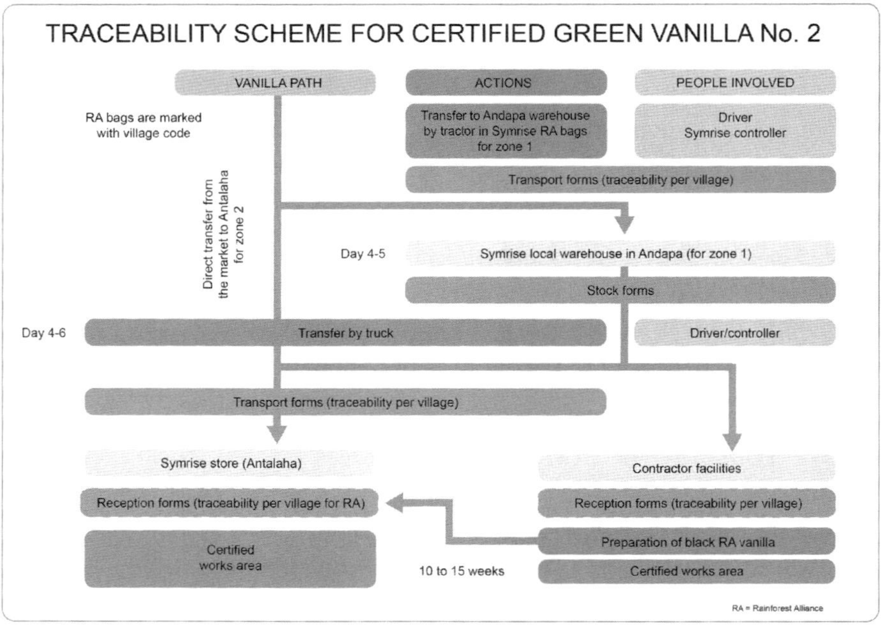

Fig. 13 Vanilla traceability II. *Source*: Symrise AG (2013)

aimed at improving farmers' living conditions and incomes, bringing social improvement. Symrise has also made an industry-wide commitment to a comprehensive multi-stakeholder dialogue for achieving a sustainability standard for vanilla. The Sustainable Vanilla Initiative (SVI) was created by a group of trading companies, food producers, flavoring manufacturers, and vanilla exporters in 2012. Symrise has been a member since its inception and is committed to its success.

Symrise hopes to achieve a series of interconnected objectives through its commitment to farmers and their communities in Madagascar. The company would like to secure the long-term availability of sustainably sourced, premium quality vanilla. In doing this, Symrise is establishing its position as a reliable supplier of premium vanilla flavorings and increasing the independence of its supplier network from a volatile raw materials market that is often hard to predict. This means dependable prices for customers and consumers as well as stable margins for Symrise.

This means that Symrise can continue investing in sustainable local initiatives. The plan is to systematically expand the mechanism for protecting the unique ecosystem and bring about a change in the economic conditions, health care, and social welfare of the farmers and their families. Symrise intends to support the farmers, implement exemplary agricultural practices directly and help develop an agricultural standard for vanilla over the medium and long term.

It is the goal of Symrise to further develop its current strategy, creating shared value for farmers, customers, and Symrise alike.

Bibliography

CITES. (2013). *Convention on international trade in endangered species*. http://www.cites.org/. Accessed June 11, 2013.

Deutscher Nachhaltigkeitspreis. (2013). http://www.deutscher-nachhaltigkeitspreis.de/1328-0-Startseite_2013.html. Accessed June 11, 2013.

GIZ. (2013). *Gesellschaft für Internationale Zusammenarbeit*. http://www.giz.de/en/. Accessed June 11, 2013.

GRI. (2013). *Global reporting initiative*. https://www.globalreporting.org/languages/german/Pages/default.aspx. Accessed June 11, 2013.

SEDEX. (2013). http://www.sedexglobal.com/. Accessed June 11, 2013.

Symrise. (2013). http://www.symrise.com/uploads/media/SYM_Unternehmensbericht_de_130311_01.pdf. Accessed June 11, 2013.

UN. (2013). *UN global compact – Communication on progress*. http://www.unglobalcompact.org/COP/. Accessed June 11, 2013.

Siemens: Managing Sustainability Along the Value Chain to Benefit Our Customers

Ralf Pfitzner and Matthias Lutz

1 Introduction

1.1 About Siemens

Siemens is a globally operating technology company with core activities in the fields of energy, healthcare, industry, and infrastructure. For 166 years, Siemens has stood for innovative strength, a passion for technology, sustainability, responsibility, and an uncompromising commitment to quality and excellence. In fiscal year 2014, our approximately 343,000 employees generated a revenue of about 71.9 billion euros from continuing operations and income from continuing operations of about 5.4 billion euros. We have business activities in nearly all countries of the world and operate more than 290 major production and manufacturing plants worldwide.

Sustainability in its different aspects has at all times been part of the company, although it was not always termed "sustainability." For example, as early as 1873, Werner von Siemens developed a technology to eliminate ash from factory emissions which today would be considered as "environmental technology." The foundation of the Siemens Health Insurance Fund in 1908 is an example of addressing the social dimension of sustainability. Furthermore, already in 1971 Siemens set up a companywide Environmental Protection Office. Today, Sustainability is a

R. Pfitzner (✉)
Siemens AG, Corporate Development Strategy, Sustainability, Wittelsbacherplatz 2, 80333 Munich, Germany
e-mail: ralf.pfitzner@siemens.com

M. Lutz
Siemens AG, Plant Data Services, Werner-von-Siemens-Str. 65, 91052 Erlangen, Germany
e-mail: matthias.lutz@siemens.com

M. D'heur (ed.), *Sustainable Value Chain Management*, CSR, Sustainability, Ethics & Governance, DOI 10.1007/978-3-319-12142-0_8

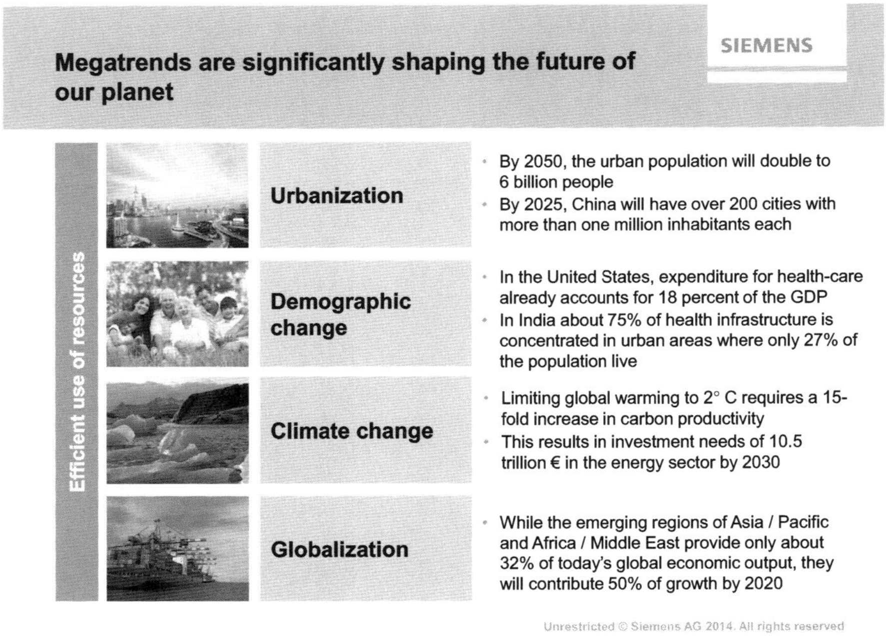

Fig. 1 Megatrends are shaping the future of our planet (Siemens 2014a)

responsibility of the Managing Board and a guiding principle within our company as it is of strategic importance for our businesses.

1.2 Megatrends

Our company, our customers, and our markets are subject to both long-term trends and short-term economic developments (Fig. 1).

Global megatrends are long-term developments that are expected to have an impact on all humans. Demographic change, urbanization, climate change, and globalization are megatrends that entail major challenges for policymakers, entrepreneurs, and scientists around the world. At the same time, however, they offer tremendous business opportunities.

Demographic change includes two major trends: the world's population continues to grow steadily – from 7.2 billion in 2013 to 9.6 billion in 2050 (UN 2013), and it continues to get older. Together, these two trends will challenge the ability of future healthcare systems to make healthcare available to everyone.

Urbanization refers to the growing number of densely-populated metropolitan centers around the world. This trend intensifies the existing demand for sustainable and energy-efficient infrastructures for buildings, transportation systems, energy, and water.

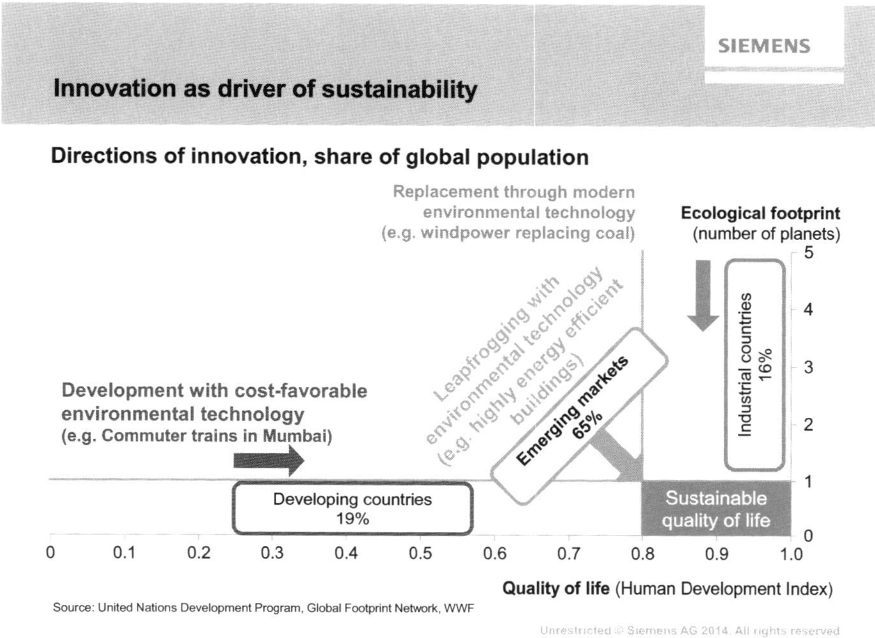

Fig. 2 Ecological footprint and human development index (Siemens 2014e citing UNDP 2014; Global Footprint Network)

We view **climate change** as a given fact and believe that reducing greenhouse gas emissions is vital to counteract the increasingly drastic effects on our ecosystem. There is serious need for innovative technologies to increase efficiency and reduce the emissions related to power generation and consumption.

Globalization refers to the increasing integration of the world's economies, politics, culture, and other areas of life. Globalization leads to increased competitive pressure and demand for economical, timely-to-market, high-quality products and solutions (Siemens 2013a, p. 170).

The global challenges with regard to sustainability are often expressed in the global footprint and the Human Development Index (HDI) as Fig. 2 shows.

According to the Global Footprint Network, humanity today uses the equivalent of 1.5 planets every year. This means that today it takes the earth 1 year and 6 months to regenerate what we use in a year. The footprint concept is based on an "accounting system that tracks, on the demand side, how much land and water area a human population uses to provide all it takes from nature. This includes the areas for producing the resource it consumes, the space for accommodating its buildings and roads, and the ecosystems for absorbing its waste emissions such as carbon dioxide" (Global Footprint Network 2014).

Humans are dependent on nonrenewable resources, especially fossil fuels, and the resulting greenhouse gas emissions are having serious effects on climate. Many developing countries and some emerging countries still are below the HDI (covering life expectancy, education, and income) of 0.8 which is considered as "high human development." To achieve a "sustainable quality of life," we believe that technology and innovation are major parts of the solution. Business can make a difference for example by increasing energy and resource efficiency in order to fulfill needs of people, to provide access to healthcare, energy in rural areas, and mobility in an effective and efficient manner, as well as limiting consumption to the equivalent of one earth to ensure a sustainable future for the Earth and for future generations.

2 Our Understanding and Organization of Sustainability

Siemens has defined sustainability as acting responsibly on behalf of future generations to achieve economic, environmental, and social progress.

We are aware of the associated high standards and the possibility of conflicting goals in balancing all three dimensions of sustainability. Nevertheless, the aim to create sustainable added value remains a key element of our corporate strategy. We are convinced that sustainability, in this sense, is also a business opportunity, and one that is worth seizing. For our value chain, this means for example that we do not compromise on sustainability standards or on compliance – neither for our suppliers nor in our own operations.

"One Siemens," our framework for sustainable value creation and capital-efficient growth, addresses this business opportunity with its three strategic directions (Siemens 2014b).

First, "focus on innovation-driven growth markets." The products and solutions in our Environmental Portfolio and the innovation power of Siemens play a central role in contributing to environmental and climate protection while also strengthening our standing in the innovation-driven growth markets that we focus on.

Second, "get closer to our customers." An intense customer focus and a competitive, globally balanced, and localized network of suppliers supports us in getting closer to our customers all over the world.

Third, "use the power of Siemens." Excellent employees are one of Siemens' vital strengths as they play a key role in our success and are the true power of Siemens. Leveraging the power of Siemens also means strictly adhering to clear principles of integrity – something we also expect of our partners and suppliers.

As these examples show, sustainability is not embellishment at Siemens – it's a central theme of our corporate strategy. In our Sustainability Program we focus on targets and activities in three areas: "Business opportunities," "Walk the talk," and "Stakeholder engagement." In the first area, we turn our approach to sustainability into concrete business opportunities. "Walk the talk" means we are committed to embedding sustainability throughout our organization and operations. In the third area, we focus on collaboration with all relevant stakeholders.

The latter is essential as many global challenges cannot be resolved by single businesses alone. Therefore Siemens partners with business organizations like the

World Business Council for Sustainable Development (WBCSD), e.g., in the area of energy efficiency in buildings, with C40, the Cities Climate Leadership Group in the area of sustainable urban development and in various other initiatives. Furthermore, Siemens has launched a global Siemens Integrity Initiative that supports organizations and projects fighting corruption and fraud through Collective Action, education, and training with over US$100 million. The initiative focuses on supporting projects that have a clear impact on the business environment, can demonstrate objective and measurable results, and have the potential to be scaled up and replicated. The Siemens Initiative is part of the comprehensive settlement between the World Bank Group and Siemens AG that was announced on July 2, 2009 (Siemens 2014e).

2.1 Organization of Sustainability

It has often been said but it remains true – a successful implementation of sustainability starts at the top of the organization. At Siemens, the central position within the company's organization, in our programs and measures we execute illustrates the importance of sustainability. Efficient sustainability management is a company-wide task that requires a clear organizational structure and a thorough anchoring of sustainability in our corporate culture. All our sustainability activities are steered by the Chief Sustainability Officer. Dr. Roland Busch, member of our Managing Board, currently holds this position. In order to coordinate and manage our sustainability activities proficiently, we established the Siemens Sustainability Board, the Sustainability Office, and the Siemens Sustainability Advisory Board.

The Siemens Sustainability Board, chaired by the Chief Sustainability Officer, is the central steering committee for sustainability at Siemens. In its regular meetings it directs our sustainability program as part of our sustainability strategy and adopts appropriate measures and initiatives. Our Chief Sustainability Officer also manages the Sustainability Office, which is responsible for enhancing sustainability even further at Siemens and for coordinating the sustainability program and other company-wide programs and measures. To help us maintain an objective perspective on our sustainability challenges and performance, we have also created the Siemens Sustainability Advisory Board, composed of eight eminent figures in science and industry from a range of disciplines and regions of the world. The Board meets at least twice a year, and through professional exchanges and practical initiatives has already contributed to the further development of our sustainability program. Furthermore, assigned Sustainability Managers in the Divisions and regional units ensure that sustainability measures are implemented throughout the Company.

The following chapters illustrate how we manage sustainability along the value chain based on our organizational setup: with our suppliers, in our own operations, and with our customers in order to mitigate risks and exploit opportunities.

3 Sustainable Supply Chain Management

3.1 Key Facts

The primary goal of all Supply Chain Management (SCM) activities is to ensure the availability and quality of the materials required to serve our customers. In order to achieve this goal, we require a globally balanced, locally anchored, and close network with our supplier base for optimal exploitation of the innovativeness of our suppliers.

The supply chain management parts of this chapter (Sects. 3.1–3.3) refer to the "source" process according to the SCOR model; EHS management and program (Sect. 4) refers to the "make" process of the model.

In fiscal year 2013, Siemens' purchasing volume amounted to approximately 38 billion euros. With direct responsibility for material costs, Supply Chain Management thus accounts for over half our value creation in sales. We procure from some 90,000 suppliers in over 150 countries and constantly increase the share of sourcing from emerging markets.

The challenges faced by SCM have grown considerably in recent years. Markets are more volatile, and not only from a financial point of view – the global economy is increasingly shifting away from the current industrialized countries to developing countries. In addition, the volume of global trade has more than doubled between 1990 and 2010 alone and the number of multinational companies (new global players from developing countries) is rising. The competition for innovations and scarce resources is consequently much tougher.

Sustainability in the supply chain has become a key requirement, mainly driven by customers and investors. In addition, we are facing a growing influence of sustainability-related legislations within the supply chain (e.g., rule on conflict minerals pursuant to Dodd–Frank-Act).

Although we were fortunately not confronted with significant issues in our supply chain in recent years, controversies around labor practice at the electronics supplier Foxconn in China or more recent issues in the textile industry's supply chain (The Economist 2013) show that it is essential to ensure sustainability standards in the supply chain and to mitigate respective risks. Sustainability requirements are therefore an integral part of all our relevant supplier management processes – such as supplier selection, supplier qualification and evaluation, and supplier development.

3.2 Risk-Based Approach

We expect all our suppliers to make a clear commitment to the principles of sustainability. Our requirements – such as respect for the basic rights of employees and environmental protection – are defined in the "Code of Conduct for Siemens

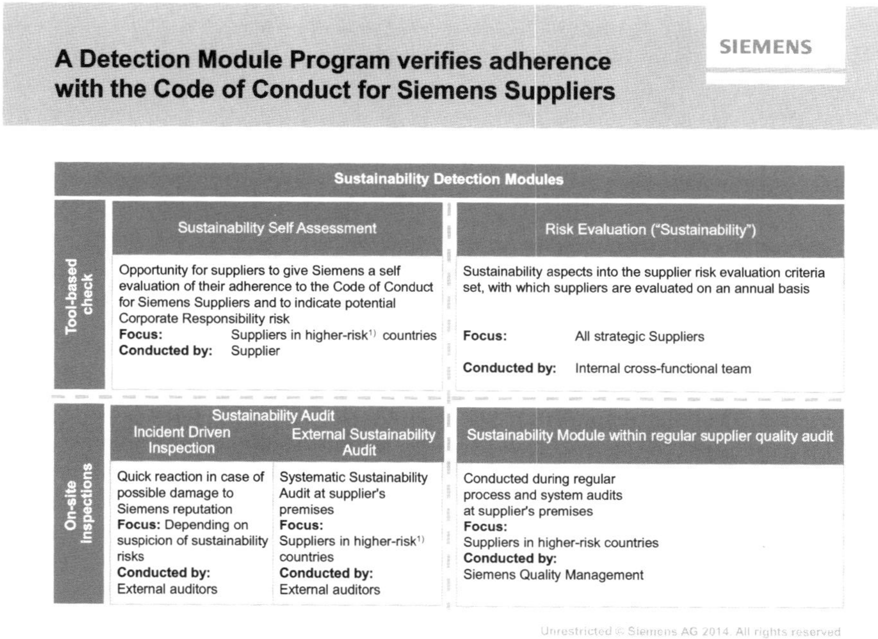

Fig. 3 System of sustainability detection modules in the supply chain (Siemens 2014c). Based on Organization for Economic Cooperation and Development and TI/CPI Transparency International/Corruption Perception Index

Suppliers," which is based on the ten principles of the UN Global Compact and reflects the content of our Siemens Business Conduct Guidelines.

Under the relevant provisions in our procurement contracts and our conditions of purchase, all Siemens suppliers must meet these requirements and also promote compliance with them in their own supply chain.

Because the supplier network is very large and widely spread, it is not possible for us to examine all suppliers to the same extent by auditing them on site. We have therefore established a risk-based system (Sustainability Detection Modules) of appropriate processes which enables us to systematically identify potential risks in our supply chain. It consists of sustainability self-assessments by suppliers, risk evaluation conducted by our purchasing department, sustainability questions within supplier quality audits, and sustainability audits by external auditors. Objectives, focus, and agents are shown in Fig. 3.

Within the last years, we rolled-out our supplier qualification process and continually increased the number of self-assessments. The suppliers in scope are mainly from non-OECD countries with a purchasing volume of more than 50,000 euros. We initially covered mainly our existing suppliers but more recently started to focus on new suppliers.

In addition, sustainability-related questions are part of all regular quality audits. If sustainability risks are identified, we specifically audit the suppliers in question

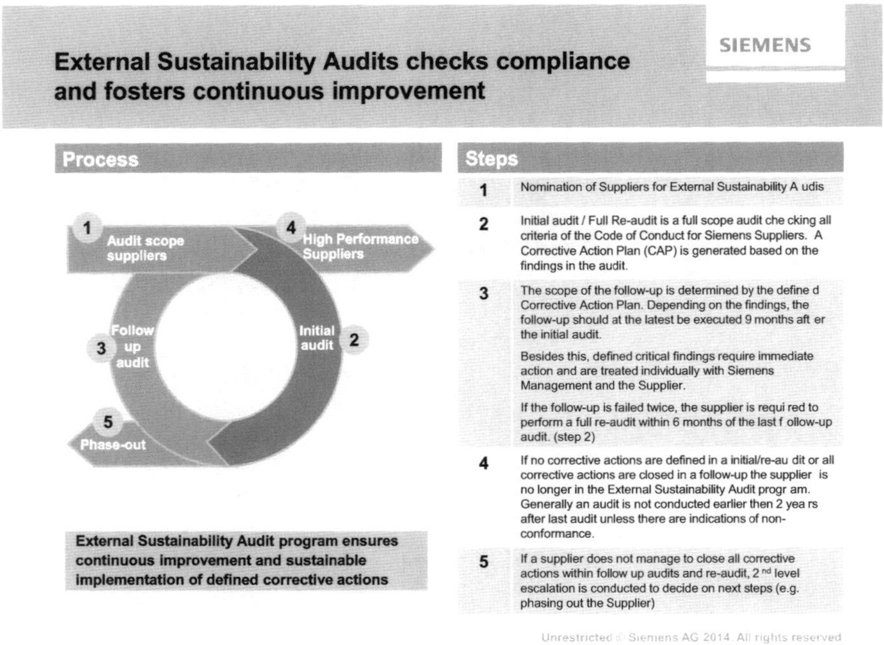

Fig. 4 System of external sustainability audits (Siemens 2014c)

on site. Figure 4 explains the respective process steps. We pay special attention to inspecting suppliers in emerging countries where we increasingly purchase products as part of our Global Value Sourcing program. If deviations from our requirements are identified, they have to be resolved by the suppliers within a reasonable period of time. In the event of serious deviations or unwillingness to implement measures for improvement, we exclude suppliers from any business with Siemens. In everything we do, we are guided by the principles of developing our suppliers in close partnership and building up their competencies for the long term. Furthermore, we conduct follow-up audits, which entail revisiting the sites to establish whether the agreed measures have actually been implemented. Deviations identified in the audits mainly relate to structural deficiencies in management systems and the lack of specific processes and guidelines at the supplier. This includes, for instance, measures to effectively prevent corruption and bribery and to rule out child labor.

3.3 *Know-How Transfer and Competence Building*

Our suppliers' commitment to complying with our sustainability principles is most effective when it is based on their own convictions. We are therefore increasingly committed to building up our suppliers' competence and intensifying knowledge transfers related to sustainability. This is why we have developed an internet-based information and training platform, which is available free of charge to all suppliers (Siemens 2014d). On top of that, sustainability is an integral part of the company-wide training programs for buyers. All employees with purchasing responsibility are obligated to take part in intranet-based training on the subject of "Sustainability in the Supply Chain."

4 Managing Our Own Operations

4.1 EHS Management Approach and Program

Siemens has a comprehensive EHS (Environmental Protection, Health Management and Safety) management system. The process requirements of this management system help our operating units to comply with applicable laws, regulations, and customer requirements, to satisfy our corporate requirements properly, and to achieve our Siemens-wide environmental targets. The environmental protection management system requires that our environmentally relevant production sites and offices must implement an environmental management system which fulfills the requirements of the internationally recognized ISO 14001 standard and also our own internal standard, known as "Specifications on environmentally compatible product and system design." This internal standard defines requirements to reduce the environmental impact of our products and systems during production, use and disposal phase, and is subsequently an integral part of our business processes.

The management system includes a number of effective and complementary environmental programs as well as a set of Siemens-wide environmental targets. We conduct regular internal reviews of our environmental performance and progress, in order to create a cycle of continual improvement.

Our commitment to continual improvement caused two environmental protection programs to come into being in fiscal year 2012: "Serve the Environment" for industrial environmental protection and "Product Eco Excellence" for product-related environmental protection. They are designed to mitigate the environmental impact of Siemens business activities and our products, to fulfill growing international requirements with regard to environmental protection, to increase customer benefits, and to proactively strengthen our position as a sustainable company.

4.2 Energy and Resource Efficiency at Our Own Sites

Our industrial environmental protection efforts focus on optimizing energy and resource efficiency at our sites. With the "Serve the Environment" program we are committed to the following Siemens-wide main targets:

- To continue our systematic effort to improve energy efficiency, and thereby achieve corresponding improvement in our carbon dioxide efficiency;
- To improve the waste efficiency each year by 1 % until 2014;
- To reduce waste for disposal each year by 1 % until 2014.

Furthermore, Siemens continues to manage water-related risks. In locations where there are particular water risks (for example as a result of aridity, high waste-water loads, or poorly developed technical infrastructure), the local sites need to define targets matched to local conditions and, in meeting those targets, effectively reduce risks and negative impacts on the environment.

We measure progress towards achieving our "Serve the Environment" program targets by aggregating the results of measures implemented locally at our sites. We calculate environmental performance on a portfolio-adjusted basis. This approach enables us to survey and compare our environmental performance over time, regardless of acquisitions and disposals.

In order to use energy efficiently for our own operations and to increase our carbon efficiency from energy use, the "Siemens Energy Efficiency Program" (EEP) was developed already in 2007 in collaboration with two Siemens divisions. According to the process shown in Fig. 5, energy analyses within the EEP framework were conducted and measures implemented at more than 100 major manufacturing sites.

With innovative concepts like the Green Building Initiative and the Energy Efficiency Program, Siemens Real Estate (SRE), which manages our properties, optimizes resource allocation, and simultaneously makes buildings more energy efficient.

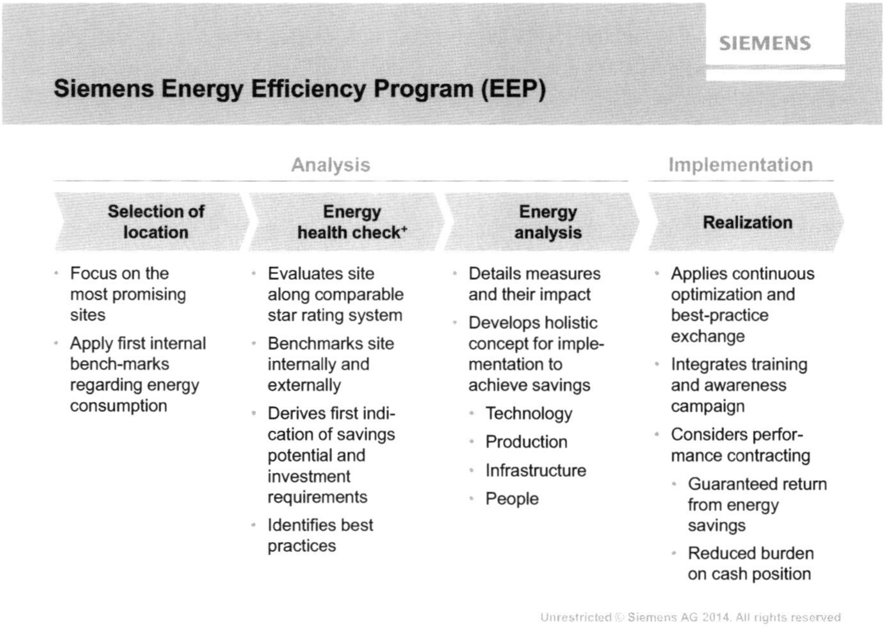

Fig. 5 Siemens energy efficiency program

Savings	
Energy cost:	€/a 3,700,000
Savings:	€/a 674,400
Investment:	€ 4 Mill.
Amortization:	4 years
IRR:	19.3%
CO_2 Emission:	t/a 2,300

Technical solution
Building Management System
Gas-combined heat power
Optimization of hydraulic systems
Ventilation system with 165,000 m^3/h with heat recovery system
Installation of a radiant warmer

Fig. 6 Financial savings and energy saving measures implemented at Siemens Krefeld Railway Engineering plant

New buildings as well as important existing constructions have to meet systematic sustainability criteria and are certified according to the international Green Building Standard LEED (Leadership in Energy and Environmental Design). Through energy-focused renovations of the building services technology and structural measures at more than 15 production locations, annual energy cost, for example, was reduced by over 4.4 million euros. Most measures were implemented under an energy performance contracting model with Siemens' Building Technologies Division and resulted in a sustainable reduction of annual CO_2 emissions by more than 16,000 tonnes (SRE 2014).

One example of successful implementation of energy saving measures is the Siemens Krefeld Railway Engineering plant, one of Siemens' most important centers in the railway industry. Each year more than 450 railcars leave the factory. In Krefeld-Uerdingen, Siemens Mobility employs around 2,200 people who develop and manufacture these rail vehicles and electrical components. Regional trains such as the Desiro and high-speed trains such as the Velaro are marked "Made in Krefeld."

A detailed analysis for the Krefeld site within the Energy Efficiency Program resulted in a package of seven efficiency measures that were implemented, leading to annual savings of approximately 675,000 euros, equalling a payback of the investment within 4 years (see Fig. 6).

Heating System

1. The most important efficiency measure consisted in building a gas-fired cogeneration plant for heat and power generation. Thusly, the power plant was implemented in parallel to the existing four boilers.
2. In addition, the hydraulics of the heat distribution were optimized. Network pumps are now controlled by a new building management system which allows for a demand-driven operation through frequency inverters.

(continued)

3. Another efficiency measure consisted in the installation of ceiling heating elements. Due to the good accessibility of the ceilings in the involved buildings and no overhead cranes, the installation of this type of heater was logical. The infrared radiant heat is, in no time, ascertained by the employees as pleasant and is expended with minimum heat energy. These dark heaters are supplied with natural gas. The direct firing with gas as the primary energy carrier also reduces the heat losses in the heating network.

Ventilation System

4. Optimization of the ventilation system involved the renewal of the complete ventilation control center (100,000 m^3/h) and retrofitting the air-handling unit in the assembly hall with a heat recovery system in form of a rotating air-to-air heat exchanger and modern control technique. As a result, the air supply temperature is now at a constant 18 °C.

Building Management System

5. This efficiency measure consisted in the creation and expansion of the building management system. For this purpose, the measures described, as well as the existing automation stations were linked to a new, joint management system.
6. The installation of an energy monitoring system provides an insight into the plant's energy consumption and allows for an optimal adjustment to the installation. The measure also included the installation of energy meters and circuit-entering of the building management system.
7. In addition, a new monitoring platform was set up. The new Energy Monitoring and Control system (EMC) ensures, through continuous recording and analysis of energy consumption, not only further energy savings but also performance control of the efficiency measures. In order to make these achievements transparent, understandable for everyone and specific to the customer requirements, Siemens developed the Green Building Monitor™. The Green Building Monitor™ is a communications means, which allows staff and guests at the entrance area of a building to get information on the energy and media consumption in this building and to motivate them with specific advice to actively contribute to an increase in energy efficiency. The implementation of these EEP measures in the Krefeld facility led to cost savings of 15 % and a reduction of CO_2 emissions by 20 % (SRE 2012).

4.3 Product-Related Environmental Protection

Siemens' products, solutions, and services are the "bridge" between our operations and our customers. The major focus on product-related environmental protection therefore is to improve the overall environmental performance of our products and solutions. We define mandatory requirements in our internal environmental standard to reduce the environmental impact of our products and systems during the product development, production, use, and disposal phases.

The "Product Eco Excellence" program supports our businesses to fulfil these requirements. Additionally, the program aims to better prepare the operating units for future regulatory and customer requirements, to strengthen environmental communication, and to broaden environmental awareness amongst our employees. The main elements of the program are:

- Being committed to continuously improve transparency regarding declarable substances, particularly in purchased parts and components. To gain transparency, we provide a list of declarable substances (LoDS), comprising substances that are restricted in use due to regional or application-specific regulations or due to potential health and environmental risks posed by these substances themselves and in the manufacture, use, and disposal of products containing them. We strive for an improved basis for assessing the environmental impact of our products and ensuring that our customers' requirements in the respective target markets are met. This also supports closing material cycles (cradle to cradle) which is becoming an increasingly important topic as global market demands.
- To develop a methodology for better assessing risks such as environmental, toxicological, and future availability risks associated with substances and materials used. The results are the basis for substitution decisions within product development. We have developed the methodology and will roll it out as part of the environmental program. We intend to verify the potential of the methodology using pilot projects.
- To establish a harmonized procedure for determining the "ecological footprint" of our products whose coverage we want to further increase. In order to determine and evaluate the "ecological footprint" of our products and systems, we have adopted the requirements of international life-cycle assessment (LCA) standards ISO 14040 and ISO 14044. The assessment results are the basis of our environmental product declarations (EPD) which are part of an internal harmonized process.

Though widespread demand by our customers does not yet exist, we see an increase of requests for sustainability-related information on products as well as for our overall sustainability performance. The main drivers are our customer's sustainability requirements in tendering processes as they intend to reduce their own carbon footprint and improve their energy and resource efficiency. In this context, life-cycle assessments and environmental product declarations are tools we increasingly use for evaluating products along the life cycle and to communicate this information to our customers and interested stakeholders.

5 Contributing to Our Customer's Success: The Siemens Environmental Portfolio

Siemens develops and manufactures mostly investment goods which last many decades. Hence, the main lever regarding overall energy and resource efficiency for many of our products and solutions is efficiency in the use phase. Though today there is hardly any premium to be achieved because a product is "green," it becomes a business opportunity once we support our customers in reducing their operational as well as total cost of ownership with energy and resource efficient products, solutions, and services.

Apart from the individual product level where EPDs are available, we address these demands with the Siemens Environmental Portfolio. Here we bundle all those products, solutions, and services that make particular contributions to environmental and climate protection. Our aim is to achieve a threefold benefit: First for our customers who improve their competitiveness as a result of lower energy costs and higher productivity, second for future generations, and third for Siemens itself, by developing attractive markets and growing profitably.

Quantifiable benefits for our customers can be illustrated with many examples. For example at the "Taipeh 101," which is 508 m high and therefore one of the tallest buildings on earth, a Siemens building automation system lead to energy savings of 18 %. Besides this, the driverless metro in Nuremberg enables a 50 % capacity increase with 15 % energy savings and thus highlights customer benefits in the area of mobility. Also in industry, the usage of optimized drive technologies can cut energy consumption by up to 70 %, leading to significant energy cost savings and short amortization periods. This is also illustrated by Fig. 7.

5.1 Clear Criteria for the Siemens Environmental Portfolio

Key features of a product in the environmental portfolio are energy efficiency, renewables, and environmental technologies. As no global standards exist for the definition of "green/sustainable products" and being credible is very important to us, Siemens defined an internal standard already several years ago. Inclusions to the Environmental Portfolio are made in accordance with strict processes on the basis of the following criteria:

Energy Efficiency This applies to products, solutions, and services that offer significantly better energy efficiency than a comparable solution. The condition is an increase in energy efficiency of at least 20 % or a reduction of at least 100,000 metric tons of CO_2 in the use phase in a given year of all installed products, solutions, and services combined.

Renewables This criterion covers technologies such as wind turbines and solutions for hydropower as well as smart grid applications like smart meters or smart control mechanisms for energy distribution networks.

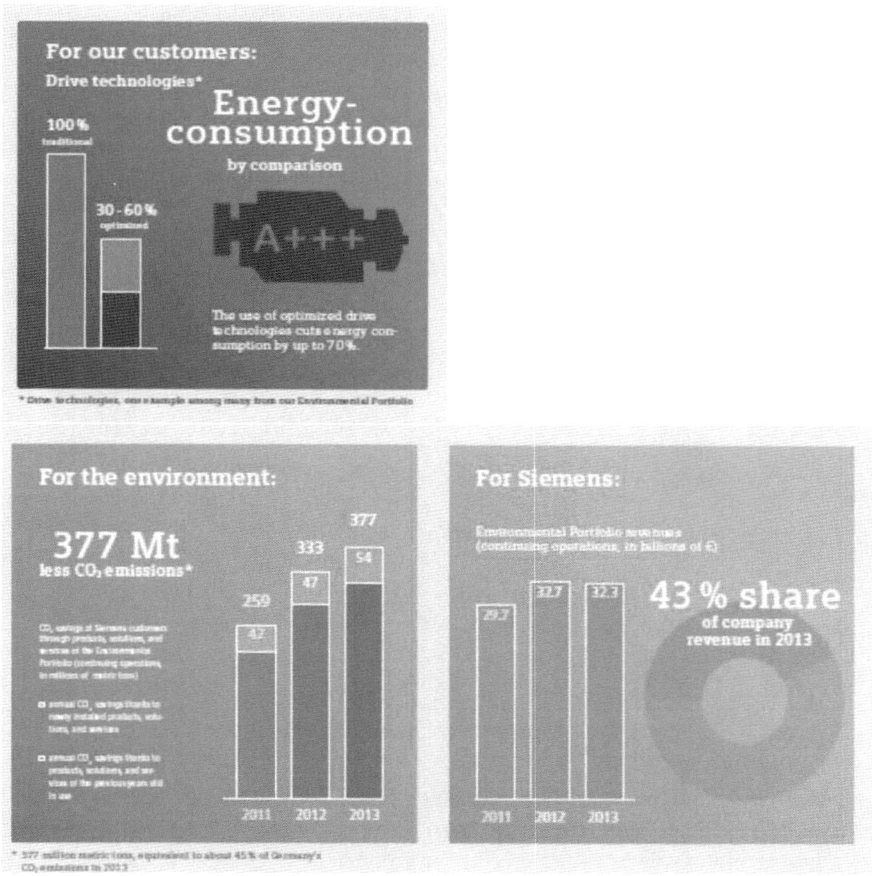

Fig. 7 Threefold benefit with the Environmental Portfolio – energy savings for customers, CO_2 reduction for the environment and growth opportunities for Siemens (Siemens 2014f)

Environmental Technologies The focus is on technologies for pollution control, water, and waste water treatment or recycling. Solutions from the healthcare sector can also qualify if major effects for the patient (noise, X-ray radiation) are reduced by at least 25 %.

Primarily the use phase is being considered – which means that the positive effects must be noticed by the customer. Every year the entire Siemens portfolio is reviewed for possible classification in the Environmental Portfolio on the criteria outlined above. The elements undergo a multiphase check in the appropriate Siemens Division and in the Corporate Sustainability Office before being admitted.

As Fig. 8 shows, in fiscal year 2013, nearly three quarters of the solutions in our Environmental Portfolio related to energy efficiency and underline Siemens' continued strategic focus on technologies in this field. Energy efficiency is not only relevant in the consumption of energy, where for example Siemens industrial motors used in conjunction with variable speed drive technology can reduce energy

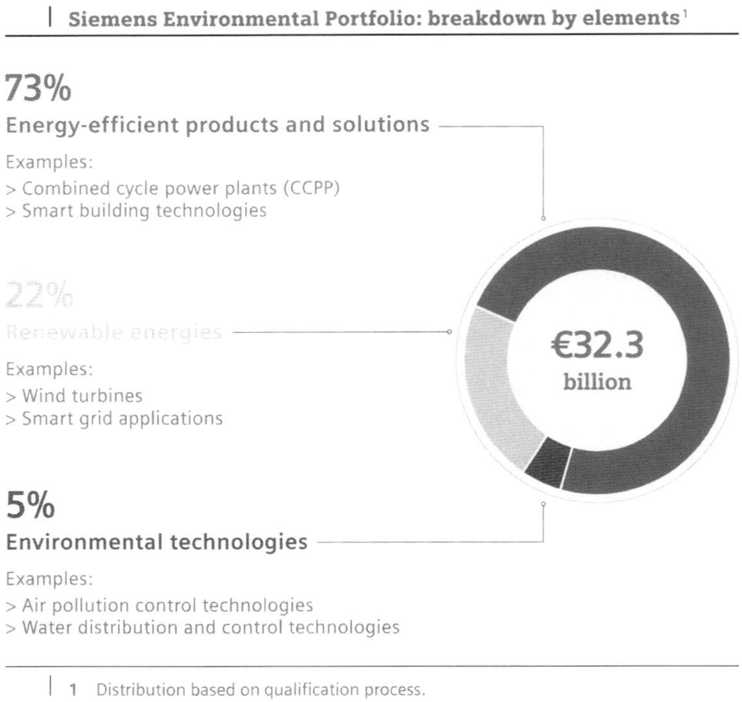

Fig. 8 Breakdown of Environmental Portfolio by elements (Siemens 2013b, p. 19)

consumption of up to 70 %. The Environmental Portfolio also offers solutions which enable efficient generation of energy, for example through our combined cycle power plants. Renewable energies account for almost a quarter of revenue generated by the Environmental Portfolio. Siemens is a leading company in this field, with technological innovations such as gearless 6 MW turbines for the generation of wind power. Environmental technologies comprise a broad range of services and solutions related to water and air pollution control, as well as products of the healthcare sector where an environmental impact reduction is achieved by reducing noise, radiation, or weight.

5.2 *Technology Fields, Impact and Growth*

The Siemens Environmental Portfolio includes ten technology areas along the entire value chain of electrification: renewable energies, fossil power generation, power transmission and distribution, smart grids, energy storage, mobility, industry solutions, building technologies, healthcare, and water. Figure 9 illustrates some innovative products, solutions, and services along the value chain of electrification.

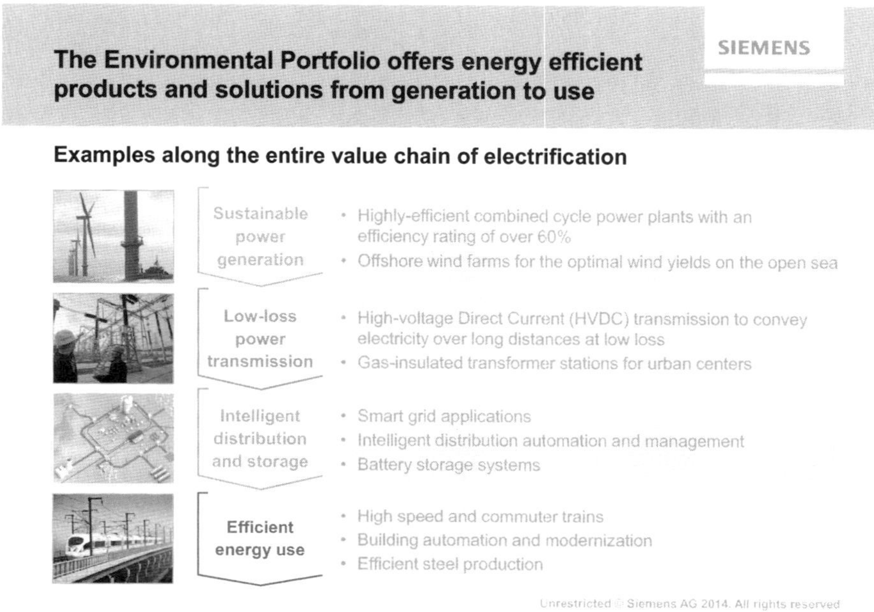

Fig. 9 Innovative technologies along the value chain of electrification in the Environmental Portfolio (Siemens 2014g)

Innovation in these technology areas continuously improves energy and resource efficiency across the entire energy conversion chain: from power generation (e.g., combined cycle gas fired power plants with efficiency >60 % or the aforementioned 6 MW gearless wind power turbines to increase competitiveness of renewable energy technologies), transmission (e.g., high voltage direct current power transmission with low losses) to smart-grid technologies and energy-efficient consumption. Increasing the energy efficiency of the demand side still has a lot of potential. Given the fact that just under 2/3 of industrial power requirements are allocated to electrical drives, 40 % of potential savings in the drive train are to be found in system optimization by means of speed control and energy recovery with inverters and energy saving motors. Other examples refer to building technologies or for example intelligent and resource efficient transport systems.

The impact of energy efficiency is primary energy and fuel cost savings for our customers. Besides this, it is a positive contribution to the environment and society – many technologies needed to stay on a 2 °C global warming track are already available.

Taking together all elements of the Environmental Portfolio which were installed at customer locations since the beginning of fiscal year 2002 and remain in use today, we have reduced customer carbon dioxide emissions by 377 million metric tons in fiscal year 2013 (Fig. 7), which is the equivalent of the combined annual emissions of the following 12 cities: Berlin, Cape Town, London, Los Angeles, Melbourne, Mexico City, Moscow, New York City, São Paulo, Seoul, Singapore and Tokyo – or roughly 40 % of Germany's annual carbon emissions.

The carbon dioxide savings demonstrate that we not only enable our customers in becoming more energy efficient but also turning the challenges of the megatrend climate change into an opportunity. Furthermore, it is an opportunity to grow profitably – revenue from products and solutions of the Environmental Portfolio reached 32.3 billion euros or 43 % of Siemens total revenue in fiscal year 2013 (Fig. 7).

6 Conclusion and Outlook

Managing sustainability along the entire value chain is a necessity today in order to mitigate risks, fulfill customer demands and act as a responsible business. It is therefore increasingly an integral part of our business rather than a separate activity.

External rankings, ratings and awards prove that we are on the right track with embedding sustainability into our strategy and operations. For example, Siemens has been part of the widely respected Dow Jones Sustainability Index for 15 consecutive years. Within this index, we were ranked as "Industry Leader in 2014 for Industrial Conglomerates for the seventh time in a row, and as Industry Group Leader for Capital Goods" for the third time. We also earned high ratings on a number of other indexes and rankings, including those created by the prestigious CDP. Siemens had one of the best scores in the world for the seventh time in a row.

We expect sustainability-related customer demands to increase in the future – in terms of product performance (driven by the need for increased energy and resource efficiency), increased transparency along the entire supply chain (e.g., information on substances and carbon footprint of products) as well as related to our sustainability management approach and fulfillment of essential standards (compliance, environmental protection, occupational health and safety).

This article focused on sustainability of our supply chain and the benefits this brings with it for our customers. Sustainability, though, goes way beyond these aspects. It has become an important feature for recruiting and for employee engagement; today's leaders care about career development but also about the "purpose" they are working for.

Sustainability is an ongoing journey where Siemens as well as people involved continuously learn and grow. Key drivers are the stakeholders: customers, shareholders and employees. The implementation into all business processes requires time and effort and it is an ongoing task to balance the "three P's – people, profit, planet" in daily decision making. Being aware of possible conflicts, staying credible and let communication follow content are possible recommendations we would make to other organizations that start their journey.

One should always bear in mind that sustainable value creation for a company requires long-term thinking. As already Werner von Siemens, the founder of the company said (Werner von Siemens 1884): "I won't sell the future for momentary profit."

Bibliography

Global Footprint Network. (2014). *World Footprint*. http://www.footprintnetwork.org/en/index.php/GFN/page/world_footprint/. Accessed 2 May 2014.

Werner von Siemens. (1884). *Letter of Werner von Siemens to his brother Bruder Carl*, 29.12.1884.

Siemens. (2013a). *Siemens annual report 2013*. www.siemens.com/annual/13/en/download/

Siemens. (2013b). *Additional sustainability information to the siemens annual report 2013*. www.siemens.com/annual/13/en/download/

Siemens. (2014a). *Company presentation – "Sustainability at Siemens"*.

Siemens. (2014b). *One Siemens*. www.siemens.com/one-siemens. Accessed 2 May 2014.

Siemens. (2014c). *Sustainability in the supply chain*. https://w9.siemens.com/cms/supply-chain-management/en/sustainability/detection/Pages/detection.aspx. Accessed 2 May 2014.

Siemens (2014d). *Web-based Training* https://w9.siemens.com/cms/supply-chain-management/en/sustainability/supplier-development/web-based/Pages/trainings.aspx. Accessed 2 May 2014.

Siemens. (2014e). *Siemens integrity initiative*. www.siemens.com/integrity-initiative. Accessed 2 May 2014.

Siemens. (2014f). *Leading you to energy efficiency – Siemens environmental portfolio*. https://www.siemens.com/sustainability/pool/umweltportfolio/siemens_environmental_portfolio.pdf. Accessed 2 May 2014.

Siemens. (2014g). *Measuring impact – A portfolio approach*. Presentation at the PE International Symposium 2014, Ralf Pfitzner, Stuttgart 2014.

SRE. (2012). *The EEP Book: A showcase of building-related energy efficiency projects in manufacturing facilities – a collaboration of Siemens Real Estate and Building Technologies*.

SRE (2014) *Siemens real estate*. http://www.siemens.com/about/en/businesses/siemens_real_estate.htm. Accessed 2 May 2014.

The Economist. (2013). *Bangladesh's clothing industry: Bursting at the seams*. http://www.economist.com/news/business/21588393-workers-continue-die-unsafe-factories-industry-keeps-booming-bursting-seams. Accessed 2 May 2014.

UNDP (2014) *Human Development Index (HDI)*. http://hdr.undp.org/en/statistics/hdi. Accessed 02. May 2014.

UN. (2013). United Nations, Department of Economic and Social Affairs, Population Division, 2013.

WWF. (2012). *Living planet report 2012*. http://wwf.panda.org/about_our_earth/all_publications/living_planet_report/2012_lpr/. Accessed 2 May 2014.

Opportunities Through Positive Impact Investing and Finance Embedded in Banking Value Chains

Karen Wendt

1 Sustainability in the Banking Industry: Status Quo

In the last few years and even today financial institutions are in the spotlight for a number of reasons. They are criticized for the financed emissions they bury in their portfolios. They are animadverted of short selling their leverage when it comes to asking clients to act fully in accordance with human rights. Controversial projects carry reputational, environmental, and social risk for financial institutions. The banking crisis, triggered by unsustainable mortgage sales and a breakdown of the supporting value chain (in this case of mortgage backed securities), has shown that banks need to revisit the way they operate, to ensure sustainable value creation for their clients.

Integrating sustainability into the whole value chain of a financial institution is an ongoing and challenging task. "Environment" and "society" are context factors that financial institutions have to take into account, in a world where information travels by a mouse click. For a long time, banks have addressed sustainability in the value chain mainly with a focus on internal processes, i.e., internal supply and purchasing systems concentrated on the financial institution's own consumption of materials and energy. Likewise, financial institutions and investors often have an active community investment programme in which employees volunteer to support a local community or raise funds for a foundation to alleviate poverty in a certain country. Community investment programmes may be praised for the specific objective they address, but they do not give much information about the sustainability of the core activities of the financial institution or the investors themselves. These activities qualify for modern philanthropy, but not for sustainable finance and investing.

K. Wendt (✉)
Responsible Investment Banking, Ährenfeldstr. 5, 82194 Gröbenzell, Germany
e-mail: info@responsible-investmentbanking.com

M. D'heur (ed.), *Sustainable Value Chain Management*, CSR, Sustainability, Ethics & Governance, DOI 10.1007/978-3-319-12142-0_9

While it is good that financial institutions use green energy to heat their buildings in winter and/or cool the buildings in summer, this does not mean that they have sustainable lending practices, products, processes, or investments, or that trading, marketing, and selling strategies follow sustainable patterns. Neither does it imply that financial institutions are concentrating on their main purpose: creating shared value or positive impacts through investment and finance. It does not mean that they minimize their negative impacts through robust environmental and social risk management strategies, policies, and procedures. It just means they finance the economy and support required investments that are associated with economic projections.

Further issues, for which asset managers and financial institutions are under severe public debate, are human rights as well as financed emissions. In terms of banks' and investment funds' relations with human rights we witnessed an astonishingly successful complaints procedure against the trustee of the Norwegian pension fund. The complaints procedure has been managed by the Norwegian National Contact Point (NCP). Despite the trustee being one of the first signatories to the UN Principles for Responsible Investment, it has been penalized for investing in the Korean Steel company Posco in spite of this company's human rights violations. On 27 May 2013, the Norwegian NCP published its final statement, concluding that NBIM violated the OECD Guidelines for Multinational Enterprises firstly by refusing to cooperate with the NCP and secondly by lacking a strategy to identify and address human rights impacts. This ruling has been a wake-up call for the financial industry and illustrates the importance of human rights for banks, investors and investment funds very clearly. Banks need to identify human rights violations in their clients' value chains to make sure the violations do not occur as a result of their financings.

Financial institutions increasingly understand that the public scrutiny puts them in a very difficult situation, where the business model of financial institutions and investors is questioned. Stakeholders increasingly demand more sustainability as well as transparency of banks' activities.

A well-known example of public action as a result of unsustainable international finance was the "Cut your Card" campaign against a major US financial institution in 2003. The campaign began when civil society and the Rainforest Action Network (RAN) criticized the bank in question for destroying the rainforest through their financing activities. RAN started a television campaign showing video clips of destroyed rainforests asking the public: "Do you really know where your money is currently?" Even though people put their money into a bank account, they normally are not aware which kind of loan this money will be used for.

2 The Challenge of Integrating Sustainability into the Banking Value Chain

Integrating sustainability into the whole value chain of a financial institution is an ongoing and rather challenging task. It is important that the sustainability strategy of an organization touches the core of its activities and is integrated consistently into the entire value chain. For a bank and an investor a sustainability strategy should address not only all of their day to day activities, i.e., the core business but also evaluate the sustainability strategy for the future based on megatrends and financial and social innovation.

If this does not happen, the organization can be accused of "Greenwashing," which is, in fact, what a lot of financial institutions are being criticized for at the moment. They are criticized for the financial emissions that result owing to their respective investment and finance portfolios, as these emissions would not exist without financing clients and their projects. Furthermore, banks are condemned of short selling their leverage when it comes to encouraging or requiring clients to adhere to human rights. Likewise some controversial projects brought a lot of both reputational, environmental, and social risk to financial institutions and their respective portfolios, which are risks that banks want to address in their own interest.

To mitigate the challenges banks began in a first step to analyze their own internal sphere of influence. Purchasing, product development, internal organizational systems, and procedures are part of this. In a second step they started to map out the context factors which matter for them, for clients or projects they want to finance. All impacts that banks affect by doing business ("inside-out" factors) as well as any external factors that impact the bank and the business model ("outside-in" factors) are called context factors. Important context factors include clients, clients' suppliers and direct product suppliers to the bank, main stakeholders (civil society, politics and government, regulatory bodies, international good governance organizations, multinational organizations, business organizations), as well as the environment. In addition the mapping typically includes technological, environmental, and organizational circumstances under which banks operate. These factors are often referred to as PESTO factors as they encompass factors of politics (P), environment (E), society (S), technology (T), and organizational learning (O).

2.1 *Context Factors and Their Contribution to the Risk Matrix*

When it comes to context factors the picture gets more complex. Important context factors stem form politics, environment, society, technology, and organizations. They influence the profitability and the risk profiles of transactions, loans, and investments and are difficult to control. Identifying and managing these factors in particular environmental and social, require that financial institutions work with experts for environmental or social due diligence.

Context factors create high interdependence and complexity for financial institutions. Systems theory provides a useful methodological basis for pinpointing complexity (see, e.g., Forrester 1977). According to this theory, companies are viewed as open subsystems, which interact with its superordinate system (i.e. its corporate environment). The complexity of such superordinate system stands for the complexity of the relevant business context (Jeschke 2015).

According to Sargut and McGarth (2011) the following four parameters characterize the design of a system:

1. Multiplicity: How many elements define the system: elements are either decision subjects (stakeholders) or decision objects (products that are subject to stakeholders' interaction).
2. Interdependency: To what an extent are system elements intertwined in a bundle of interrelating cause/effect schemes?
3. Diversity: To what extent are system elements and the nature of their interrelations similar or dissimilar?
4. Dynamics: To what an extent are both, the set of relevant elements as well as their interrelations, subject to change over time? To what extent – and to what magnitude – have stakeholders or relationship patterns changed within a given time period, and to what extent can they be expected to do so in the future?

The complexity of the system increases with growing multiplicity, interdependency, diversity, and dynamism of its constituting system elements (Jeschke 2015.). At the same time, each of the four complexity parameters claims its own specific management response: High multiplicity calls for a widened environmental scanning approach, typically reaching beyond directly interacting market partners (ibid). With a high degree of elements' interdependency, causal analysis will be required to grasp the "big picture." Otherwise, management focuses attention on symptoms, not causes (ibid). Highly diverse systems challenge corporate analysis to appreciate the particularities of the individual system elements. Flexibility and coordinative capacities are required to translate diversity into business implications (ibid). Finally, with increasing environmental dynamics, some continuous monitoring with early warning signals is required to continuously update on the business environment.

According to Jeschke, the combinations of these parameters leads to a set of 16 types of system design, as illustrated in Fig. 1.

Due to their limited degree of complexity, simple situations (*1* in Fig. 1) – as they are quite common in our everyday life – can be interpreted with intuition. Complicated situations (*9* and *13* in Fig. 1), in contrast, will require support of technical solutions.

One trigger of the banking crisis were financing products (for examples housing mortgages), which were sold to customers who could not afford to pay the mortgage or loan. At the time, banks believed that a very good management of the value chain was in place, which would allow them to minimize risks and at the same time generate business and loans for which they can take an interest rate. Clients' payment difficulties were not an issue as long as property prices went up, because payment arrears could be mitigated by increasing the mortgage loan. Alternatively,

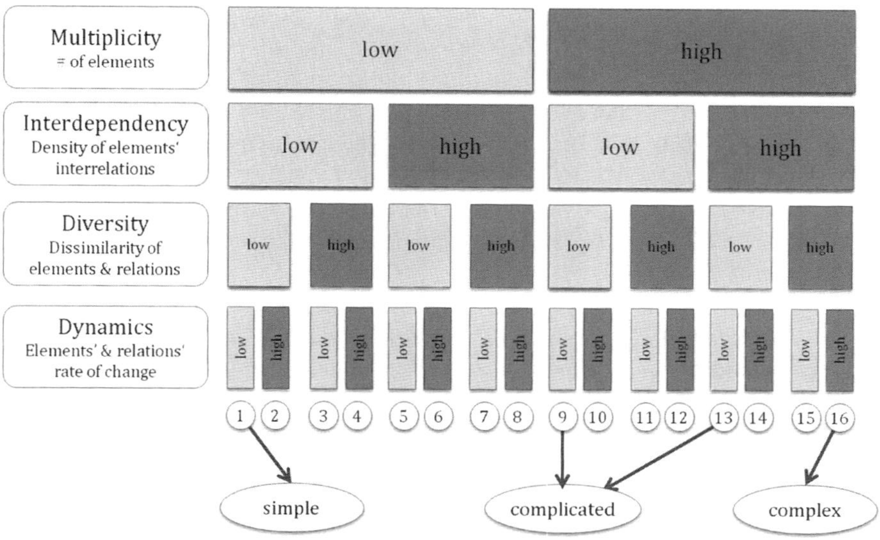

Fig. 1 Types of system design (Jeschke 2015)

property could be sold and the mortgage redeemed. Problems only developed when property prices declined, the demand in the housing market experienced a downturn or loans were no longer covered by the value of the property. As a consequence the product value chain, created around mortgages – i.e., the mortgage backed security market – collapsed and brought banks into a vulnerable position. Not only banks' reputation was affected, but also their liquidity, top line revenue as well as profit and loss statement. The banking crisis has therefore made it obvious that banks need to revisit their value creation processes to ensure sustainable value to clients and use sophisticated technological tool for mapping the context factors and their impacts.

Similarly, financial institutions need to measure and understand the magnitude of their loan and investment portfolios for the emission of greenhouse gases and start to manage the risk accordingly. Climate change brings about increased risks of financing the controversial technologies. For instance, investments in carbon intensive technologies cause contributing to climate change, which continues to be a pressing issue. Sustainable models need to capture the dynamics of carbon capture methodologies and analyze the long term lock in costs for coal.

2.2 *The Impact of Climate Change in Finance*

In 2010, nearly 200 nations agreed that global warming must be limited to 2 °C in order to avoid the worst consequences of climate change, such as a decline in water availability by 50 % by 2060 in many regions, which would happen if we carry on with our current consumption trends. This scenario is described in the Worldbank Turn Down the Heat report and could cause the large-scale displacement of populations, an increase in epidemic disease, rising sea levels, and extreme heat waves, according to the report, and the potentially exceed the assimilation capacity of many societies and natural systems. Climate change comes at a significant cost for people and society. To provide an example of the costs of climate adaptation risk, one may consider the costs for reconstruction after the Typhoon Haiyan hit the Philippines in 2013 which are estimated at a total of US$15 billion, according to The Economist (2013). Similarly, the damage attributable to the 2005 Hurricane Katrina alone has amounted to more than US$0.1 trillion in 2012 (Venugopal 2015).

These figures challenge the models of insurance companies worldwide and provide evidence that climate-friendly markets are needed and conventional value chain management ought to be overhauled to include analysis of those external context factors. In fact, the developed world needs to take serious steps to adjust its means of production and consumption. The mismanagement of public goods such as water, emissions, fisheries, and other ecosystems services cannot be allowed to continue.

"Are the worlds' financial markets carrying a carbon bubble?", asked the 2012 Carbon Tracker Initiative's Report, to illustratate the problem and in 2013, the Carbon Tracker followed suit with its report on "wasted capital and stranded assets" (Carbon Tracker 2012, 2013). Climate change has developed into a risk to nature and humanity, and likewise presents a huge risk to the financial community and insurance markets. Despite this, current value chain management fuels climate change, increases climate adaptation risks, and subsequently even threatens humanity. While the reaction of policymakers to the challenge appears to be slow, especially given the short time which remains to change course, public banks are moving away from financing coal. Analysts of the largest commercial banks question the business rationale for further investment in coal.

2.3 How Can Banks Cope?

Banks have to reinvent themselves and integrate environmental, social, and governance considerations more systematically into their value chains. This entails implementation into strategy, products, and conducting impact, risk, and context factor analysis for all investments and finance activities. It is a challenging process, if there are no systems and procedures for understanding and managing environmental, social, and governance issues in place. Therefore for instance Equator banks require an environmental and social management system in place to ensure the due management of all ESG issues. Likewise Financial institutions should have Environmental, Social, and Governance Risk Management Systems (ESG) in place to monitor their own business, the markets, as well as their own value chains and the value chains of their clients. This includes risk maps, best in class mitigation strategies on ESG and measuring the success of the mitigation strategy by comparing risks and impacts before and after implementation of the mitigation strategy.

This should include a risk map including the context factors illustrated in Fig. 1, as well as best in class risk mitigation and management strategies which are offered by a number of Soft Law Standards and Good Practice Guidance Notes like the Equator Principles, the IFC Performance Standards, the UN Guiding Principles for Human Rights in Business, the UN Principles for Responsible Investment, the OECD MNE Guidelines, ISO 26000, just to name a few. The tool box can be complemented by a number of ISO certifications which help the financiers and clients to implement good practice in the field of ESG. The criteria formulated by the Global Reporting Initiative (GRI) are likewise a helpful guide in implementing good practice Environmental and Social Governance. These approaches in first place help to address risks identified, however companies and banks can likewise use them for screening and identification of positive impacts. As a rule of thumb we can assume that companies and banks which are unable to identiy risks and negative impacts are likewise unable to identify opportunites that stem from the sphere of environmental and social context factor analysis.

The application of the aforementioned principles and guidelines creates the fundament for identifying positive impacts and not only risks and negative impacts. Impact analysis is very similar for positive impacts as it is for negative impacts. Screening can be used for discovering opportunities and positive impacts as well, which then in a third step can be integrated in a similar way into the project or product profile though the use of an Action Plan. The methodologies are very much the same for identification and mitigation of negative impacts as for positive impacts. For negative impacts it is important to get them mitigated through the combination of a screening process and the implementation of an Action Plan. For positive impact creation the screening process will purposely look at identifying potential positive impacts and add measures and objectives to the profile of a transaction and embed them into the deal profile as well through an Action Plan. Thereby the market value of the company will be increased. This increased value then can be materialized when selling a project company for instance in an IPO process, as the company will be more resilient against downside scenarios and have a robust profile with regard to acceptance and social licence to operate in the market place.

In both cases – mitigation of negative impacts as positive impact discovery – it is important to note, that the impacts need to be monitored and the Action Plan therefore needs to define not only objectives but likewise key performance indicators for each objective to make the impact measurable. This guarantees that the Action Plan will be enforceable, and will be eligable for monitoring and reporting.

3 Embedding Sustainable Practices in the Banking Value Chain

Many banks have started to deal with sustainability a decade ago. They have signed the Equator Principles and address sustainability in sector lending policies. They use Environmental, Social, and Governance (ESG) principles when lending. To address ESG risks banks must look into their clients' behaviors, systems, market strategies, supply chains, and production mechanisms. Further steps on common standards and approaches have been taken by the Equator Principles III, which were launched in June 2013. They embrace not only risk management perspectives, but also include an explicit commitment to human rights in finance through executing rigorous due diligence. They also take up a leading role in addressing climate change independently from risk considerations. This progress is important and helps to streamline sustainability in an important area of banking. It broadens the focus of risk management towards people and climate change. As these commitments are brand new, best practices do not yet exist. Nevertheless, they need to be established by focusing on issues like management systems, collaboration, shared procedures, shared learning, policies, frameworks, and most importantly – a robust system for stakeholder engagement. The Equator Principles III now embrace human rights in finance through executing strong due diligence and to accept a leading role in addressing climate change. The question is: Is this enough or is there more to be done? The answer will depend on how far we will be willing and prepared to go.

Notwithstanding the important steps that are being taken so far by implementing ESG principles through sustainable lending and investment practices, I believe there are some additional areas in which banks need to thoroughly address sustainability.

In general, the definition of sustainability, which financial institutions usually use, includes the objective to meet the needs of the organization and stakeholders through protecting, sustaining, and enhancing human, natural, and financial capital inside and outside of the organization. This is a very broad definition which leaves a lot of flexibility for interpretation. Some financial institutions have a strategy or at least an extensive plan regarding how sustainability can be put into practice using objective key performance indicators (KPIs) and processes throughout the entire value chain. In an ideal scenario the strategies tend to seek evidence using balanced score cards, thus translating vision into action and allowing for organizational learning to take place. They may have committees to discuss, define, project, and steer sustainability throughout their value chain. Sometimes they even use platforms where they share and resolve issues in collaboration with their peers. An example of this is the Biodiversity Initiative which the Equator Principles Financial Institutions Association is pursuing with industry associations like ICMM and IEPICA, the ORSE Platform in France, the Equator Principles Forum, the Thun Group of Banks, and UNEPFI.

However, most institutions today do not have a strategic approach which addresses all their activities. There needs to be an implementation plan to ensure that the strategy will lead to specific actions.

The next chapter will address the implementation of ESG in the lending practices of banking and explain how banks can scrutinize the supply chains of their borrowers for human rights, social and environmental issues using the World Bank's guidelines and the International Finance Corporations Performance Standards. In doing so, banks can pursue a variety of routes: Many use sector policies and Equator Principles for their lending portfolios, but this will not support them in advisory business, treasury activities, trading, IPOs, equity investment, or hedge business with their clients. In order to fully assess their entire portfolio banks need to integrate ESG issues into the "Know Your Customer Check." However, this does not address the liquidity management, internal and external capital allocation business, which finally needs to be steered using a value chart and key performance indicators. Addressing human rights and financed emissions are further urgent and very new topics towards people orientation and the creation of positive impact through finance.

Their Environmental and Social Management System of a Financial Institution has to manage all these internal as well as external factors.

3.1 *Addressing ESG Risks and Opportunities in Banking*

To determine the aspects that matter in banking value chains and develop the right approach for each context factor, a number of action fields need to be reviewed and thought through (see Fig. 2). The question is: What are the context factors that pose the most relevant risk to the financial institution? (It is worth mapping those context factors and likewise review how they are changing over time). Also, it is important to understand, which context factors have the potential to force a financial institution to abandon a transaction.

To illustrate how banks determine in which areas they are investing their sustainability efforts, some of the applicable focus areas are explained below.

Operational Ecology Banks have started to look into sustainability in a number of ways. To protect their reputation many banks have evaluated their purchase procedures and internal processes to see where they can save CO_2 by looking into green buildings, CO_2 neutral business trips, saving paper, electricity, and heating.

Sustainable Product Development Following the financial crisis many banks started to analyze their financial product development procedures, to make sure that products are transparent, understandable, and fair to the customer.

Principles for Responsible Investment in Asset Management – In addition, many banks have signed the Principles for Responsible Investments (PRI) to make sure they are able to offer funds to their clients that exclude sectors such as weapons, stray bombs, or child labor.

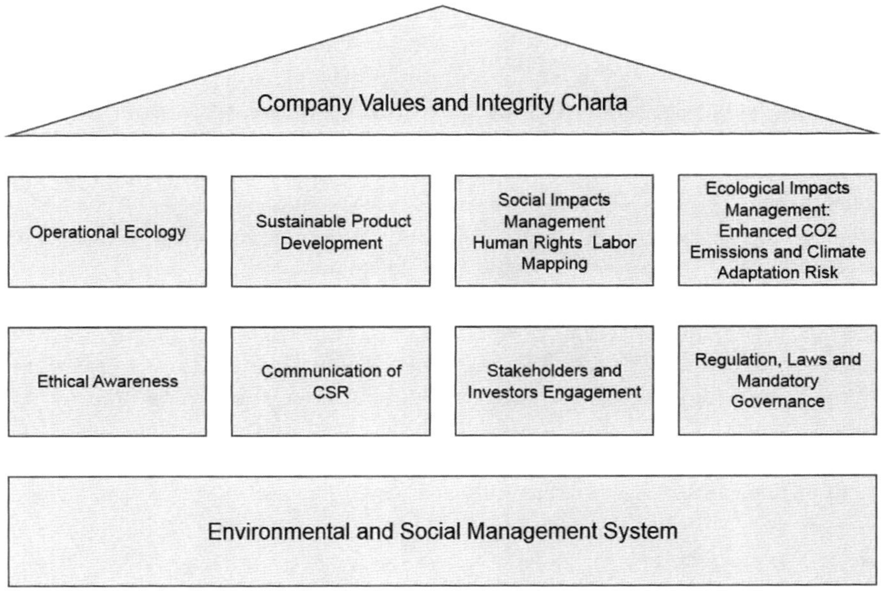

Fig. 2 Action fields for ESG in financial institutions (Karen Wendt, 2014)

Foresight and a Sustainability Strategy A bank's strategy is very often supported by raising awareness about their respective corporate social responsibility efforts as well as communication about the matter.

Whereas operational ecology, communication and awareness raising, as well as product development are mainly focused on internal procedures, environmental and social impact management in finance through Equator Principles and in investing through the PRI can, in fact, be regarded as a supply chain issue. Institutions will normally apply exclusion lists to make sure that funds or companies with adverse practices are identified and excluded from the service portfolio. They therefore service the external market perspective. Impact identification procedures are very much the same for negative as for positive impacts. They normally occur in the sphere of influence of the finance or investment transaction meaning if the transaction would not there the positive or negative impact would not occur. Therefore impact identification whether negative (focus on risk management) or positive (focus in opportunity creation) use the same logics and often the same context factors. Just to give an example. The creation of local jobs in a developing region can be a positive impact, but what if labor is not locally sourced or the investment is even using forced labor or child labor in its supply chain. The example shows two characteristics: First, impacts for a certain value can be positive or negative depending on the way of implementation and the methodology and the mapping for positive and negative impacts is quite similar.

3.1.1 Sustainable Lending and Investment Practices

To address the ESG risk associated with their lending portfolios banks can pursue a variety of methods for the integration into their value chains:

- By using sector policies and adhering to the Equator Principles on the lending side
- By measuring the climate impact of their lending and investment portfolios
- By calculating the financed emissions embedded in their activity,
- By integrating the Principles for Responsible Investment (PRI) for Asset management.
- By implementing ESG issues into the "Know Your Customer Check" and reviews of their product development processes.

Adding People Focus to Risk Management: Addressing human rights and the social effects of climate change are further urgent issues for top management where there is a trend to development towards the creation of positive impacts through investment and finance.

3.2 Addressing ESG Risk in Project Lending: "The Equator Principles"

In the same year of the "Cut your Card" campaign, other banks had reached a comparable tipping point with civil society campaigns. The public, grumblingly disagreeing as to where banks invested their money, and clients, worrying about their money and whether it was safe, forced banks to create a first framework on managing environmental and social risk in project finance and beyond. To address environmental, social, and governance (ESG) risks, many banks have committed themselves to the "Equator Principles," a risk management framework for determining, assessing, and managing environmental and social risk in structured lending. The Equator Principles are primarily intended to provide a minimum standard for due diligence, developing a management and mitigation strategy and measuring it to support responsible risk decision-making and best in class impact management. Even today they are the most effective and internationally accepted voluntary framework for managing environmental and social risk in project lending. They are the basis for the management of non-technical risks in international lending; its annex provides a useful overview of these risks and their mitigation. The EPs involve a voluntary commitment by the signatories, to ensure that financial institutions conduct due diligence procedures, and that clients and borrowers analyze and manage the impact of their projects in accordance with the World Bank environmental and social standards and particularly with the International Finance Corporation's Performance Standards (Cochard 2015). The latter covers issues such as forced population displacement, respect for biodiversity, and human rights. The institutions who have signed this document commit themselves to due diligence on the projects they finance, to analyze the social and environmental risks and impacts of the projects, and to ensure that the borrowers develop action plans to reduce their impacts as much as possible and offset problems that cannot be avoided.

For example, according to the Equator Principles, financial institutions have to investigate into extra-financial risks. ESG risks while project specific, can normally be classified into risk categories. High risk issues vary widely, but include some of the following issues listed b: These risks are likewise described in the illustrative list of project risks available in the Annex to the Equator Principles (www.equator-principles.com).

- Multiple emissions at or close to regulatory limits
- Large-scale constructions with an extensive temporary or migrant workforce
- Workers health and safety as well as systems performance
- Significant retrenchment, migrant labor, forced labor, child labor, discrimination, lack of rights for collective bargaining
- Extensive contaminated soil and risk of soil or water contamination
- Unsustainable demand of water resources
- Involuntary resettlement or economic displacement of populations
- Potentially significant adverse impacts on vulnerable or endangered species and/cultural heritage or habitats
- Impacts on cultural heritage or a monument due to for instance increased traffic access
- adverse NGO attention with local community grievances
- Impacts on ecosystems services
- Impact on indigenous people

Other important factors that are part of the risk profile assessment of EPs include:

- Environmental Impact Assessment exemptions
- Extended authorizing processes
- Lack of stakeholder engagement on siting decisions
- Lack of analysis on legacy issues for brownfield sites
- Limited capacity and enforcement of national regulatory requirements, as they can impact on the quality of regulatory controls that are defined within decision documents and permits used to manage ESG project risks
- Lack of cohesion or inconsistencies with national development plans or strategic assessments
- Government led resettlement

Usually, in large scale projects, initial due diligence takes the form of a comparative analysis of the prepared project documentation with the World Bank Safeguard Guidelines and the IFC Performance Standards, on which the Equator Principles are based. It normally also includes a site visit to assess the potential Environmental, Health, and Safety (EHS) and social risks. This is often the first time that the client and the lender ESG group liaise in detail on the ESG project risks (Cousins 2015). Requests for relevant ESG project documentation routinely had to include: EIAs, Social Impact Assessments (SIAs), risk assessments, details of Environmental and Social Management System (ESMS) arrangements, feasibility studies, engineering reports and designs, soil investigations, information on air and water quality modeling, health and safety performance data, monitoring, expropriation plans, and information on stakeholder engagement for the project (ibid).

What is equally important is the discussion with client EHS representatives in order to assess the management capacity within the client's organization and its contractors, and how ESG risk management is organized and monitored. Moreover, one needs to understand the management procedures of contractors which the client has put in place to ensure the application of the World Bank standards throughout the supply chain. The subsequent analysis includes issues such as (Cousins 2015):

- The role of the EHS manager (if they have one)
- The role of the community liaison officer (if they have one)
- Senior management involvement in ESG issues
- The status of the company's (and contractors') environmental, health and safety, human resources
- Safety performance and monitoring
- The level of engagement with local communities and how concerns or complaints are managed
- Potential supply chain issues including the role of contractors and sub-contractors in managing ESG risks while carrying out project activities
- Mapping of the project supply chain and identify labor rights or human rights issues as well as environmental risks
- Monitoring and reporting arrangements

3.3 *Sector Policies Used in Lending*

Most other activities of banks other than structured activities could probably never be subjected to the due diligence required under the Equator Principles. This is because they do not fulfil two necessary conditions (Cochard 2015).

On the one hand, the Equator Principles, as they exist today, are based on the assumption that the use of the funds is precisely known and is linked to the construction or expansion of an industrial asset or infrastructure (existence of an impact study and a plan for the management of residual impacts).

On the other hand, financial institution and their clients ought to have the necessary leverage (for example, when the bank is financing equipment used in the construction of a large project). The question is whether the client in question has access to the impact studies of the entire project and whether he can influence its characteristics. For financial institutions – the question is therefore whether they can enforce an Equator Principles Action Plan. Project financing is traditionally subject to significant due diligence and to tailored legal documentation owing to the risks involved for the bank (reimbursement is solely based on the project's cash flow, without guarantee from the developer) and an Equator Principles Action Plan to mitigate and manage any environmental and social risks and impacts. However, the same requirements are not usually applicable to other methods of financing and their social acceptability may be unsecure. Imagine, for example, an individual client accepting a property loan from a bank on the condition that a maximum temperature level within the property is respected, or a car loan is provided on the condition of certain eco driving commitments (Cochard 2015). What happens, where early reimbursement of the loan is demanded if these conditions are not met? What seems natural for project financing within the framework of the Equator Principles is not obviously applicable in other cases (ibid).

Incorporating the principles adopted by the UN Human Rights Council in 2011 the OECD's key principles for multinational companies stress the obligation to "seek ways to prevent or mitigate adverse human rights impacts that are directly linked to their business operations, products, or services by a business relationship, even if they do not contribute to those impacts." This obligation is based on the performance of reasonable due diligence (Cochard 2015).

As part of their CSR policy financial institutions cannot ignore major issues of public concern like climate change or net loss in biodiversity, which have been identified as real threats by science and are a direct consequence of our modern life and way of doing business. Each financial institution needs to decide and determine its own financing and investment policies. Other than the Equator Principles no level playing field exists for sector policies. As a financial institution does not have the same leverage as in project finance, sector policies use exclusion criteria.

Both the Equator Principles and sector CSR policies contribute to banks' management of credit and reputational risk related to environmental and social impacts of the activities they finance. Banks that develop sector policies are therefore generally looking to synchronize as many as possible of their requirements regarding the two approaches, while they acknowledge that their leverage for action differs. The general idea is that a bank does not end up financing a project under one method which it would not finance under another, even if the nature of the potential due diligence depends largely on the financing method used (Cochard 2015).

3.4 The New Investment Rectangle: Liquidity, Risk, Return, and Impact of Funds

Banks and investors have greater freedom in determining where to allocate money, but likewise more responsibility. With the proliferation of the internet across the globe, information can be accessed very easily. Misconduct, discrepancies between commitments and actions of governments, private sector corporations, or banks quickly become visible to the skilled, information-filtering community. This makes it necessary for any company, but in particular for investment banks and investors, to strategically rethink their business models as they are under immediate and permanent scrutiny from stakeholders. Some of these are powerful enough to influence the profitability of their investments, but also their reputation and model of operation. They can provide banks and investors with a different matrix of context factors to their investments and lending practices in a very short time, and so can different political and regulatory frameworks. One could argue that this leads investment banks to even more short-termism in order to get rid of the risks that context factors may pose. This, however, may be short sight rather foresight, owing to the fact that with increased short-termism, the cross-selling opportunities and customer retention and loyalty vanish. Moreover, short-termism and risk avoidance through short turnover periods do not make a financial institution or an investor immune to reputational damage.

Negative impacts can still be traced back to the structuring or underwriting institution or investment company in our current information age. Confidentiality is no shelter against revelation. As long as investors and banks engage in long-term business, one can always argue that using environmental and social foresight only in long-term business but not in the short-term (where the risk is passed on quickly or immediately) is arbitrary. Doing so does not make an institution immune from reputation risk like the consistent application of environmental and social consideration and best practice does.

The degree of interconnectedness and links between context factors will provide more complex situations for decision making and it will be important to understand the key context factors that can make or break a single deal, or even an institution as a whole. Coincidently, the role of fiduciary agreements and duties of investors are currently being redefined and now focus increasingly on investors' responsibility towards society and their fundamental ethical norms, as the case of the Norwegian pension fund illustrates. This has the potential to be a game changer in the discussion of the principal-agent problem. Banks need models to calculate the so called extra-financial risks and impacts that context factors pose to investment and finance in monetary and use it as a basis for terms and need to use mitigation of extra-financial risks and impacts as a decision making tool.

The investment triangle of liquidity, risks, and return is now complemented by a forth component: the impact of use of funds. Clients want to know and institutions need to show how they use clients' money. This entails the due application of environmental and social governance when investing clients' money. Otherwise, clients are migrating funds to more responsible institutions, and subsequently remind banks that they, in fact, hold the role of the principal. In fact, fund managers and wealth managers are the agents for their investees, the shareholders, and therefore can influence the investment target on its sustainability strategy for instance by using an engagement strategy for ESG, if the client is at risk due to the sector it is operating in or due to its own track record with civil society.

As a result, a rectangle of investment and finance comes into existence, which includes liquidity, risks, return, and the impact of funds (Fig. 3).

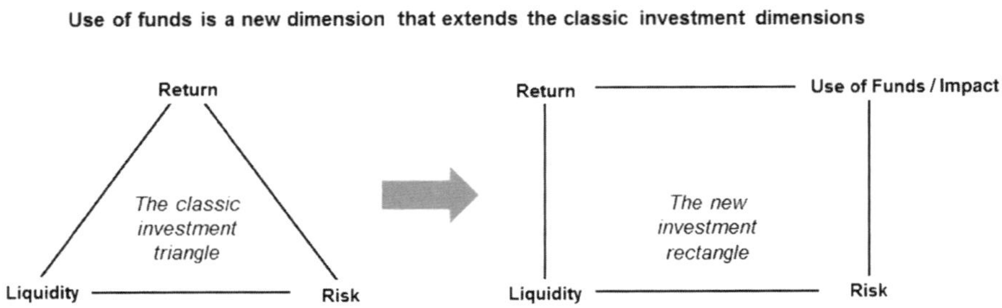

Fig. 3 The new investment rectangle (Wendt, own Graphic, 2014)

4 A New Paradigm in Investment and Banking Is Emerging

4.1 Investors: The New Drivers of Sustainable Development

One thing is needed to fundamentally change markets: leverage. Many banks and companies use CSR, but many of them do not put it at the core of their respective strategies. This may become increasingly dangerous, because sustainability and CSR are at the edge of growing into key success factors. CSR, governance, and transparency and reporting are attracting increasing focus from potential institutional investors. The industry increasingly embraces the concept of impact. While conventional wisdom states that ESG is a necessary cost centre that reduces reputational, operational, or credit risk, ESG can likewise be a framework for profit creation and strategic direction. ESG due diligence can add value to investors throughout the investment process, from selection to exit, for example for Initial Public Offerings (IPO). This is demonstrated for instance by the Aloe Investment Fund. The industry increasingly embraces the concept of value creation through ESG.

Additionally, in alignment with Non-Governmental Organizations (NGOs), investors increasingly ask for transparency and evidence of integration of ESG into the entire value chain and company operations. Banks are no exception to the rule. In 2013 a group of financial investors responsible for a portfolio of US$3.3 trillion urged 1,900 companies from 44 countries to join the United Nations Global Compact and to comply with its ten principles. Investors (and shareholders) are subsequently taking up matters of climate change, biodiversity, ecosystems services, and access to drinking water. Similarly, the Carbon Disclosure Project has more than 700 members and more than US$87 trillion in funds. This project, which requires companies to become active regarding climate, ecosystems services, and biodiversity, may be the most potent new player in the field of sustainability and has the power and leverage to change the rules of the game. Within this process, rating agencies often play the role of an enforcement agent and are catalysts in the process of positive impact creation.

4.2 Positive Impact Investing and Financing

A new far-sighted and proactive approach of value chain management in investment and banking, is looking at opportunities: the creation of shared value and positive impact. Value chains of production and consumption have to change in public and private sector alike, to collectively help fostering climate-friendly solutions, products, and, above all, climate-friendly markets.

Positive impact investment and finance likewise put the focus on supply chains, but also adds the extension and transformation of supply chains and consequently includes elements which do not show up directly in the social reporting of investment and banking. While social responsible investment (SRI) uses screening to avoid the exposure of the investment portfolios that are harmful for society or environment, Impact investing intentionally aims to create a positive, measurable impact through profitable businesses.

A new and proactive component of value chain management in investment and banking consider the opportunity side of value chain management and particularly the creation of shared value and positive impacts. Value chains of production and consumption will have to change and innovative means will have to be put forward in order to allow public and private sectors to collectively pursue climate-friendly solutions, products and, above all, climate-friendly markets. Some market players like EMPEA the Emerging Markets Private Equity Association does acknowledge the approach and has build an impact investment council to identify new opportunities and to provide relevant research.

Accessing finance for climate-friendly projects can be challenging today owing to the limited track record of these markets and their currently evolving state. As a result, the private sector has limited awareness and discomfort. Yet, there are the first promising structures which have been put in place by clients and financiers like green bonds and climate bonds.

Global megatrends will force society, like the Green Bonds which business and banking to extend value creation beyond financial goals in order to take environmental and social solutions on board. This applies in particular to population growth, climate change and adaptation, ecosystems services, human rights and labor, fresh water, and access to food and agricultural services. In all of these areas, the financial system is called to duty, as political solutions start to play a role only slowly and timidly. The individual versatility of financial institutions and their clients, as well as mobile venture capital and equity funds, will be the key elements for economic success. They will determine our future and the extent as well as circumstances under which economic attainment can take place.

Positive impact investment and finance goes in fact one step beyond shared value. Shared value means asking a company to concentrate its focus on maximizing economic and societal value by investing capital and other resources and by leveraging core activities and partnerships for the mutual benefit of people and shareholders. It is comparable to the concept of amalgamated value where financial, environmental, and social performance are calculated and blended in one indicator. The underlying meaning is that companies create business and societal value when they take a broad and long-term view of their business activities.

Positive impact investment and finance also puts focus on supply chains, but adds elements of their extension and transformation. It subsequently also includes features which currently do not show up in the social reporting of investment and banking and do not fit under current standards and schemes. They draw on cross functional and cross sectorial cooperation and the creation of shared knowledge.

Positive impact investment and finance puts the creation of positive impact at the very centre of strategy, product development, technological innovation, and supply chain transformation. A number of products currently emerge in this field; while some of them are still small, many have the potential to become mainstream. Hence positive impact investment is a natural evolution of to socially responsible investment (SRI), which uses screenings to avoid portfolio exposure to socially or environmentally harmful investments, whereas positive impact investing intentionally seeks to create a positive, measurable impact through profitable businesses. Like SRI, impact investing is also a process by which investment managers screen, evaluate, and monitor investments for environmental and social factors. The difference is just the intention: scouting for positive impacts rather than concentrate on exclusion lists. In an ideal world one would combine the risk management aspect of negative impact avoidance with the positive impact scouting aspect and create a new asset class on impact investing.

Some examples for positive impact illustrate how these investments pay off for the company and the investor alike, by providing a new and sustainable way of value creation.

Social Impact bonds are created by a public commitment to pay a group of private sector investors for their social success or positive social impact outputs and outcomes as measured by Key Performance Indicators. Impact investing differentiates between outputs (direct performance criteria as number of people getting vaccination) and outcomes (direct and indirect results – reduction in diseases due to vaccination). Outcomes are more diverse due to the set of context factors to be considered, less direct and therefore often more difficult to define and measure than outputs. A broader discussion of this issue can be fond on the EUROSIF page.

The first Social Impact Bond was implemented in the United Kingdom to reduce prison recidivism. Similar pay for performance models exists in ecosystems services and for customers at the bottom of the pyramid.

Climate Bonds and Green Bonds can also be described as being on the rise as they are rapidly creating a new market. The Climate Bond Initiative estimates that the number of outstanding climate-themed bonds doubled between 2012 and 2013.

The International Energy Agency (IEA) estimates that, on current trajectories, the world is, in the words of IA Chief Economist Fatih Birol, “barrelling” towards 6–7 °C warming, and that this would have “catastrophic” impacts.

The IEA also estimates that, worldwide, $1 trillion of investment in energy, transport and building sectors are required each year – above business as usual – to reduce energy-related carbon emissions in line with a 2 °C global warming scenario.[1]

Climate Scientists now recognizes that 2 °C warming is now very likely, leading to significant adaptation pressures. According to the UN Environment Programme, adaptation and the sustainable management of natural resources such as forests, fisheries, agriculture and water will require an average additional annual investment of $1.3 trillion out to 2050.

In order to meet the IEA's $1 trillion target, the challenge is not to creating new capital, but by shifting a portion of existing investment into low-carbon development.

Public sector balance sheets are severely constrained and are likely to remain so. The bulk of the money is going to have to come from the private sector, in particular from the $83 trillion of assets under management by institutional investors.[2] If structured correctly, the good news is that the $1 trillion required is investment not cost. Investment in high capital expenditure projects can deliver stable returns over a long period using a thematic bond market.

A thematic market is a labelled bond market where use of proceeds are specifically devoted to a particular purpose, in this case climate change and environmental problems.

Many investors – for example those representing USD23 trillion of assets under management that signed 2013 declarations[3] about the urgent need to address climate change – express interest in green bonds, subject to their meeting existing risk and yield requirements. That interest in *equivalence* has been the key driver in sustained issuance and oversubscriptions of thematic green bonds in 2013 and 2014.

From 2007 to 2012, the market grew slowly with only a small spike in 2010 but in mid to late 2012 three French provinces, Ile-de-France, Provence-Alpes-Côte d'Azur and Nord-Pas de Calais, issued green bonds that were heavily oversubscribed – this increased the market interested in thematic bonds. In 2013 the IFC issued a $1 billion (benchmark size) green bond in February and shortly after the EIB issued a 650 million euros Climate Awareness Bond, which it then tapped again to make it a 900 million euros. The size of these bonds were a turning point in the market (up to that point, few bonds reached $200 million) and stimulated interest from both banks and investors.

The sector is currently mostly fuelled by public sector issuance such as the Ministry of Railways in China, the World Bank and some development banks. However, Climate Bonds transform existing capital supply chains and allow big institutional investors access to climate funding. Besides exponential population growth from 2.5 billion people to 12.5 billion people from 1950 to 2050 and rising urbanization with increased need for public goods like water, food, ecosystems and

---

[1] International Energy Agency, ETP World Energy Outlook (2012).

[2] OECD (2014).

[3] http://globalinvestorcoalition.org/.

health services, education, and climate change there is another megatrend: natural resources become scarce. Consumers demand more sustainable produced goods (i.e., those produced without child labor, with decent working conditions and without damage to nature and biodiversity), making the access to sustainable resources a new development. At the moment there are not as many sustainable produced goods as the market demands. Taking sustainable production to the level of local farming in Africa, Asia, and South America is something that companies and banks in cooperation with development banks and development agencies and Development Finance Institutions can aim for.

As companies are forced to take up responsibility for the production of goods they buy, there is the need to go further upstream into the value chain to control a sustainable way of production by local producers and to keep access to these local resources (Kleiterp 2015). When a company only buys its goods in one harbor such as Accra or Mombasa, it will lose access to scare resources. Using development cooperation funds to improve sustainability and efficiency deep in the value chain (for example with small farmers) combines the interest of companies in developed and developing countries (ibid). One can foresee that in the future more funds from development cooperation will be used for value chain finance as this combines the pressure to use funds in the national interest and to create positive impact in developing countries. However development corporations and banks now need to find new ways to support these trends in their financing and investment strategies. Some of the examples on positive impact investing provide evidence on the effectiveness and profitability of such an approach.

Positive Impact Investment and Finance has the potential to align interests of customers, shareholders, stakeholders, investors, and banks themselves towards a universally-shared objective, overcoming the classical trade-offs and dilemmas faced by banks and investors.

5 Conclusions

Most environmental and social risks in finance come from the sphere of context factors, which are mostly interrelated and increase the complexity of the system. They usually stem from politics, environment, social system, technology, and organization. Those coming from environment, social system, and politics are the most difficult to assess and manage. They affect the value chain of financial products, their cash flows, cash flow calculations, as well as company values, liquidity, and fungibility of assets. They can not only harm the reputation of financial institutions, investors, and products, but also their intrinsic financial value. In order to identify positive impacts banks, investors and fund managers will need a similar mind set in screening context factors as for assessment and identification of negative impacts. Banks need to focus on climate change for creation of positive impacts and on climate adaptation to avoid risks. Renewable resources and resource efficiency, human rights and preservation of nature, biodiversity, and ecosystems services are not only risk management issues but can be used to create positive impacts. In order to reduce social tension the financial industry will need to help in addressing poverty and income distribution.

While the Equator Principles, Sector policies, the Principles for Responsible Investment are best practice frameworks to address risks, the opportunity side of creating positive impacts for the population and environment is still not given enough attention, but this will be necessary for the future economic, ecologic and social equation in investment and finance. We are loosing a lot of potential if we continue to overlook the upside potential. Positive Impact Investing and finance can create new customers, alleviate poverty, and become mainstream, especially because megatrends are currently shifting to more integrated value chains and international cooperation.

Bibliography

Carbon Tracker. (2012). *Unburnable carbon – Are the world's financial markets carrying a carbon bubble?* http://www.carbontracker.org/site/wp-content/uploads/downloads/2011/07/Unburnable-Carbon-Full-rev2.pdf. Accessed 22 May 2014.

Carbon Tracker. (2013). *Unburnable carbon 2013: Wasted capital and stranded assets*. http://www.carbontracker.org/site/wastedcapital. Accessed 22 May 2014.

Cochard, E. (2015). Translating standards into successful implementation: Sector policies and equator principles. In K. Wendt (Ed.), *Responsible Investment banking*. Springer International.

Cousins, D. (2015). Implementing environmental and social risk management on the ground: Interfaces between clients, investment banks, multi-laterals, consultants and contractors: A case study from the EBRD. In K. Wendt (Ed.), *Responsible Investment banking*. Springer International.

Economist. (2013, November 16). Worse than Hell. http://www.economist.com/news/asia/21589916-one-strongest-storms-ever-recorded-has-devastated-parts-philippines-and-relief. Accessed 30 May 2014.

Equator Principles. (2013). *The equator principles*. http://www.equator-principles.com/resources/equator_principles_III.pdf. Accessed 22 May 2014.

Forrester, J. W. (1977). *Industrial dynamics* (9th ed.). Cambridge, MA: MIT.

Jeschke, B. (2015). Managing assets in a complex environment: An innovative approach to sustainable decision-making. In K. Wendt (Ed.), *Responsible investment banking*. Cham: Springer.

NPC Norway. (2013). *Norwegian bank investment management violates OECD guidelines*. http://www.responsiblebusiness.no/en/2013/05/27/norwegian-bank-investment-management-violates-oecd-guidelines/. Accessed 22 May 2014.

Kleiterp, N. (2015). The new development cooperation: The importance of the private sector. In K. Wendt (Ed.), *Responsible investment banking*. Springer International.

Sargut, G., & McGarth, R. G. (2011). Learning to live with complexity. *Harvard Business Review, 9*, 1–10.

Venugopal, S. (2015). Mobilising private sector climate investment: Public-private financial innovations. In K. Wendt (Ed.), *Responsible investment banking*. Springer International.

UN Guiding Principles for Business and Human Rights. http://www.business-humanrights.org/SpecialRepPortal/Home/Protect-Respect-Remedy-Framework/GuidingPrinciples

Henkel: Sustainability in the Value Chain: From Philosophy to Practice

Frank Roland Schröder, Dirk Holbach, and Thomas Müller-Kirschbaum

1 Sustainability and Resource Efficiency: One of the Challenges of the Future

For decades, the discussion about the underdeveloped countries on our planet's southern hemisphere has influenced public debate. Increasingly, however, we are witnessing the striking transformation of former developing countries into dynamic economies which goes hand-in-hand with an intensification of global trade (UNDP 2013).

From a socioeconomic standpoint, this is a positive development but it puts additional pressure on the earth's limited resources. In 2008, average global resource consumption was 1.8 times higher than the biocapacity of the earth[1] – up from 1.5 – (in 2007). To achieve a balance between the earth's biocapacity and our resource consumption over the medium and long term, it is essential that resource consumption and added value are decoupled in the production process. The technology to achieve this goal is already available. Ernst Ulrich von Weizsäcker, for instance, identifies a number of ways that resource efficiency can be improved by a factor of 5 using different technologies (von Weizsäcker et al. 2010).

The United Nations Environment Programme (UNEP) highlights the strong correlation between economic growth, quality of life, and resource consumption. The European Union also dedicates a great deal of attention to resource efficiency. In the "Europe 2020 Strategy," the European Commission underlines the important role of resource efficiency in strengthening sustainable growth (European Commission 2010).

---

[1] Footprint Network (2013)/biocapacity, also called biological capacity, is the capacity of an ecosystem to provide useful biological materials and absorb waste produced by humans.

F.R. Schröder (✉) • D. Holbach • T. Müller-Kirschbaum
Henkel AG & Co. KGaA, Henkelstraße 67, 40589 Düsseldorf, Germany
e-mail: roland.schroeder@henkel.com; dirk.holbach@henkel.com; thomas.mueller-kirschbaum@henkel.com

M. D'heur (ed.), *Sustainable Value Chain Management*, CSR, Sustainability, Ethics & Governance, DOI 10.1007/978-3-319-12142-0_10

To assess the efficiency and effectiveness of resource efficiency measures, a quantitative analysis is needed. In the wake of the 2008 financial crisis, Joseph Stiglitz and Amartya Sen sharply criticized the concept of the gross domestic product, a purely monetary indicator, which gives rise to the question: which methods can be used to measure both resource consumption and added value? One method is the Human Development Index of the United Nations (UNDP 2013), which has been compiled since 1990, while another is the Happy Planet Index of the New Economics Foundation (New Economic Foundation 2012).

2 Triple Bottom Line and Life Cycle: A Holistic Approach to the Sustainability Strategy

The term *sustainability* was originally used in the context of forest management and relates to the long-term use of natural resources. The concept encouraged the abandonment of measures designed for short-term gains, but which also depleted the resources in the process. A modern interpretation of the term *sustainability*, which also emphasized the dimension of time, was introduced in the Brundtland report. According to this interpretation, sustainable development "... meets the needs of the present without compromising the ability of future generations to meet their own needs..." (United Nations 1987).

The concept of the *triple bottom line* (Elkington 1998), sometimes also referred to as the three pillars of sustainability, is often used to illustrate sustainability. To fully exploit the concept of sustainable development, equal priority must be given to each of these three basic elements of sustainability, i.e., economic, environmental, and social aspects. Favoring one element at the expense of the others limits the potential of sustainable development. There may be a certain legitimacy to focusing purely on environmental issues. However, if social or economic aspects are adversely affected, this development can no longer be considered sustainable. In this respect, strengthening one element at the expense of the complementing aspects limits the overall sustainability performance.

Consequently, sustainability should give equal consideration to value (which includes the economic and social factors) and footprint (which expresses environmental factors). For example, resource efficiency can be improved by increasing the value created while keeping the footprint the same. Conversely, there is an increase in resource efficiency if the footprint is reduced, while the value can be held constant. A strategy that primarily aims to reduce the footprint while accepting restrictions in use is called a *sufficiency strategy*.

In 2010, Henkel decided to increase resource efficiency as part of its sustainability strategy by improving the relationship between the value it creates and the footprint of its products and technologies by a factor of three within 20 years (Henkel AG & Co KGaA 2011). One way to reach this goal would be to increase the added value by 50 % and simultaneously lower resource consumption by 50 %. The quotient of 1.5 and 0.5 yields the target factor of three.

From an operational standpoint, it is advisable to break the concept of value and footprint down into additional categories. The footprint can be divided into categories such as "Raw Materials and Waste," "Energy and Climate," and "Water." These three categories play a key role in the resource use and in the impact assessment, both of which are important elements in the *life cycle assessment* [LCA] methodology.

The discussion over value has not yet reached this level, but categories are also useful here. In addition to performance, which is extremely important for the value of a product or a service, two other categories that could be considered are "Health and Safety for Consumers and Employees" and "Social Progress." The tools for systematically capturing, evaluating, and quantifying data have not been fully developed in these categories. While the LCA is already well-established for footprints, the tools used to measure social progress are still in an early stage.

3 From the Sustainability Strategy to Day-to-Day Operations

Based on the considerations outlined above, the question is: how can the sustainability strategy, as a guiding principle, be incorporated into day-to-day operations?

To improve the sustainability profile of processes and products, it is important to analyze the entire value chain since correlations between the individual phases in the value chain often exist. This prevents specific processes, which are critical in environmental terms, from being shifted outside of the boundaries of consideration. In an ideal scenario, this analysis does not just include the primary effects, but also secondary effects. Such a process requires an exact analysis of the relevant processes and intensive communication with all partners in the value chain. This applies, in particular, to partners who lack either information or interest (e.g., consumers of household laundry products) (Müller-Kirschbaum and Schröder 2012).

By linking the categories that describe sustainability with the individual phases in the value chain, one can create a matrix. Six sustainability categories and six value-chain phases produce a matrix with 36 fields of action. The number of fields is arbitrary and can be changed at any time without affecting the basic concept. What is important is defining a space that encompasses all of the possible fields for improvement. This gives a transparent overview of what has been accomplished and the deficiencies that still exist. The matrix (see Fig. 1) was developed by Henkel and the CSCP[2] and presented to the various stakeholders (CSCP 2012).

The individual fields of action naturally contribute in different ways to optimizing the sustainability profile. Fields of action that have a significant impact on value or the footprint within the value chain are called *hot spots*.

[2] CSCP stands for the Collaborating Centre on Sustainable Consumption and Production, Wuppertal, Germany.

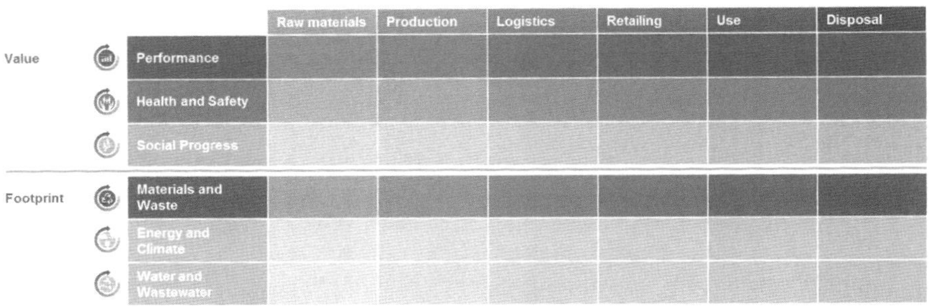

Fig. 1 Sustainability matrix. *Source*: Henkel (2013)

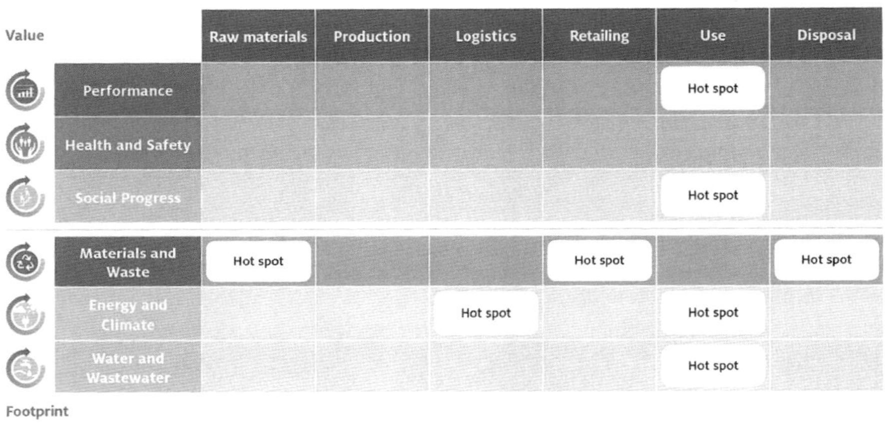

Fig. 2 Hot spots for an automatic laundry detergent in the sustainability matrix. *Source*: Henkel (unpublished)

For the footprint, *hot spots* are identified based on life cycle assessment [LCA] methods. Studies of laundry detergents show that, across the entire life cycle (including raw material production, production of the finished products, transport and distribution, as well as the use and disposal), the highest energy consumption – and thus generally also the highest carbon emissions – occurs during use (A.I.S.E. 2001) as a result of the large amount of energy needed for heating water. In this example, where the use phase has been identified as the hot spot, the most effective way to reduce energy consumption and carbon emissions is therefore to lower the washing temperature.

For the value categories similar methods are not as well established. Stakeholder surveys, which can be conducted in a variety of different ways, are thus important instruments (CSCP 2012).

The *hot spots* identified generally apply for one product category (e.g., automatic laundry detergent). They can be transferred into the matrix to indicate areas of potential action (see Fig. 2). The *hot spot analysis* therefore helps to identify areas of high concern and facilitates the planning of effective measures to improve the sustainability profile of a product within a given category.

4 Application of the Sustainability Matrix in Day-to-Day Operations

Incorporating the strategy into the company's day-to-day operations is of central importance. However, even if the *hot spot analysis* identifies those fields of action which have the greatest effect on the sustainability profile, it is advisable to take measures that do not directly impact *hot spots,* if these measures can be implemented quickly and with little effort.

The following section describes how the sustainability matrix approach is integrated into day-to-day operations using selected examples from the areas of laundry detergents and household cleaners. The examples focus on products, including their development and production. However, the matrix can be used to analyze internal and external processes, as well.

4.1 The Role of Internal Company Measures in Optimizing the Sustainability Profile

As explained above, the matrix encompasses the entire value chain. Methodologically speaking, it makes sense to start with the company's internal process steps, in what is known as the *sphere of control*.

The question is whether it is possible to increase value and reduce the footprint at the same time. If an index value of 100 is arbitrarily assigned to the washing performance of a powder laundry detergent based on 1975 development standards, then today's products would have an index value of 140 (Scheidgen and Meier 2013). At the same time, modern powder laundry detergents also shorten washing times, which reduces the amount of time needed for housework and thus contributes to the category of *social progress*.

In addition to this increase in value, which primarily affects the use phase, the footprint has also been significantly reduced. Modern powder laundry detergents produce better results at a washing temperature of 20 °C than powder laundry detergents from 1975 at a washing temperature of 40 °C. Simultaneously, the recommended dosage of powder products has been lowered from 275 to 90 g (Scheidgen and Meier 2013). These improvements have been achieved through a number of innovations, including the use of modern enzymes and innovative polymers as well as advances in production technology.

Reducing the recommended dosage, however, doesn't just have a primary effect, i.e., reducing the resources used for the manufacturing of products. In addition, these measures entail a range of secondary effects that further enhance the sustainability profile. Less packaging material, for example, is needed for compact laundry detergents; in other words, the specific volume of packaging per wash cycle is smaller. Other measures, including the use of modern packaging materials as well as improved production technologies lower the material input still further, independent of the reduction in recommended dosage mentioned above. Compact products and less packaging also result in a reduced logistical effort, which lowers the material and energy needed for transport and storage. In turn, this cuts the emissions produced during transport.

This example shows how measures taken in product development, targeting the value and the footprint, stimulate a number of primary and secondary effects along the entire value chain that make it possible to increase the value of a product while simultaneously reducing its footprint.

There are similar examples in other product categories. For instance, in the case of automatic dishwashing detergents, the number of functions has been increased considerably (low-temperature performance, improved drying, better short-wash cycle due to improved solubility) while the weight of the dishwashing tabs has simultaneously decreased. This is another example how the value of a product can be increased while keeping the resource requirements, and thus the footprint, almost constant.

For the sustainability strategy to be successfully implemented, it must be integrated into the individual phases of product development process. Basic aspects, that help improve the sustainability profile, are already identified in the concept phase of product development. In the subsequent phases of the innovation process, these qualitative analyses are supplemented by semiquantitative and ultimately also quantitative calculations on the basis of *hot spots* and individual fields of action. All phases of the value chain are, of course, included in the analyses. It is key to conduct the analysis across all system boundaries taking into account the relevant parameters.

The concept of increasing value and simultaneously reducing the footprint can also be applied to internal processes such as the production of laundry and home care products and their logistics. Hence, all new production systems and processes are assessed and optimized based on sustainability criteria. Existing systems are also systematically included in these analyses. From 2011 to 2012, Henkel successfully reduced CO_2 emissions and water consumption (CO_2: 652,000 to 651,000 metric tons and water 7.9 to 7.7 million m^3) despite a slight increase in the production volume (from 7.55 to 7.59 million metric tons) (Henkel 2013). The concept can incorporate individual measures that primarily involve the energy efficiency of single machines and technical equipment, the process steps for heat recovery, and the recirculation and reuse of process and cooling water.

One particularly important step for the Henkel Laundry & Home Care business sector is ISO 50001 energy management certification. Fifteen of 28 production sites were certified to meet this standard in 2011 and 2012, all other sites are scheduled to follow in 2013. In parallel, the first building in a completely new process facility at the Körösladany site in Hungary achieved LEED certification ("Leadership in Energy and Environmental Design") in 2012.

To make energy consumption transparent, a specially designed real-time "Global Energy Metering" system is being developed and implemented. This system allows for the analysis and optimization of data from important consumers in the production facilities over time. These measures are supported by instruments applied systematically including a plant-specific "Sustainability Scorecard" that tracks all relevant variables over the long term as well as a "Scouting Tool" that regularly assesses the use of renewable energies. It takes into account the local availability of different forms of renewable energy, investments in systems and equipment, and demand at the production facilities under constantly changing economic conditions. The company works continuously on improving its performance and establishing new standards in sustainability with the help of a "Best Operating Practice Sustainability" program.

In terms of optimized logistics, the warehouse for finished goods have a direct impact on the carbon footprint. Henkel Laundry & Home Care favors a cross-border central warehousing concept for nearby production sites which enables the company to optimize the logistics network with respect to specific greenhouse gas emissions (per metric ton-km) and costs. By concentrating on a small number of logistics centers, the level of automation can be increased, reaching 100 % in some cases. This positively affects energy performance. Combined with other optimization steps (e.g., optimized lighting and heating), energy consumption can be reduced by up to 40 % compared to conventional warehouses.

All of these measures have led to impressive progress in reducing the operational footprint without compromising the performance of operations. The accomplishments also extend to occupational safety. The successful implementation of the company-wide "Zero Accidents Program" lowered the number of reportable occupational accidents to 1.0 per million hours worked by 2012.[3] In 2012, the average for the entire European soaps, detergents, and maintenance products industry was 7.9 accidents per million hours worked (A.I.S.E. 2013a).

Henkel's progress over a 10-year period is summarized in Fig. 3.

4.2 How Partners in the Value Chain Contribute to the Sustainability Profile

In addition to the parameters focusing on products and their internal production processes, Henkel depends on intensive collaboration with external partners in the value chain to further improve the sustainability profile of products.

[3] Lost time accidents per million hours worked.

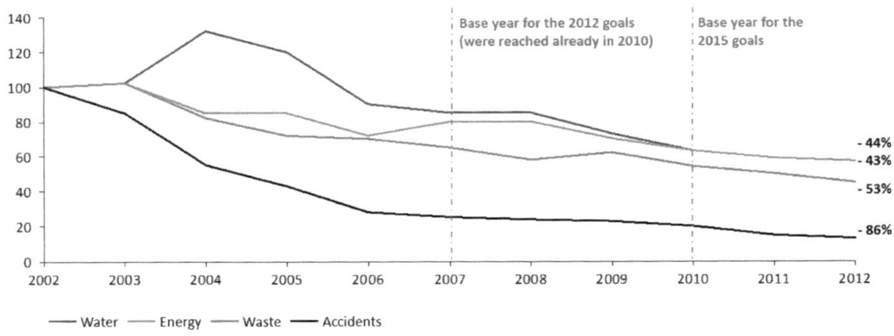

Fig. 3 Changes in the environmental indicators (per metric ton of production volume) and occupational accidents (per million hours worked) between 2002 and 2012. *Source*: Henkel (2013)

Suppliers of ingredients also help increase value through innovation. The choice of raw materials used affects the sustainability profile of the ingredients and thus also the final products made from them.

Ingredients based on renewable raw materials play an important role in laundry detergents and household cleaners (Schröder 2012). In terms of production volumes, surfactants have the greatest importance. In the context of renewable raw materials, palm oil and domestic plant-based oils are of secondary importance for surfactant production due to their chemical structures causing technical limitations. On the other hand, palm kernel oil and coconut oil are the most important renewable raw materials in surfactant production in terms of quantity. By a sequence of chemical reactions, these two oils yield glycerin and surfactants with chain lengths ranging from 12 to 14 C atoms, which are beneficial for the washing and cleaning processes as well as cosmetic applications. Examples include fatty alcohol sulfates, fatty alcohol ether sulfates, and fatty alcohol ethoxylates. These surfactants are generally a combination of renewable raw materials and components based on mineral oil as well as inorganic chemicals. Even the alkyl polyglucosides, which comprise 100 % renewable raw materials, are not fully natural but based on chemical conversion of natural plant ingredients.

Tropical plant oils – specifically palm oil and palm kernel oil – are increasingly the subject of public debate. In 2012, the total plant oil market amounted to 169 million metric tons with a clear focus on palm oil (market share 35 %) (U.S. Foreign Agricultural Service 2014). The market shares of palm kernel oil and coconut oil were 4 % and 2 % respectively. While the contribution that the palm oil sector makes to economic development is widely acknowledged, deforestation of rain forests, clearing of high conservation value area [HCVA], and depletion of high carbon stock area, such as forests and peat land, caused by new plantations are the subject of intense debate in this context (Forum Waschen 2013). Displacement of indigenous animal populations and the negative impact on the biodiversity is of similar concern. To protect forests and the animals that live in them, the Roundtable on Sustainable Palm Oil [RSPO] developed a certification system to foster the development of sustainable palm oil (RSPO 2013). One important component of sustainable palm oil is increasing the productivity of the palm oil plantations. Poorly managed plantations, often owned by small farmers, only produce approximately 1–2 metric tons of palm oil per hectare. In comparison, well-managed plantations can yield as many as 8 metric tons of palm oil per hectare. An improvement in productivity and the intensified use of fallow land reduces the need for newly cleared land and thus decreases the pressure on the rain forest (WWF 2009). In addition to that, a high productivity also contributes positively to economic development.

Overall, the use of certified, sustainably produced vegetable oil contributes positively to sustainability in the area of raw materials. From the perspective of the sustainability matrix, the improvement in productivity, resulting in a decreased pressure on natural habitats, contributes to the category of "Social Progress" category while an increase in occupational safety in the palm oil industry positively affects the "Health and Safety" category.

The analysis is not limited to the selection of ingredients and raw materials used for their production. It also includes the production processes of suppliers and contract manufacturers. Henkel's pilot projects designed to capture key sustainability parameters such as energy, water, and wastewater were launched in 2011 and are currently being expanded. The ultimate goal is for this core data, including the figures for all contract manufacturers, to be checked on a monthly basis. These measures are designed to help further improve the sustainability profile across the entire value chain.

4.3 The Importance of the Use Phase in the Value Chain and the Role of the Consumer

Though laundry accounts for only 6 % of the total household electricity consumption in Germany (Bürger 2009), energy consumption over the entire life cycle is highest during the use phase for automatic laundry and dishwashing detergents. Lowering the washing temperature doesn't just reduce the need for primary energy. Depending on the primary energy sources used, the CO_2 emissions can also be reduced. So too can solid waste (cinder and ash) if solid primary energy sources like coal are burned.

The industry's effort to develop products that deliver the desired results even at low washing temperature have already been discussed. However, average washing temperatures in the individual European countries still vary considerably and are much higher than modern laundry detergents require (A.I.S.E. 2013b)

At an optimum washing temperature, that also takes hygiene concerns into account, the consumer can considerably reduce the footprint while getting the same value. Using less energy generally saves consumers money. Normally, however, the consumer is not aware of these related impacts, which is why consumer information and education is so important to sustainable consumption.

4.4 *The Importance of the Disposal Phase for the Sustainability Profile*

Laundry detergents and household cleaners make their way into the aquatic environment as a natural consequence of their intended use. As a result, the industry began looking at the environmental impacts of its products and their ingredients at an early stage (Müller-Kirschbaum and Schröder 2012). However, environmental risks cannot be attributed solely to the actual ingredients disposed of via waste water. Energy generation and transport also produce harmful emissions that can either directly or indirectly end up in the aquatic environment. In this respect, the measures discussed here that reduce the dosage and lower the washing temperatures also directly or indirectly contribute to reducing environmental impact.

5 Evaluation and Outlook

The world's growing population and rising global consumption are putting more and more pressure on our planet's limited resources. Which is why resource efficiency is one of the top priorities on the global political agenda.

To cope with these challenges and develop adequate solutions, the concept of sustainability can serve as a guiding principle because it examines economic, environmental, and social elements at the same time. It is essential to maintain the balance between these three elements. A quantitative evaluation allows to assess whether the measures that have been undertaken already are efficient and effective. Even though initial algorithms for quantitative evaluation have been introduced, more research is needed because the inclusion of relevant *hot spots* typically depends on stakeholder acceptance especially in the value section of the sustainability matrix.

Incorporating sustainability into the company, however, requires a clear strategy that, in an ideal scenario, is rigorously implemented in operationally relevant measures. To get a holistic picture, correlations between the various measures need to be taken into account. These relationships can intensify effects: compact products, for example, result in savings with respect to raw materials but require a reduced effort in logistics, as well. But negative correlations are also possible. Lowering washing temperatures, for instance, reduces energy consumption but also can have adverse effects on hygiene in some cases and thus negatively affects the "Health and Safety" category. The examples in the area of laundry detergents and household cleaners show how the sustainability profile can be improved within one's own sphere of control and as well sphere of influence (this also includes upstream suppliers for a manufacturing company). The full potential, however, can only be realized if all stakeholders along the entire value chain (sphere of interest) are integrated.

It is important that the sustainability assessment is integrated into the innovation process. At Henkel, this was accomplished in a multistage process, the Sustainability#Master®, which aims to create a quantitative evaluation with standardized instruments based on a qualitative and semiquantitative sustainability assessment. To optimize the sustainability profiles of products and processes, Henkel strives to increase value while simultaneously reducing its footprint since this deems to be the best approach to allow a better living while safeguarding the limited resources of our earth.

Bibliography

A.I.S.E. (2001). *The life cycle assessment of European clothes laundering*. Brussels: A.I.S.E..

A.I.S.E. (2013a). *A.I.S.E. activity and sustainability report 2012–2013*. Brussels: A.I.S.E..

A.I.S.E. (2013b). *Low temperature washing initiative, substantiation dossier*. Brussels: A.I.S.E..

Bürger, V. (2009). *Identifikation, Quantifizierung und Systematisierung technischer und verhaltensbedingter Stromsparpotenziale privater Haushalte*. Freiburg: Öko Institut.

CSCP. (2012). *The sustainable consumption index: A product development tool*. Wuppertal: CSCP.

Elkington, J. (1998). *Cannibals with forks – The triple bottom line of 21st century business*. Hoboken: Wiley.

European Commission. (2010, March 3). *Communication from the commission – Europe 2020: A strategy for smart, sustainable and inclusive growth*. COM (2010) 2020 final, Brussels.

Footprint Network. (2013). http://www.footprintnetwork.org/. Accessed 04 May 2014.

Forum Waschen. (2013). *Fakten zur Verwendung von Palmkernöl in Wasch-, Pflege- und Reinigungsmitteln in Deutschland* (Status 28 March 2013). http://forum-waschen.de/. Accessed 4 May 2014.

Henkel. (2011). *Sustainability report 2011*. Düsseldorf: Henkel.

Henkel. (2013). *Sustainability report 2012*. Düsseldorf: Henkel.

Müller-Kirschbaum, T., & Schröder, F. R. (2012). *Nachhaltiger Konsum: Ressourceneffizienz durch Innovation in Die Neue Führungskunst – Unternehmen nachhaltig führen*, Symposion, Düsseldorf.

New Economic Foundation. (2012). *The happy planet index: 2012 report – A global index of sustainable well-being*. London: New Economic Foundation.

RSPO. (2013). *Roundtable on sustainable palm oil*. http://www.rspo.org. Accessed 04 May 2014.

Scheidgen, A., & Meier, F. (2013). Of powders and pouches – Detergent innovations, trends and challenges. *SOFW Journal, 139*(6), 2–8.

Schröder, F. R. (2012). Tenside auf Basis von Nachwachsenden Rohstoffen für Konsumentenprodukte. *Euro Cosmetics*, 19–23.

United Nations. (1987). *Report on the world commission on environment and development, "Our Common Future"* presided over by Gro Harlem Brundtland. http://www.un-documents.net/ocf-cf.htm. Accessed 11 June 2013.

UNDP. (2013). *Human development report*. http://hdr.undp.org/en/2013-report.

U.S. Foreign Agricultural Service. (2014). http://www.fas.usda.gov. Accessed 04 May 2014.

von Weizsäcker, E. U., Hargroves, K., & Smith, M. (2010). *Der Faktor Fünf – Die Formel für nachhaltiges Wachstum*. Munich: Droemer.

WWF. (2009). *Sustainable oil palm development on degraded land in Kalimantan*. London: World Wildlife Fund.

VAUDE: Sustainable Value Creation as a Corporate Mission Statement for Small and Medium-Sized Companies

Antje von Dewitz

1 Sustainable Value Chain Management in Small and Medium-Sized Enterprises

Many small and medium-sized companies in Germany are family-owned. VAUDE Sport GmbH & Co. KG, which was founded by Albrecht von Dewitz in 1974 and has been run by his daughter Antje von Dewitz since 2009 onwards, also falls into this category of companies that place a greater emphasis on maintaining a sustainable way of thinking and working than the average company. Former Economic Affairs Minister Rainer Brüderle once even described family-owned companies as "the cornerstone of the sustainable development of our social market economy" (2011).

Owners of family-owned companies have made an enormous contribution to keeping jobs safe, even during the financial crisis, because of the close connection they have with their own company.

Motivated by intrinsically personal reasons, family-owned companies are often interested in concepts that are not aimed at short-term success, but aimed at developing sustainable and therefore durable and future-oriented economic and value systems. This approach is based on stability, mutual trust and a strong sense of responsibility. The VAUDE motto, "Dedicated to making (y)our world a better place," is a good example of this sense of responsibility. It is not an advertising slogan, but rather a promise based on an essential inner drive. The reason for this is a special feeling of responsibility for the company's specific area of activity: a supplier of outdoor equipment products, whose success depends on customers enjoying themselves and relaxing in a natural environment, needs to preserve and protect that very environment.

A. von Dewitz (✉)
VAUDE Sport GmbH & Co. KG, VAUDE Straße 2, 88069 Tettnang, Germany
e-mail: info@vaude.com

M. D'heur (ed.), *Sustainable Value Chain Management*, CSR, Sustainability, Ethics & Governance, DOI 10.1007/978-3-319-12142-0_11

The sustainable policy embedded in many family-owned companies often takes the form of a Stakeholder Value Approach: it is assumed that the interests of all relevant stakeholders will be taken into account in all decisions made for managing a company. Stakeholders comprise employees and owners, suppliers, customers, creditors, government, and society. Nowadays, social and ecological indicators also have a central role in all areas of the value chain in addition to traditional (mostly financial) KPIs.

2 Sustainability in the VAUDE Global Value Chain

VAUDE has production sites in Germany, China, and Vietnam. About a third of all products are manufactured at company-owned production facilities and the remainder are outsourced. The central logistics hub is located at the company headquarters in Tettnang, in southern Germany. Ever since it was founded, VAUDE has made strenuous efforts to make social and ecological improvements in all areas of its value chain in order to operate as sustainably as possible – extending beyond national borders and the company itself. The company set up the Ecology recycling network in 1994, for example. This involved developing recyclable products (all components made of polyester), setting up the appropriate network for returns via trade and a partnership with the recycling industry for returned goods to be recycled into new components. However, it became necessary to introduce a new system because the number of items returned by end-users was too low to be viable in ecological terms. VAUDE has since been working on an alternative approach with the FairWertung e. V. umbrella association, a national network of over 100 nonprofit organizations. It has developed new standards for collecting clothing and has committed itself to greater transparency and responsibility regarding the collection and recycling of used clothing since 1994.

Since 2009, VAUDE has also followed a consistent sustainability strategy that includes all areas of the company in an integrated forward-thinking approach. VAUDE was the first company in the outdoor sector to introduce an environmental management system, with the aim of achieving sustainability through an even balance between economic, ecological, and social factors. This strategy applies to the entire product lifecycle and to all decisions made at the company headquarters, such as using electricity from renewable sources and recycled paper only.

The company head office in Tettnang is entirely climate neutral. This means that all unavoidable greenhouse gas emissions generated by the company are offset through a cooperation with the Swiss foundation myclimate. VAUDE is a pioneer for this approach in the outdoor sector. The offsetting process works as follows: greenhouse gases such as CO_2 are generated via the consumption of energy such as electricity, gas, and fuels as well as via the consumption of materials for production, paper, office materials, computer resources, and packaging materials. All waste produced by the site, business travel, and logistics are also included. A team of myclimate experts uses data recorded for all of these areas to produce a complete

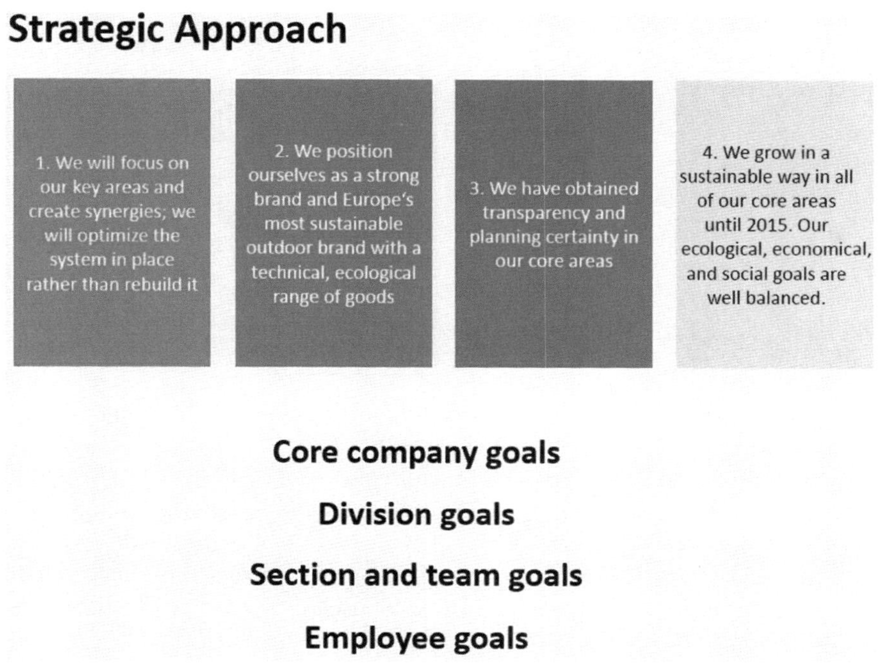

Fig. 1 Division of corporate strategy into individual objectives at all hierarchical levels. *Source*: VAUDE (2013)

Corporate Carbon Footprint, which since 2012 has covered all areas of the company. This not only provides options for reducing emissions, but also records precise details of unavoidable greenhouse gas emissions that can be offset through a myclimate climate protection project. VAUDE buys offsetting units, whereby the same amount of CO_2 generated at the company headquarters is reduced elsewhere. This involves a project in China that helps to reduce climate-damaging emissions by replacing traditional coal stoves in rural households with clean gas stoves that run on biomass. This improves air quality inside the houses and the health of the families living in them.

2.1 A Strategic and Methodical Approach

Sustainability can only be achieved through consistent interaction of individual measures and objectives. VAUDE follows an overarching sustainability strategy for minimizing unnecessary resource consumption that can be divided into four parts. The individual objectives at a corporate, departmental, divisional, and employee level are subordinated to these four parts. Figure 1 illustrates this strategic approach and shows that, unlike many other companies, VAUDE does not see sustainability merely as part of its corporate philosophy and communications strategy, but predominantly as a global priority.

The VAUDE sustainability strategy is based on three distinct approaches:

- Application of the most stringent recognized standards available in the textile sector;
- Application of external expertise across a wide range of areas;
- Total social transparency and external controls for achieving desired objectives.

2.2 A Systematic Approach

All corporate areas of activity and responsibility relating to environmental protection and CSR are systematically embedded in the VAUDE sustainability strategy, which is known as the VAUDE ecosystem. This is also based on the VAUDE vision, its positioning and values. The company aims to be "Europe's most sustainable outdoor brand by 2015," and is therefore positioning itself with an equal focus on products and CSR. (This is evident from the homepage of the VAUDE website, for example: product and CSR sections are given equal status). The diagram below illustrates the systematic approach in more detail (Fig. 2).

This vision, as well as the company's corresponding positioning and value system, give rise to numerous tangible components that, in their entirety, combine to form the overall VAUDE ecosystem:

This represents an integrated and systematic view of the individual areas of the value chain as they effect the environment. This occurs at three levels: company headquarters in Tettnang; products and suppliers; and society:

1. The first level concerns resource-saving business practices in administration and production and the use of renewable energy and various investments made to preserve biodiversity.
2. The second level concerns innovative, environmentally friendly materials and production procedures and recycling.
3. The social level is incorporated in the value chain mainly through an active environmental policy via industry associations, joint environmental projects with competitors, and cooperation with associations, universities, and colleges. The system is subject to regular independent checks to ensure that it works properly and retains its certification. In 2008, VAUDE was the first company in the outdoor sector to have an environmental management system certified according to the European Eco-Management and Audit Scheme (EMAS) and ISO 14001 (International Organization for Standardization) international standards. An independent expert has assessed the implementation of the environmental program every year since then. The assessment includes both product objectives and measures implemented at the company headquarters, such as strict procedures for sorting waste in all departments.

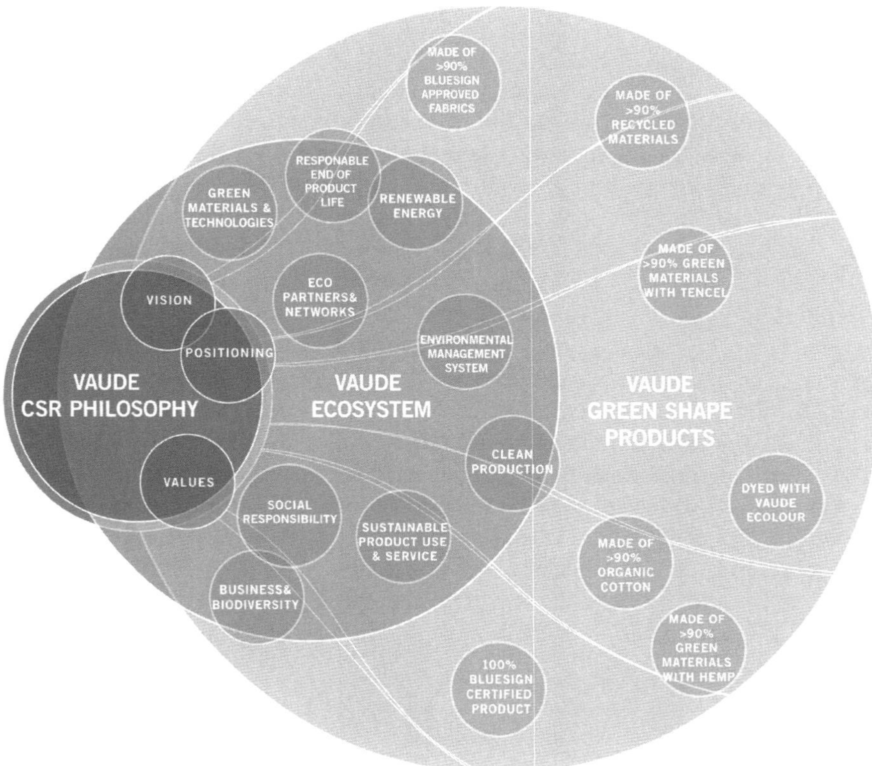

Fig. 2 The VAUDE ecosystem. *Source*: VAUDE (2013)

A major emphasis is placed on developing and producing environmentally friendly products. In particular, the manufacture of Green Shape products pays specific attention to the following: the use of environmentally friendly materials and technology; low-pollution production; environmentally friendly use and maintenance throughout the entire product lifecycle; recyclable materials; the use of renewable energy; promotion of biodiversity; an environmentally responsible approach to administration and logistics; CSR; and collaboration with sustainable partners and networks.

Green Shape products accounted for 78 % of the 2014 summer clothing collection. VAUDE would like to increase this figure to 80 % for its clothing collection and 100 % for its children's collection as part of its aim to be "Europe's most sustainable outdoor brand by 2015." This is a realistic goal, given that there are hardly any environmentally friendly products like this on the global market or, if there are, they are extremely expensive. Nevertheless, VAUDE hopes to set a trend through gradually increased demand and encourage a new way of thinking in favor of sustainability within the sector.

3 Green Shape in the Global Product Lifecycle

Expectations toward the outdoor sector are high: products must be functional, of premium quality and long-lasting, offer protection, and facilitate maximum physical performance. These expectations require innovations in the areas of materials and product technology; and although these innovations meet consumer requirements, they also have an impact on the environment. This impact can be either direct – through the consumption of resources or chemical substances used in production, the release of hazardous substances during the production process – or indirect through the consumption of resources by administrative processes. VAUDE therefore aims to harmonize two central aspects throughout the lifecycle of its products: the product's capability and the product's impact on the environment. Its motto of "performance meets ecology" describes the company's overarching and omnipresent product philosophy.

VAUDE applies its environmental management system to all points along the product lifecycle to ensure that the global value chain is structured sustainably, end-to-end. The following details show the steps taken by VAUDE in terms of sustainability in the Green Shape product lifecycle. The restricted scope of this article means that we cannot address all of the relevant points here, but the examples chosen should provide a clear picture of the VAUDE approach to environmental management.

3.1 Raw Materials Sourcing

VAUDE always seeks to use environmentally friendly materials and technology, with an additional focus on using sustainable raw materials. The company uses organic cotton for manufacturing items in its Green Shape collection, for example. This cotton is cultivated according to stringent ecological guidelines and without the use of pesticides, chemical fertilizers, or genetically modified seeds. This reduces the amount of hazardous substances when collecting, which are harmful not only to the end-user but also to the workforce involved in harvesting the cotton. This is a considerable step forward, as around 25 % of all insecticides and 11 % of all pesticides used around the world are used in traditional cotton production, although, according to figures issued by the Pesticide Action Network Germany, cotton accounts for only 2.4 % of all global agricultural land use.

The cultivation of organic cotton also uses considerably less water than the traditional cotton growing process, and natural crop rotation maintains natural soil fertility. Social factors are also considered at this stage of raw materials sourcing: organic cotton farmers are mostly organized into cooperatives. This means that they obtain a higher price for their organic cotton and are less affected by raw materials speculation on the global markets.

VAUDE products made with a 90 % minimum of organic cotton have the VAUDE Green Shape label. At this point, 10 % of the product must be supplemented with other fibers such as elastane, which makes products easier to maintain and more durable.

In addition to organic cotton, the company also uses the natural fiber Tencel, which is obtained from sustainably managed wood: it is produced in a closed production cycle, where emissions are kept to a minimum. All solvents and water used are recovered and recycled. No additional chemical components are required to ensure that clothing items have a high degree of functionality, because the material is very comfortable due to its surface qualities and minimizes the growth of bacteria caused by sweat.

VAUDE has also closely examined the downy feather it uses as a raw material. Used for filling sleeping bags and down jackets, downy feather is an entirely natural product that is both renewable and biodegradable. Obtaining it as a waste product from the food industry makes it efficient from an ecological and economical viewpoint. However, geese and ducks are not always cared for and fed according to animal welfare standards and for this reason, cannot be described as sustainable. VAUDE therefore only uses down from slaughtered birds that have not been bred for the French specialty foie gras and requires its suppliers to provide confirmation of this. The supplier for the down clothing is also a bluesign® system partner and the down used is bluesign® approved. However, it has become difficult to maintain complete control over how the birds are handled and treated across the global down market.

This is the reason why the company decided to work on a cooperation with an independent organization with a view to auditing the down supply chain end-to-end, starting from the goose farm to the slaughterhouse, the trader's guarantee of origin, and finally, the processing of the downy feather by our suppliers. This process is designed to provide total transparency for the entire supply chain.

VAUDE also uses various recycled basic raw materials to close the product lifecycle at various points – polyester, for example. New products are also made out of used PET bottles: they are collected, cleaned, ground down, and made into granules, which are then melted down and spun into thread. This produces premium quality polyester and saves up to 50 % of the energy used in the manufacturing process and around half of CO_2 emissions.

3.2 *Obtaining Raw Materials and Manufacturing*

The bluesign® standard, the most stringent global standard for textile products, has been used by VAUDE since 2001 when it obtains its raw materials (fabric production process) and manufactures its Green Shape products. All of the bluesign® standard's maximum values for hazardous substances are lower than the legal limits and much lower than other common textile standards (e.g., Ökotex 100). The bluesign® process also screens all environmental aspects of production (such as the use of materials, energy and water consumption, wastewater solutions, exhaust systems, noise pollution, waste management, and the handling of hazardous substances). The screening process is conducted via intelligent Input Stream Management, which instead of random tests for hazardous substances at the final product stage focuses instead on continuous improvement through long-term collaboration. This ensures that products do not contain any hazardous substances.

This integrated, long-term, and therefore sustainable process helps to identify and implement opportunities for improvement. This achieves the highest levels of resource productivity and the lowest levels of hazardous substances throughout the entire production chain.

A distinction is made between two different labels to ensure that compliance with the bluesign® standard is measurable: bluesign® approved fabric and bluesign® product. A product can bear the first label if at least 90 % of its textile surface is bluesign® certified. This mainly relates to inner and outer materials and all printed materials. The remaining 10 % of the textile surface is not allowed to come in direct contact with the skin during use and must adhere to stringent limits for the management of hazardous substances. The second label, bluesign® product, means that at least 95 % of the textile surface is certified. At least 30 % of all other components – such as zips, buttons, and embroidery – must also be bluesign® certified. None of the other product components are allowed to be in direct contact with the skin and must also comply with the criteria and limits of a bluesign® Restricted Substance List.

56 % of the 2013 Bike & Mountain Sports Summer Collection was bluesign® certified. VAUDE has set itself the goal of producing 100 % of its clothing collection and the entirety of its children's collection according to the bluesign® standard by 2015. VAUDE will publish standards across the board to ensure that end-users are also provided with the utmost transparency on compliance. For example, bluesign® approved fabrics and bluesign® products will be identified by hang-tags.

VAUDE's specific aim is to achieve compliance with the above standards, for example by using its own dyeing system called ecolour. This environmentally friendly dyeing procedure involves the direct addition of dye pigments while the thread is being spun. The traditional floating dye process on the other hand involves spinning undyed thread and then coloring it in various stages of dyeing and washing. Compared to the traditional dyeing process, the ecolour process cuts CO_2 emissions by more than half during the dyeing process and reduces water consumption by 89 %, because a considerable number of washing processes can be eliminated. The ecolour process also renders the dye pigments washed out in the traditional process obsolete. The ecolour process consumes 63 % less energy overall. All VAUDE products dyed using the ecolour process therefore have the VAUDE Green Shape label. However, it is not possible to dye all products with ecolour, as it only works for a certain number of colors and threads. The company is still looking into dyeing processes that are as environmentally friendly as ecolour, but that are also suitable for other materials.

VAUDE has also firmly rejected the use of PTFE membranes in all of its products, as perfluorinated and polyfluorinated chemicals (PFC) are required for their manufacture. These are hazardous to people and the environment during the manufacturing process in the country of production, during washing of the final product and at the time when the product is disposed. Although products with PTFE membranes sell better on the market due to the strong marketing presence of the biggest manufacturer of PTFE membranes, Goretex, VAUDE decided in favor of using the PTFE-free alternatives Sympatex (from polyetherester) and Ceplex (from polyurethane) and has taken on a pioneering role with regard to environmental impact and sustainability. VAUDE has also set itself the goal of ensuring its production is totally free of perfluoroctanoic acid (PFOA) by the end of 2014, as well as the goal of stopping the use of PFCs as soon as it is technically feasible, which is a market challenge in terms of water-repellent materials. PFOA can enter the environment or the human body as pollution or as a degradation product of fluorochemicals.

The social component also plays a significant role in the global value chain in terms of manufacturing. Through its voluntary membership of the FAIR WEAR Foundation (FWF), VAUDE is pursuing its goal of ensuring fair working conditions at all production facilities throughout the world where its products are made. Being an independent nonprofit organization, the FWF is a multi-stakeholder initiative supported by many different NGOs, unions, and associations in the textile industry. As a member, VAUDE has made a commitment to comply with its strict Code of Labor Practices. This Code is applied both to VAUDE's organizational conditions at headquarters in Tettnang and to its employment practices at its production plants, most of which are in Asia. The Code contains the following criteria (Fair Wear Foundation 2013):

- No child labor
- No forced labor
- Safe and healthy working conditions
- Proper employment contracts
- Payment of a minimum living wage
- Freedom of association and the right to collective negotiating
- No employee discrimination
- No excessive working hours.

The FWF is an independent organization and regularly screens for compliance with its Code. This is done through factory inspections, interviews with employees on and off the factory premises, and the involvement of local organizations such as unions, business federations, and human rights groups. This ensures a long-term collaboration that is profitable and sustainable for all concerned.

3.3 *Packaging, Transport, and Logistics*

Some 95 % of VAUDE products manufactured and sold in Europe are transported to Hamburg by container ship. From here they are taken to Ulm by rail and then by truck to the company headquarters: this is the most sustainable delivery option from the ecological viewpoint. This is why VAUDE makes very little use of air transport and only uses it in exceptional circumstances. Because VAUDE has its own warehouse in Tettnang, it has considerable influence over the transport of its products to reach customers, who in this case are specialist retailers. The latest technology ensures that processes are optimized to the highest degree: items are dispatched by HGV from the central warehouse in optimum numbers per pack and all packing boxes are also made of recycled paper.

3.4 *Sales and Marketing*

Environmental commitment can only really be successful if a company is also economically successful. It is therefore important to raise end-user awareness at the point of purchase: only informed and conscious customers will take responsibility at the point of purchase and will also be prepared to pay a comparatively higher price for a sustainably produced product.

The indirect environmental aspect of sales as an interface with the specialist retail sector and end-users therefore has significant relevance for the environment and the company has considerable influence here. VAUDE is the first company in this sector to be successful in positioning the issue of sustainability in the retail environment in separate sales areas and window displays. Similar to organic sections in food shops, this has drawn much greater attention to the issue of sustainability and Green Shape products from target groups who are already sensitized to the issue, as well as customers who have been indifferent in the past. VAUDE helps specialist retailers to achieve this with a package of measures: in the first instance, the Green Shape products themselves, as well as suggestions made by VAUDE for in-store decor and information materials. Acting as ambassadors, VAUDE sales teams make a major contribution to raising awareness of sustainability issues in the specialist retail trade and therefore amongst customers too. For the same reason, throughout 2013 VAUDE held regular sales meetings where it trained over 2,800 sales assistants from all over the world on its sustainability policy. This is a win–win situation for all concerned: brand, retail sector, end-users, and the environment.

VAUDE also positions itself clearly as a sustainable company in its marketing activity. This shapes all corporate communications from the inside out. The relevance for the environment cannot be underestimated here, as a committed company sets an example and acts as an ambassador to other companies, customers, and outdoor enthusiasts. An important factor for VAUDE is to ensure that there is absolutely no greenwashing in its corporate marketing activity. It does not portray its brand image as greener than it actually is by special spending on advertising. This means that the company avoids over-publicizing its individual sustainability activities and successes. Instead, VAUDE communicates its commitment in an all-embracing and transparent approach – on its own website, for example. Its main tool is the annual VAUDE sustainability report, providing detailed information on progress in implementing individual objectives for sustainable value creation. Neither is the company hesitant to talk about possible problems (e.g., market-driven). This gives it credibility and makes it possible to evaluate processes properly.

3.5 Usage and Product Service

The company has very little influence over how products are used by end consumers, yet users are responsible for almost half of the ecological footprint of a product. This is why the company places great importance on supporting consumers with protecting the environment after they have purchased their VAUDE products. This support includes advice on product maintenance using eco-friendly substances, washing instructions, and the sale of environmentally friendly water-based NIKWAX solvent- and fluorocarbon-free aftercare and waterproofing products.

However, the company places the greatest importance on longevity in terms of the sustainability of their products when they are used: VAUDE uses high quality materials and the best processing technology. Quality, durability, and design are the cornerstones of its product development. People sometimes joke about this being a disadvantage in terms of sales, as nobody will buy new products if the old favorites are still working perfectly well after 10 or even 20 years. VAUDE offers its own excellent repair service to stop products being thrown away ahead of time.

3.6 End of Product Life

VAUDE uses recycled polyester obtained from a mainly chemical process to save on waste and also on additional fossil oil extraction required for producing polymers. Various polymers in the synthetic fibers are recovered and reconstituted. This produces granules, which can be used for spinning thread and weaving fabric. However, given that the company specifically focuses on closing its own product lifecycles, VAUDE insists on the highest quality and durable workmanship at the product design and material selection stages. VAUDE is also one of the first companies in the outdoor sector to collaborate with the umbrella association FairWertung e. V. – a national network of over 100 nonprofit organizations dedicated to greater transparency and responsibility with regard to collecting and recycling used clothing. FairWertung collects used clothing, shoes, and well, maintained sleeping bags and rucksacks at its community and secondhand shops or via collection bins. Consumers prefer this to the option of returning items to the specialist retailer.

VAUDE therefore passes on all unsellable product returns and samples directly to FairWertung member organizations. This ensures that used outdoor products are disposed of in a responsible and sustainable way by being fed into the recycling system.

4 Success and Being a Role Model

VAUDE has set itself the target of being "Europe's most sustainable outdoor brand by 2015 in objectively measurable terms." The company is taking this mission forward in its industry sector and has already achieved considerable progress:

Since 2001, VAUDE has worked in line with the most stringent environmental standard available – bluesign®, which controls the textile value chain end-to-end. VAUDE was also the first outdoor company to be eco-certified by EMAS. The VAUDE Green Shape guarantee identifies products that are made in a particularly environmentally-friendly and resource-friendly way. The company headquarters in Tettnang and all of the products manufactured there have also been climate neutral since 2012.

The company has been recognized for its commitment to sustainability many times and sees this as confirmation of its success, as it pursues its ambitious goal for 2015. In 2011, for example, the company received the German Sustainability Award (TOP 3) in the category "Germany's Most Sustainable Strategies for the Future" and the Eco Responsibility Award for being the most sustainable company at the ISPO sports equipment trade fair in Munich.

VAUDE won the VFS – Verbund Service und Fahrrad e. V – Ethical Award in 2012, and CEO Dr. Antje von Dewitz received the B.A.U.M. environment award for her commitment and was received by the German Federal President Joachim Gauck.

As early as 2001, VAUDE was also certified as a family-friendly company by Beruf & Familie GmbH on behalf of the Hertie Foundation. A re-audit was carried out in 2004 and helped the company to provide new services under a family-friendly, human resources policy. It assessed the systems already in place, identified individual development potential, and helped to devise practical, needs-oriented steps for the future as part of an overall strategy. Company agreements are now in place for flexible working hours, equal opportunities, child care, parental/caregiver leave, family services, workflows, information, and communications to bring personal and professional goals into harmony. These agreements have already been implemented, and management will work with employees across all roles and departments to take these agreements forward as part of a continuous process. By providing the support for a healthy work–life balance, employee motivation and satisfaction levels are increased, which is good for sustainable corporate growth.

Expertise, team spirit, self-reliance, and high levels of personal initiative define the VAUDE way of doing things. Its equal opportunities culture and day-care center are a good example of this approach. The equal opportunities policy is embedded at all levels of the company: men and women are treated equally, full- and part-time employees have the same opportunities for promotion and career development, and nonprofessional skills and social commitment are taken into consideration when filling vacancies.

The VAUDE day-care center was set up in 2001 to make it easier for as many parents as possible, both within VAUDE and outside, to achieve a work–life balance and provide children with an educational environment. Without traveling times to an external day-care center and a host of other organizational benefits, the VAUDE on-site child-care center takes the pressure off the busy routine for the whole family. VAUDE received the "Freedom and Responsibility" award under the patronage of the Federal President for the sense of social responsibility it has shown in setting up its own day-care center. The birth rate among VAUDE employees has tripled since the center was set up.

In 2002, the Baden-Württemberg Ministry for Economic Affairs awarded VAUDE third prize in its "Equal opportunities for men and women in business" competition for its commitment. VAUDE also received the "Career and Family Audit Certificate" for its family-friendly initiatives from the Minister for Family Affairs Renate Schmidt and the Minister for Economic Affairs Wolfgang Clement in 2005. VAUDE reached the final round of the "Success Factor Family" business competition organized by the Ministry for Family Affairs for the third time in 2012, and is officially one of the most family-friendly companies in Germany.

5 Conclusion

VAUDE has taken on a pioneering role in the outdoor sector through its commitment to sustainability. Although this route may be difficult, requiring more effort and resulting in higher costs, the company has come to the unequivocal conclusion that it is well worthwhile.

The employees of this family-owned company are highly motivated and engaged with the company's approach to sustainability, which is to deliver premium products with an ever-decreasing ecological footprint. This is recognized by customers, who are attaching greater importance to products manufactured in an ecologically sound and fair way. VAUDE has also raised awareness of ecological issues via corporate communications that carry the message of sustainability as an all-consuming issue. The outlook for the company is also positive from an economic viewpoint: for some years, turnover has increased faster than the average rate for the sector.

Bibliography

Brüderle, R. (2011, Januar 2). Familienunternehmen stärken die Marktwirtschaft. *Handelsblatt.* http://www.handelsblatt.com/unternehmen/mittelstand/grusswort-rainer-bruederle-familienunternehmenstaerken-die-marktwirtschaft/3752806.html. Accessed 9 May 2014.

Fair Wear Foundation. (2013). *Labour standards.* http://www.fairwear.org/488/labour-standards. Accessed 9 May 2014.

International Paper: Creating Value Through Sustainably Managed Natural Resources

Teri Shanahan and James McDonald

1 Sustainability at International Paper: An Overview

International Paper is a Fortune 105 manufacturer of paper and packaging products, headquartered in Memphis, Tennesse, USA. International Paper has been incorporated since 1898 and produces uncoated free sheet (standard copy and printing papers), packaging for consumer products (like coffee cups and boxes that protect cosmetics and other goods), corrugated boxes, and the fluff pulp used in hygiene absorbent products like diapers – everyday items that rely on fiber from forests around the world.

At International Paper sustainable practices are not only a company commitment, they are a way of doing business that reaches back more than a century. IP former Chairman and CEO John Faraci explains how fiber demand is a net positive for global forests: "When people use wood products, they are helping to sustain billions of acres of forests. Forests are a key part of why Earth can support life, and at International Paper we are passionate about sustaining forests." In many ways, sustainability was an International Paper practice before it became a buzzword. As David Kiser, Vice President, Environment, Health, Safety & Sustainability explains, "International Paper's reputation is built upon more than 100 years of what people would describe as sustainability. We have long been a steward of natural resources like forests and water." Today, International Paper operates in more than 24 countries. Employees collectively speak more than 20 languages across a global workforce of 65,000. At its core, International Paper is a people-focused company, including our employees, our customers, people that live in communities where we operate, governments, and shareholders. These stakeholders are the reason International Paper is committed to ethical operations that demonstrate deliberate, and continuous improvements in our societal and environmental footprint. Our culture of ethical operations that respects people, cares for the natural environment, and creates value for our stakeholders is what we call "The IP Way."

T. Shanahan (✉) • J. McDonald
International Paper, 6400 Poplar Avenue, Memphis, TN 38197, USA
e-mail: Teri.Shanahan@ipaper.com

M. D'heur (ed.), *Sustainable Value Chain Management*, CSR, Sustainability, Ethics & Governance, DOI 10.1007/978-3-319-12142-0_12

1.1 It Begins in the Forests

For many people, the idea of sustainable forestry is a counterintuitive concept, and the enormous benefits of sustainable forestry are not common knowledge.

According to the Food and Agriculture Organization of the United Nations (FAO)

> Forestry is 'sandwiched' between two extreme and conflicting views: on one side are the superficial, mainly urban, perceptions of the felling of a tree or the hunting of forest wildlife as environmental offences; and on the other side are the poor practices and negative impacts on forests that generally come from outside forestry. A significant challenge for the forestry profession is to communicate and demonstrate the simple idea that one of the best ways of saving a forest is to use it (FAO 2012).

FAO describes sustainable forestry as "the stewardship and use of forests and forest lands in a way, and at a rate, that maintains their biological diversity, productivity, regeneration capacity, vitality and their potential to fulfil, now and in the future, relevant ecological, economic and social functions, at local, national and global levels, and that does not cause damage on other ecosystems." The cycle of sustainably-managed forests can go on forever, and it is visualized in Fig. 1.

Furthermore, "when sustainable forest management is practiced, the value of the natural forest can largely be maintained" (FAO 2012). This means that managed forests can continuously provide habitat, protect biodiversity, and maintain clean water and economic and social benefits in a never-ending cycle.

As long as growing, harvesting, and using wood fiber is economically viable, the sustainable forestry cycle – no matter how long – can be infinitely repeated. Harvard University Economics Professor Edward Glaeser sums it up nicely in a Boston Globe opinion piece: "When people use more paper, suppliers plant more trees. If we want bigger commercial forests, then we should use more paper, not less" (Glaeser 2007). It's an interesting quote considering that all too often the general public focuses on the felling of forests and tends to overlook the fact that trees – to sustain the cycle – must be planted – and then replanted again.

The world has about 4 billion hectares (or about 10 billion acres) of forested land; that is about 30 % of the total land mass (FAO 2012, p. 9) (Fig. 2). As shown by Fig. 3, only a small fraction – < 1 % – of the wood growing was reported to be harvested in 2005, the year of the most recent major UN FAO update (UN FAO Global Forest Resources Assessment 2012, p. 86).

Economic use of the world's forests has long been a driver for their existence, and the harvesting can be done in a way that is infinitely repeatable. When a preferred economic use of land – for agriculture, mining, or development – occurs, forests are often lost permanently; but when the trees have economic value, that value can sustain forests. Figure 4 shows that more than half the wood used by humans is for fuelwood, while about 47 % is used for industrial purposes. Of that 47 % for industrial purposes, Fig. 5 shows the various products wood provides.

While deforestation does continue to occur around the globe – largely due to the conversion of land for agriculture – FAO reports that the trend has been slowing

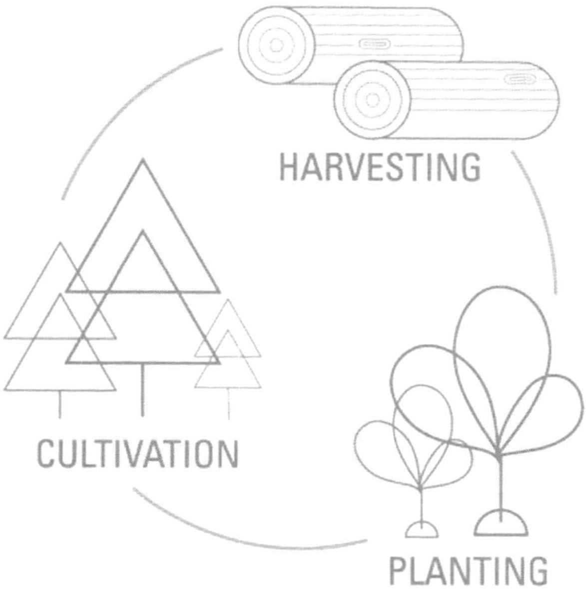

Fig. 1 Forest growing and harvesting cycle

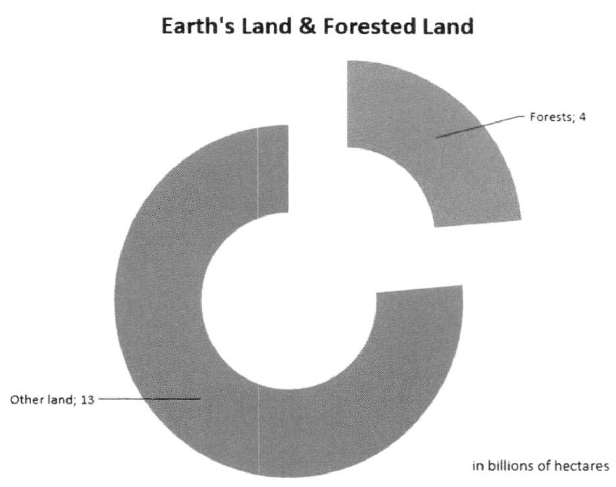

Fig. 2 Earth's forests

down (FAO 2012). This can also be seen in Fig. 6. Most temperate zone forests (where advanced economies and healthy forest products industries exist) have stabilized and some countries, such as the United States, have even reported net gains in forest cover since 2005 (FAO 2012).

From a commercial perspective, growing forests is a long-term effort. The time it takes from planting a tree until it can be used as raw material for products can be as short as 4–7 years for eucalyptus plantations in tropical climates, to 20–35 years in temperate zones such as loblolly pine in the southeastern United States, and to

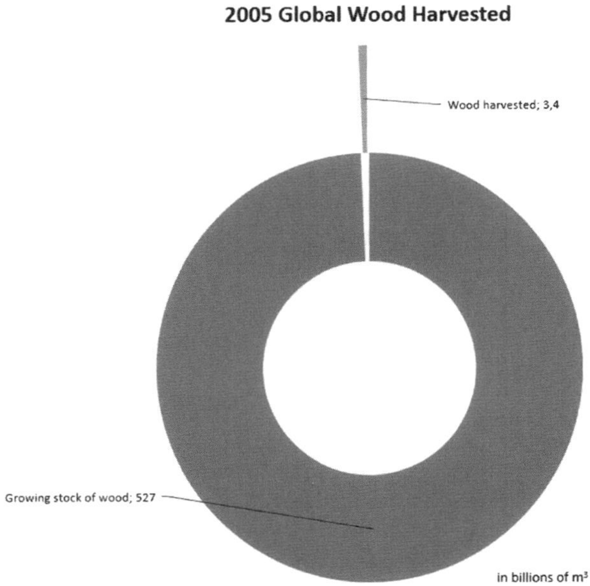

Fig. 3 Percentage of wood harvested globally

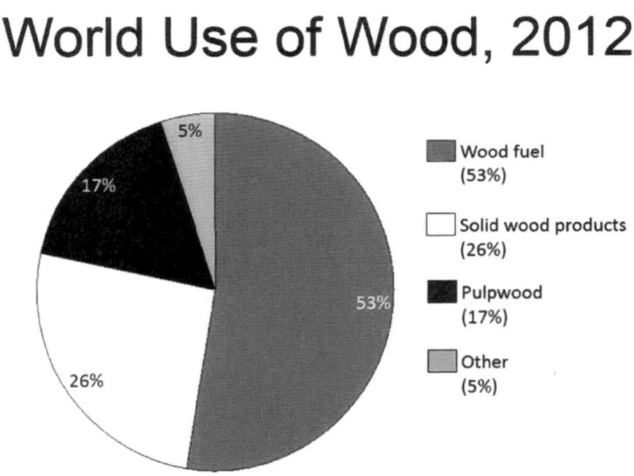

Fig. 4 How humans use wood (FAO 2014)

even more than 100 years in colder zones such as spruce and fir species in Canada and Russia.

1.2 *The International Paper Legacy*

More than 100 years ago, good citizenship at International Paper began with a focus on sustainable forestry. Our economic reliance on working forests is clear, and we have long realized that the communities where we operate value forests, too.

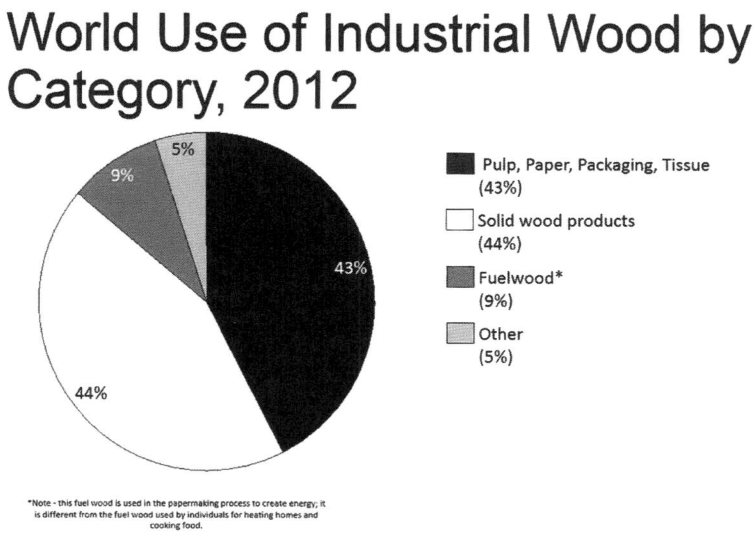

Fig. 5 The industrial use of wood (FAO 2014)

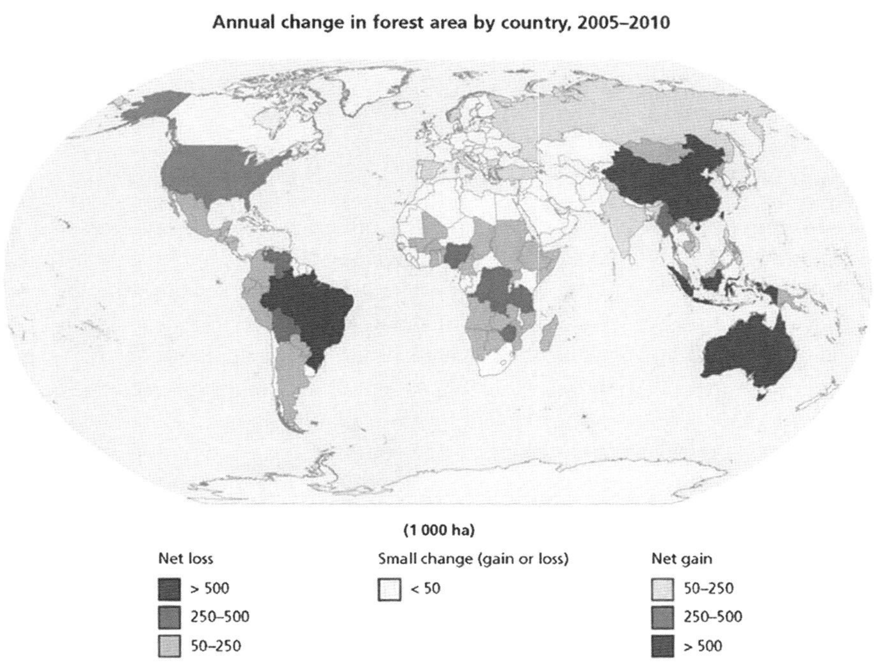

Fig. 6 State of the world's forests (FAO 2010)

Forests serve as a critical source of clean water as the roots of trees filter about one third of our drinking water. Forests also provide habitat for wildlife, and recreational opportunities for people. Managing forests responsibly matters to our stakeholders, and it matters to International Paper.

Until 2005, International Paper was the largest private landowner in the United States – and the second largest behind the U.S. federal government. As part of a larger enterprise-wide transformation plan to streamline its business and assets, the company divested nearly 7 million acres of forestland across 17 states in the U.S.

International Paper worked closely with The Conservation Fund and The Nature Conservancy to ensure these lands would maintain their unique beauty and their value for future generations. Over the years, International Paper also protected an additional 1.5 million acres across the United States – more than half the size of Yellowstone National Park – through conservation easements and donations to state and federal entities. These included the following areas:

Okefenokee Swamp, Georgia
Great Dismal Swamp, North Carolina
Appalachian Trail, Vermont
Royal Blue, Tennessee
Provincial Park, Canada
Various projects, New Hampshire
Neches River, Texas
Silverlake, Georgia
Lake Erling, Arkansas
Boggy Slough, Texas
Adirondacks, New York

Despite many changes and divestitures, healthy, abundant forests continue to represent the core of International Paper's business model. Today, International Paper buys around 70 million tons of wood each year from landowners on five continents, including 55 million tons in the U.S. Most of our wood grows in seminatural, "working" forests, although in Brazil and in India we source wood from purpose-grown plantations (Eucalyptus and Casuarina, respectively). Ensuring these forests continue to thrive is a key part of International Paper's commitment to being a responsible corporation. The company closely monitors worldwide trends and encourages suppliers to practice sustainable forestry through landowner outreach, logger training, and with its Supplier Code of Conduct (International Paper 2013). The U.S. has enjoyed a stable 750 million acres of forested land for the last 100 years, this despite the population having tripled, according to the U.S. Forest Service (2001, p. 3). Perhaps most telling, 68 % of the forests in the U.S. are working forests, so the land is expected to generate an income; the key question is whether that economic value will be generated by growing trees or via some other land use.

The companies that make forest products from these 500 million acres of working forests in the US represent (American Forests & Product Association 2012):

- 887,000 jobs
- $50 billion in payroll
- 4.5 % of the U.S. manufacturing GDP

2 Ambitious Targets Drive Sustainability Along the Value Chain

2.1 Targets that Drive Performance

International Paper's sustainability efforts are now reaching further than at any other time in the company's history. Numerous stakeholders play a role in International Paper's sustainability continuum; from its global facilities to its sourcing teams, to products, customers, communities, and suppliers. In the categories vital to International Paper's business – wood fiber, water, energy, and people – International Paper has set voluntary goals with specific, measurable targets. By establishing tough but achievable goals, we keep a constant focus on tracking metrics and improving performance year-over-year.

In 2012, International Paper announced a set of 12 sustainability goals to guide global operations through 2020 (Fig. 7). Our Senior Lead Team (including those who report directly to our CEO and Chairman) endorsed the creation of the goals to take a deliberate, continuous improvement approach and apply it to all areas of our business. These wide-ranging objectives were developed to focus and bundle our sustainability efforts. The definition of the targets was driven by goal teams and two internal councils – the Manufacturing Council and Environment, Health, Safety and Sustainability Council. Our Sustainability Steering Team, including leaders from all our businesses and our functional areas, also endorsed the goals. These leaders played a key part in establishing a strong vision for the future built on continuous improvement through extensive investigation and many robust discussions.

Setting International Paper's first voluntary public sustainability goals was an important accomplishment. Bringing together a wide range of views and expertise from a diverse set of internal stakeholders – such as employees, senior leaders, customers, and external technical experts – helped to establish strong collaboration, dialogue, and goal sharing. One year later, a common set of priorities were defined that embrace the company's legacy accomplishments and capture its sustainability vision as part of The IP Way going forward.

2020 Voluntary Sustainability Goals: 2013 Progress — INTERNATIONAL PAPER

Sustainability Area	Updated 2020 Goals	2013 Accomplishments From 2010 Baseline
FIBER CERTIFICATION	35% global increase in third-party-certified fiber volume.	Surpassed our original goal of 15%, with 22.8% increase in certified fiber volume since 2010. We reset our goal to 35%.
AIR EMISSIONS	10% reduction in pollutant emissions (SO_x, NO_x, PM) from aligning with our energy efficiency initiatives by 2020.	Achieved 14% reduction from 2010 levels.
SUPPLY CHAIN	Establish baseline supply-chain performance and implement plans to improve by 2013.	Established baseline supply-chain performance and implemented plans to improve.
ENERGY EFFICIENCY	15% improvement in purchased energy use by 2020.	3.7% gain in efficiency since 2010.
GHG EMISSIONS	20% absolute reduction in global GHG emissions (Scope 1 and 2) associated with the production of our products by 2020.	5.8% reduction since 2010.
SAFETY	Accident-free workplace.	21% decrease in LIFE events since 2010.
FIBER EFFICIENCY	Reduce fiber loss in the manufacturing process by achieving performance of less than 0.75% fiber loss.	Completed standardized measurement methodology and reporting format.
WATER QUALITY	15% reduction in mill wastewater discharges of oxygen-depleting substances (BOD) to receiving streams.	BOD levels increased in 2013 but still remained 13% below 2010 baseline.
WATER USE	Map water usage through our manufacturing locations by 2013; develop site-specific plans by 2015 to reduce use in strategic watershed areas by 2020.	Completed risk mapping and ranked mills by composite risk score.
SOLID WASTE	Reduce manufacturing waste to landfills 30% by 2020, and ultimately to zero.	International team analyzed data from mills across the globe and put forward an aggressive new goal to reduce waste.
RECYCLING	15% increase in the recovery of Old Corrugated Containers (OCC) by exploring new sources and diverting usable fiber from the landfill.	18.7% decrease in OCC.
PHILANTHROPY	Measure and report on our charitable support for education, literacy and health and human services in the communities where we operate.	Donated approximately $11.9 million to charitable organizations in 2013. Global team is working on measuring employee volunteer hours.

Fig. 7 International Paper's 2020 voluntary sustainability goals (International Paper 2012)

2.2 *Translating Sustainability Vision and Targets into Action*

To help propel success and to ensure the translation of the vision into action, International Paper appointed a series of goal project teams responsible for implementing the initiatives. Additionally, senior company leaders act as "Goal Champions" to drive the company's performance against each goal. The introduction of these teams and leaders puts International Paper in a strong position to accomplish its goals by 2020.

IP is reasonably confident in the success of some goal areas, including reducing greenhouse gas emissions, improving fiber certification, energy efficiency, and reducing NOX and SOX emissions. Some areas, like fiber efficiency, water use, water quality, and philanthropy will be more challenging and have less obvious economic benefit. IP's supply chain goal is a huge challenge, with 100,000 global suppliers, so we committed ourselves to setting a goal without knowing how we will achieve results. While we don't have all the answers, our sustainability Goal Champions will be key to developing innovative solutions.

3 Sustainability along the Value Chain

3.1 *Recovering Used Fiber for Paper Products*

Recovering wood fiber from the waste stream is a profitable business, and the market for recovered fiber has been growing for decades. Today, the world uses 55 % recovered fiber and 45 % virgin wood fiber for all pulp, paper, and packaging products (Pöyry 2012). International Paper is committed to increasing fiber recovery through recycling and other collection practices. Recovering wood fiber has both economic and environmental benefits. Recycled fiber can be a cost-effective alternative to virgin wood, so increasing the amount that is recovered may reduce costs. And, when wood fiber does not go to a land fills, we avoid the creation of methane gas, a potent greenhouse gas. In fact, when paper products decompose without oxygen (like in a landfill), methane gas is created; when paper products decompose with oxygen present, no methane is created, but only CO_2, which is a far less potent greenhouse gas. This is one demonstration of how International Paper considers sustainability throughout the entire value chain of its products.

In the United States, International Paper recovers, processes, or facilitates the sale of more than 6 million tons of fiber each year, making International Paper one of the larger recyclers of paper in the country. Related to this, International Paper has set a 2020 goal to increase recovery of Old Corrugated Containers (OCC) – often known as used cardboard boxes – by 15 % from its 2010 baseline. In 2012, International Paper put strategies in place to boost recovery rates by building up its internal capacity, working with suppliers, and acquiring new sources of materials for recovery.

Paper is already one of the most highly recycled products humans use, with 65 % being recovered in the U.S., about 71 % in Europe, and 45 % (and rising quickly) in Brazil. Some believe that using recovered fiber could "save a tree." The reality often surprises people: when a working forest loses its economic value, it can be converted to an alternative land use to continue generating income, or it can be sold for development.

"Wood fiber recovery in the U.S. is a significant environmental success story. International Paper's Fiber Sourcing and recycling teams continue to build upon that success by working with generators and suppliers to identify and capture new sources of recyclable materials. We're confident that these efforts will further increase our recovery rates in coming years," said Thomas Cleves, Vice President and General Manager, Containerboard and Recycling and Fiber Recovery Goal Champion at International Paper.

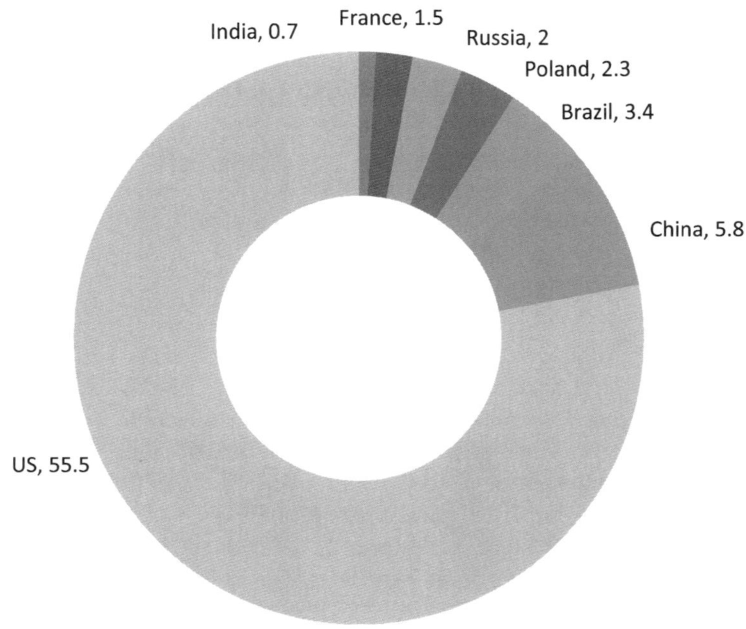

Fig. 8 Origins of International Paper's fiber

3.2 Responsible Sourcing of Primary Fiber

International Paper's large global demand for fiber is a key economic driver for the existence of millions of acres of forested land. In 2012, the company purchased nearly 70 million tons of wood fiber in the United States, Brazil, Europe, the Middle East, Africa, Russia, China, and India. This can also be seen in Fig. 8. In Brazil, International Paper sources fiber primarily from company-owned land. In Russia, the fiber is sourced from government forestland, either through third parties that have access to government leases or directly through International Paper's forestland leases with the Russian government. In the United States, International Paper's main source of fiber is from small, private landowners and from timber management companies.

International Paper is committed to sourcing fiber from responsibly managed forests, especially through the state implementation committees of the Sustainable Forestry Initiative® (SFI®) – a nonprofit that establishes and enforces forest certification standards. Working with suppliers and landowners to encourage reforestation of harvested acres, encouraging landowners to third-party certify their land to standards set by the Forest Stewardship Council (FSC®), SFI®, or the American Tree Farm System® (ATFS®) helps to ensure land is managed on a sustainable basis.

International Paper's energetic efforts working with suppliers and landowners have resulted in the exceeding its 2020 goal to increase third-party certified fiber on a global basis. This engagement with our wood suppliers and landowners is critical as they need to clearly understand why the added administrative cost of annual audits is important to IP. It can be especially challenging when speaking with a long-time (often several decades) supplier who has been operating responsibly; we help these suppliers see why certification serves a purpose. The original goal was to increase certified fiber by 15 % over the baseline of 2010, when approximately 25 % of our fiber was certified. By the end of 2012, we had achieved an increase of 18 %, bringing our total certified fiber level to nearly 22 million tons, or 32 % of our supply. In May 2014, International Paper announced its new goal, which is a 35 % increase (still using 2010 as a baseline). This will bring us to approximately 35 % certified fiber, which is about 24 million tons. This progress requires a tenacious pursuit of success, with significant landowner engagement over long periods of time. It is important to note, however, that many of our suppliers follow sustainable forestry principles long before becoming certified. In the U.S. "Best Management Practices" for forestry were developed more than 50 years ago, and have been evolving and more widely applied since then. In many cases, certification acts as a "proof point" to demonstrate long-standing practices.

International Paper will not accept fiber from endangered forests or illegally logged forests and forests. The company's extensive management practices as well as intimate knowledge of our suppliers and their landscapes are effective tools to ensure our wood is secured from well-managed forests. Our Environmental Management System requires extensive landowner and logger education and regular audits for safety and forest management. We also participate in World Wildlife Fund® for Nature's Global Forest and Trade Network (GFTN), which reaffirms this commitment. The GFTN exists to support and facilitate greater coordination of national and regional efforts to expand responsible and credibly certified forest management, including technical assistance throughout the certification process and enhanced marketing opportunities.

Using certified fiber in International Paper products serves as a proof point that the forests providing our wood are being managed responsibly and reduces the chance that illegally harvested timber ends up in the fiber purchased for operations. International Paper customers and consumers also value forest certification in helping them choose paper and packaging products from responsible sources. In 2012, 26 % of the fiber sourced to our U.S. mills was third party forest management certified (ATFS, FSC® or SFI®). Outside of our North American operations, approximately 58 % of the combined fiber sourced to International Paper mills is third-party certified (International Paper Sustainability Report 2012).

Globally, International Paper follows a certification policy that recognizes all credible third-party standards. All of the forest leases International Paper directly manages and harvests in Russia are certified according to FSC® national forest management standards, and in Brazil all the forests owned and managed by IP are certified in accordance with the Programme for Endorsement of Forest Certification (PEFC™). Approximately 90 % of IP's Brazilian's forested land is also certified to FSC standards.

3.3 Reducing Our Footprints in Energy, Emissions and Water Use

International Paper's manufacturing processes generate and consume large amounts of energy, resulting in atmospheric emissions, including greenhouse gases (GHGs). Improving our energy efficiency results in fewer emissions and reduced costs, so it makes sense that we work strenuously to reduce energy consumption and increase efficiency. One of our voluntary goals is to increase energy efficiency by 15 % by 2020, from a 2010 baseline.

Approximately 70 % of the energy required for International Paper's global mill system is met through the use of renewable, carbon-neutral biomass. Hundreds of millions of dollars are invested in modern pulp mills to be able to recover "black liquor," which is a by-product from the pulping process from parts of the tree that would otherwise go unused, as well as spent pulping chemicals. This is a major use of renewable energy. IP also burns wood residuals from harvests and from saw mills; almost no wood goes to waste. It sees additional opportunity to further reduce the purchased fuels and particularly the fossil-based fuels as well as the purchased steam and electricity portion of the company's consumption. International Paper's converting operations, distribution business, and nonmanufacturing operations also have an impact, but these entities comprise a small amount of its total energy use.

International Paper has achieved significant company-wide reductions in energy use and GHG emissions over the past decade. The company continues to evaluate and select improvement projects from across our global operations. Subsequently, in 2012, International Paper approved 66 energy efficiency projects, which we expect will result in the reduction of 3.3 billion cubic feet of fossil fuel use and a GHG emissions reduction of 270,000 metric tons each year. Energy efficiency considerations allow International Paper to make decisions regarding its fuel mix. It switched from fuel oil and coal to cleaner-burning natural gas at several U.S. mills where it has made economic sense. New acquisitions present additional opportunities for improvement.

In 2012 energy efficiency in International Paper's mills improved 2 % compared to a 2010 baseline. The company saw a year-over-year increase in efficiency between 2011 and 2012 due to increased self-generation of electricity at some specific sites. This self-generation stems from the recovery and use of "black liquor," a residue from trees during the pulping process. One-fifth of International Paper mills set new record lows for purchased energy consumption during 2012. Working with the U.S. Department of Energy (DOE) on cost-effective energy efficiency improvements, International Paper has committed to reduce energy intensity 25 % over 10 years across its U.S. facilities.

Like many industrial processes, we use energy to generate steam that powers our operations. A pulp and paper mill's "life cycle" begins with its wood yard, where supplies of wood and chips are received and stored. The wood stores must be actively managed, and this requires energy to power the heavy equipment used in the process. Next comes the pulp mill, where wood is "cooked" chemically and processed mechanically to become pulp. If producing bleached products, a bleach plant is next; this step represents another energy use. Finally come paper machines, where the pulp is spread to create a web of paper, and then energy is applied to dry the sheet. Every step in this process needs energy, and although it represents just 8 % of the total product cost, it is aggressively and minutely managed.

For example, mills in Augusta (Georgia), Franklin (Virginia), Pensacola (Florida), Riegelwood (North Carolina), and Savannah (Georgia) reduced the amount of coal and fuel oil they burned. At the Ticonderoga mill in New York – which currently burns fuel oil as a secondary source to wood residuals – International Paper is developing plans to extend a natural gas pipeline to the facility. Looking around the globe, International Paper's biomass boiler at its Mogi Guaçu plant in Brazil, a $90 million dollar investment, went through numerous rounds of testing in 2012 and became fully operational at the end of the year. The boiler will decrease global fossil fuel GHG emissions by 200,000 tons per year. In Russia, the combined heat and power (CHP) facility at the Svetogorsk Mill became fully operational at the end of the 2012. The unit will generate more than 180,000 MWh of electricity annually at efficiencies significantly higher than stand-alone power generation, displacing grid electricity and reducing emissions.

These global efforts are paying off, as 2012 was an important year for the reduction of energy emissions. International Paper was one of 20 organizations, and the only forest products company, to win the U.S. Environmental Protection Agency's inaugural Climate Leadership Award. International Paper was recognized by the EPA for leadership in achieving voluntary fossil fuel GHG reductions of more than 40 % from 2000 to 2010. This kind of success isn't easy; in fact it required $300 million invested in more than 100 energy improvement projects. International Paper's energy efficiency and GHG reduction efforts also received the American Forest & Paper Association's (AF&PA) "Leaders in Sustainability" award – one of only five companies to be recognized at the association's inaugural Better Practices, Better Planet 2020 Sustainability Awards program.

International Paper tracks global GHG emissions and reports them annually through participation in the Carbon Disclosure Project (CDP™), the European Union Emissions Trading Scheme, U.S. EPA's Mandatory Greenhouse Gas reporting rule, and various state, regional, and national reporting programs.

More than this, the company participates in the debate on climate change policy at regional, national, and international levels, and advocates a clearer regulatory framework. In 2012, International Paper engaged with the World Business Council for Sustainable Development, the AF&PA, and National Council for Air and Stream Improvement on energy issues such as the carbon neutrality of biomass.

As Donna Harman, AF&PA President and CEO said, "International Paper is an exemplary leader in sustainable business practices. Their work on energy efficiency and greenhouse gas reductions sets them apart as a leader among industry peers and in the broader manufacturing community" (International Paper 2012).

Another important focus area is the water footprint of the worldwide operations. Water is essential to International Paper's manufacturing and operating facilities around the world. It plays an important role in a number of areas in a pulp and paper mill. It is needed in the chemical "cooking" of wood to turn it into pulp, in the "slurry" of water and pulp that is applied to the paper machine, and in the creation of steam to drive our equipment. However, most of the water used in our processes is returned to the local water supply after it is fully treated. International Paper recognizes that its value goes far beyond its use in papermaking. Reducing its water footprint, especially in strategic water-stressed areas, is important for International Paper and its sustainability agenda. In 2012, the total volumes of influent and effluent per metric ton of production at International Paper's manufacturing sites did not change. This marks the third consecutive year that the company's rates have remained steady.

Therefore, the company has set goals to map water use, to evaluate opportunities to align water use with local supplies, and to further improve the quality of the water returned to local release environments. Most of the water used in operations – nearly 95 % – is returned to the environment after appropriate treatment; about 4 % remains in our products, and the balance enters the atmosphere through evaporation.

Focusing on efforts to reduce the water used in its mills, International Paper started with regions of the world where pressure on water supplies is particularly acute. As a first step, in 2012 International Paper began comprehensively mapping current water use and future water needs at all of its manufacturing sites. International Paper considers regional water stress indicators to help us select priority facilities to explore further water use reduction opportunities. Given the complex nature of regional water issues, it is challenging to identify projects and opportunities that will maximize the impact of its investments on our business, communities, and the environment. While water-scarce areas are likely to present these opportunities, International Paper remains committed to using water wisely throughout our global manufacturing operations.

International Paper works hard to ensure the water used in its global paper mill system is returned to the environment in a clean condition. Improvements in existing treatment plants and new plants, like our corrugated packaging mill in Kenitra, Morocco – the first wastewater treatment facility at a paper mill in Morocco – mean sizeable reductions in the oxygen depleting nature of International Paper's treated water. Wastewater treatment operations at integrated paper mills face unique challenges associated with both the scale of operations and individual facility constraints. This means that it is often best to focus on minimizing process material losses, which reduces the amount of oxygen depleting substances that must be treated in the first place. International Paper's existing wastewater treatment systems operate at high levels of removal efficiency. This requires significant investment in both process controls and wastewater treatment facilities to achieve these results.

As International Paper moves forward in setting targets and identifying opportunities, the company expects to see more progress relative to its 2010 baseline. Since International Paper returns the majority of water after treatment, the company has a role to play in protecting local waterways near its mills. The company committed to achieving a 15 % reduction in mill wastewater discharges of oxygen depleting substances to release environments by 2020. By the end of 2012 IP we had achieved a 27 % reduction in these substances, so we surpassed the 2020 goal. During 2014, it will evaluate whether we can sustain this new level, or whether we should set an even higher target.

3.4 Ensuring Compliance with Our Standards

International Paper is committed to excellence in environmental, health, safety, and sustainability (EHS&S) practices and performance, and continually works to better manage natural resources.

The company seeks to do business with customers, suppliers, vendors, contractors, joint-venture partners, and other business associates who share a high standard of ethical business behavior, and champion the innovative and ethical management of natural resources. IP works with suppliers of responsibly grown fiber and routinely certifies their fiber supply chain to widely recognized standards. Their commitment extends beyond forests to include manufacturing and supply chain excellence, as demonstrated by the breadth of its 2020 voluntary goals.

To fulfill these global commitments, International Paper holds leaders responsible for engaging employees in complying with applicable laws and regulations, implementing our global EHS&S management system and Global Performance Standards, and transparently reporting our EHS&S metrics and progress against our commitments. In 2014 the company is working to enable every facility and every business to see how each part of the company plays a role in its environmental performance; all employees are personally connected to making progress toward International Paper's goals through their everyday work.

During 2012, International Paper did not pay any significant fines for environmental noncompliance. However, the company did anticipate paying significant fines in 2013 arising from an August 2011 wastewater discharge exceedance at the Bogalusa, Louisiana, paper mill that resulted in a significant fish kill in the Pearl River. At the time of the incident, the mill was owned by Temple-Inland. When International Paper acquired Temple-Inland in February 2012, the company was aware of the matter and has worked closely with the Louisiana Department of Environmental Quality in developing and implementing corrective measures at the mill. Under International Paper's ownership, the Bogalusa mill has made significant progress in reducing its environmental footprint, and a comprehensive multimillion dollar improvement to the mill's wastewater treatment system is under way.

4 Engaging with Communities and Stakeholders

The impact and success of a global corporation requires more than managing inputs and outputs in the global value chain (and the data associated with it). International Paper is a company highly focused on people. We have extensive processes for developing our employees, including regular one-on-one discussions with the employee's manager as well as annual assessments and discussions about employees' potential in each business' People Council. We offer an enormous array of training and development tools for our employees, and our Leadership Institute has specific programs for several hundred employees with highest long-term potential each year.

Global corporations have a responsibility to give back to the community. As International Paper has grown and evolved for more than a century, the corporation has strengthened its commitment to customers, to employees, to shareowners, and to sustainable practices. As a result, the communities where IP operates have become, in many ways, stronger.

International Paper believes strongly in being a good neighbor and responsible corporate citizen. In 1952, the International Paper Foundation was created to support nonprofit organizations that brought about positive change in the communities where International Paper employees live and work.

Sixty years later, International Paper continues this mission through philanthropic giving and volunteer efforts around the world. To ensure a maximized impact, we have set a 2020 global sustainability goal to measure and report on our charitable work. International Paper's long history of charitable giving has allowed the company to develop local relationships, establish strategic initiatives, and encourage employee volunteerism in countless projects and programs. International Paper conducts global philanthropic activities through International Paper Foundations in the United States and Poland, the International Paper Institute in Brazil, and through business contributions and donations made through the International Paper Employee Relief Fund (ERF).

In 2012, International Paper Foundation grants, ERF donations, business contributions, and in-kind contributions totaled more than $12.3 million worldwide. International Paper targets philanthropic activities across four key focus areas: employee involvement, environmental education, health and human services, and literacy. A key program for the past several years has been International Paper's "Coins 4 Kids" effort, focused on feeding children in Kenya via school lunch programs; this has helped several thousand children stay in school.

Another important way that IP supports communities is through the economic value of the jobs our operations offer. Pulp and paper mills typically operate in areas that are abundant in forests, but not densely populated by humans. In these rural areas, mills are frequently one of the largest employers and represent a significant part of the area's tax base. IP's presence in these small towns also means that supportive businesses – hospitals, restaurants, schools, and more – can flourish. The jobs in our factories are highly sought, and we engage actively with our communities to ensure we have open dialogue and trust, including through our Community Advisory Councils.

5 Conclusion

In 2013, International Paper launched several new collaborations to help the company refine and advance sustainability efforts globally. The company launched a $7.5 million donation to, and collaboration with, the National Fish and Wildlife Foundation's "Forestland Stewards" initiative. International Paper joined the World Business Council for Sustainable Development and its Forest Solutions Group, and became a voluntary member of the U.S. Department of Energy (DOE) Better Plants program. It also become a member of the World Wildlife Fund's® Global Forest & Trade Network.

Sustaining and driving progress is a key part of International Paper's sustainability platform.

Sustainability is a journey for International Paper – one that is pursued with leadership, goal setting, measurement, ongoing transparency, and employee engagement as core elements of the company's sustainability initiatives. The company is continuing to strengthen its sustainability approach and assess the risks and opportunities around this ever-changing concept.

With a sustainability strategy that drives toward continuous improvement and embeds sustainable resources in its value chain, International Paper will deliver even greater business results, constantly staying ahead of the customer and consumer curve. As current Chairman and CEO Mark Sutton says, "At International Paper, our sustainability success – like our business success – is guided by one simple ideal: Do the right things, the right way, for the right reasons. Our 65,000 employees around the globe are united in our quest for continuous improvement, and they are the reason International Paper's best days are ahead."

Bibliography

American Forest & Product Association. (2012). *National economic impact fact sheet – United States*. http://afandpa.org/docs/default-source/default-document-library/click-here.pdf?sfvrsn=0. Accessed 28 May 2014.

FAO. (2010). *Global forest resources assessment 2010 – Maps and figures*. http://www.fao.org/forestry/fra/62219/en/. Accessed 16 May 2014.

FAO. (2012). *State of the forests*. http://www.fao.org/docrep/016/i3010e/i3010e.pdf. Accessed 1 March 2014.

FAO. (2014). *FAOSTAT*. http://faostat3.fao.org/faostat-gateway/go/to/download/F/FO/E. Accessed 27 May 2014.

Glaeser, E. (2007). A road map for environmentalism. *Boston Globe*. http://www.boston.com/news/globe/editorial_opinion/oped/articles/2007/05/21/a_road_map_for_environmentalism/. Accessed 1 March 2014.

International Paper. (2012). *2012 sustainability report – Progress*. http://www.internationalpaper.com/documents/EN/Sustainability/IPsustainability.pdf. Accessed 1 March 2014.
International Paper. (2013). *Supplier code of conduct*. http://www.internationalpaper.com/documents/EN/SupplierDiversity/SCOC.pdf. Accessed 4 April 2014.
Pöyry. (2012). *World of fibre – Are we heading from era of RCP to a future of virgin pulp?* http://www.wpauly.it/wp-content/uploads/2013/01/Poggio+2012-Poyry.sp_.pdf. Accessed 16 May 2014.
U.S. Forest Service. (2001). *U.S. forest facts historical trends*. Washington, DC. U.S. Department of Agriculture.

Independent Capital Group: The Importance of Sustainable Value Creation as an Investment Criterion

Mirjam Staub-Bisang

1 The Importance of Sustainable Value Creation as an Investment Criterion

Across all sectors, corporate strategy is being oriented toward sustainability. As a result, all industries face the question of how to integrate sustainability into design and structure of the value chain. Not only should environmental and social criteria be linked with every economic activity, but they have to be. Moreover, it goes without saying that companies and corporations need to have clear organizational structures and adopt fair commercial practices. This is what is meant by the concept "good governance," which represents the third dimension of sustainability, alongside social and environmental aspects.

As the basis for long-term economic success, sustainability is an important investment criterion for long-term investors. In financial markets, this view is currently gaining widespread acceptance. Listed companies and their suppliers are therefore well advised to focus their attention on sustainable thinking and action. In this context, we are not talking about corporate social responsibility (CSR) measures, which focus on "good corporate citizenship" – a company's awareness of its social responsibility – but rather on the integral consideration of sustainability criteria throughout the entire value chain.

M. Staub-Bisang (✉)
Independent Capital Group AG, Gottfried-Keller-Strasse 5, 8001 Zürich, Switzerland
e-mail: msb@independent-capital.com

M. D'heur (ed.), *Sustainable Value Chain Management*, CSR, Sustainability, Ethics & Governance, DOI 10.1007/978-3-319-12142-0_13

1.1 A Definition of Sustainable Investments and Investment Strategies

The concept of "sustainable investments" denotes a market of investment products and strategies as well as the process of responsible, ethical, future-oriented, and principled investment itself. Generally, a sustainable investment strategy gives real consideration to what is known as ESG (environmental, social, and governance) criteria, without dispensing with the requirements of market returns (Staub-Bisang 2011).

Investment strategies may be described as sustainable when they embrace social and environmental aspects along with economic ones. They often include ethical criteria and – notably where equity investments are concerned – corporate governance considerations as well. Sustainable investment strategies are principally applied to equity, bond, and property investments (Staub-Bisang 2011) (Fig. 1).

Common procedures when implementing sustainable investment strategies include positive and negative screening, as well as the integration of ESG criteria

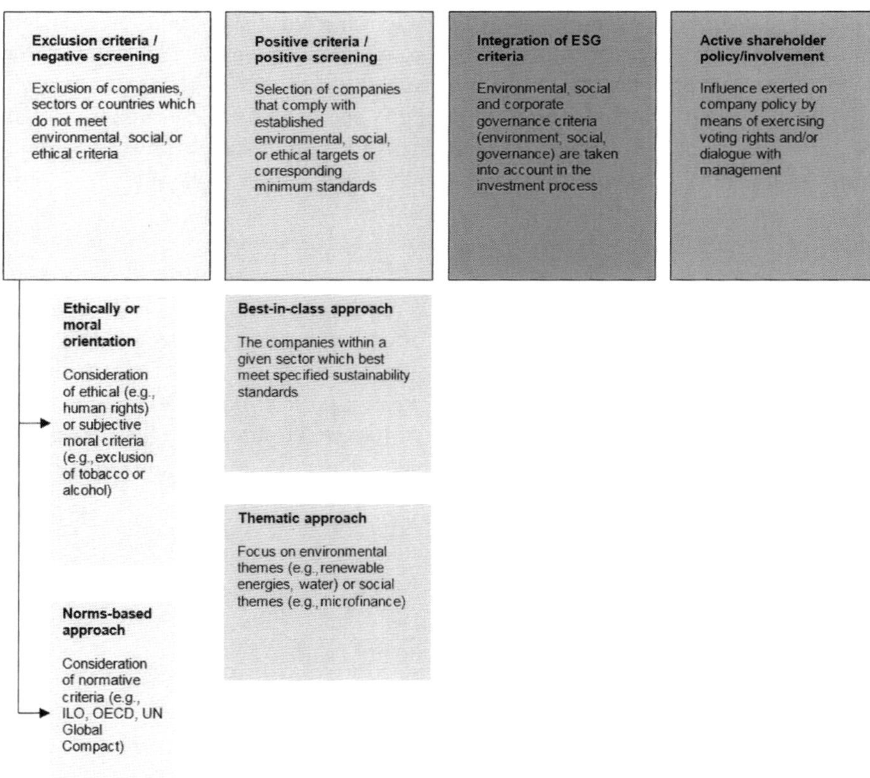

Fig. 1 Overview of sustainable investment strategies. *Source*: Staub-Bisang (2012)

and the adoption of an active shareholder policy. Positive screening denotes the selection of companies (or other investments) that fulfill certain sustainability requirements. Negative screening excludes business sectors or practices that breach environmental, social, or ethical standards from the investment universe (Staub-Bisang 2012). The decision is based on the consideration that capital should not be made available to certain sectors or companies that damage the environment or act antisocially or unethically. Exclusion criteria can be defined based either on the subjective, moral convictions of the investor or on accepted norms. Norm-based screening includes investments in any conceivable sector or company, but only if they comply with certain international standards and norms, such as those set by the OECD, the ILO (International Labour Organisation), or the UNO. The aim of negative selection is to avoid investment in certain companies or sectors, whereas the use of positive criteria places the emphasis on the conscious selection of preferred securities based on sustainability criteria.

The approach known as best-in-class, which is the basis of the Dow Jones Sustainability Index, for instance, represents one characteristic form of positive selection. A best-in-class strategy does not exclude any sectors, goods, or services, but rather subjects them to a stringent sustainability and rating process carried out by various sustainability rating agencies. Thus, under this approach; for example, "dirty" petrochemical companies are analyzed according to environmental and social criteria with the aim of finding the "cleanest" in the sector, namely the leaders in sustainability (Staub-Bisang 2012). Rating agencies evaluate companies' sustainability based not only on publicly available information such as annual or sustainability reports, but also on direct written and oral surveys and external information obtained from databases and stakeholders such as nongovernmental organizations. The sustainability analysis covers the whole value creation and supply chain as well as stakeholder relationships with a company.

The result is a sustainability rating within the control group that is based on an assessment of a particular company's activities in various social and environmental fields and a comparison with the activities of its competitors (Staub-Bisang 2011). For details of the analysis process of rating agencies, see Sect. 2.2.

A further characteristic form of positive selection, aside from to the best-in-class approach is the thematic approach, which focuses on specific sectors. The priority in this approach is the specific investment theme that has commercial potential and addresses environmental and social concerns in the context of sustainability. Thematic investments include, for instance, investments in renewable energies and water treatment and distribution, as well as resource efficiency in general; microfinance investments address the social challenge of poverty in emerging economies (Staub-Bisang 2012).

Increasingly, institutional investors who do not wish to restrict the investment universe by using positive or negative selection are adopting the ESG integration approach. This approach supplements traditional financial analysis by including ecological, social, and governance criteria (ESG) with the goal of improving long-term investment success. ESG criteria are chiefly incorporated in the investment analysis on the basis of risk considerations, such as reputation risks of companies as a result of environmentally damaging production practices or impending class actions by employees. This approach aims at having a potentially positive influence on financial performance by helping to avoid future risks and their associated costs.

Governance, in contrast, forms the major focus of an active shareholder policy, whereby shareholders aim to exert direct influence on a company's management. The simplest form of exerting influence is the exercise of voting rights at a company's annual shareholders' meeting; additionally, direct contact with the management can be sought. Under certain circumstances different investors may join forces [sometimes via third party organizations such as International Shareholder Services (ISS)] and bring joint pressure to bear so as to induce a company's management or board of directors to act on controversial issues such as, for example, excessive incentive schemes.

Generally, investors adopt a combination of these strategies. Institutional investors often integrate ESG criteria into their investment processes in combination with the active exercise of voting rights. This corresponds to the guidelines of the UN Principles of Responsible Investment (UN PRI), which has a current membership of some 1,200 institutional investors with managed assets of US$30 billion (PRI 2013a) (see chart below). One of the commitments undertaken by signatories to the UN PRI is the integration of ESG criteria into their investment processes, the mandatory disclosure of significant ESG issues, active exercise of voting rights, and reporting on activities and results (Fig. 2).

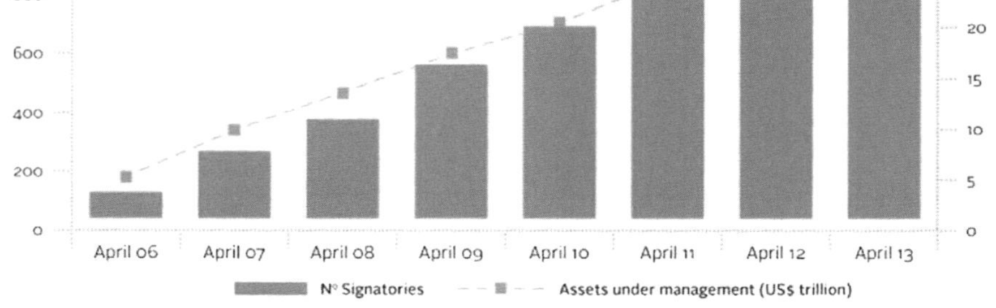

Fig. 2 Signatories of PRI. *Source*: PRI (2014)

1.2 The Market for Sustainable Investments

The global volume of sustainable investments was estimated to be over 10 trillion euros by the end of 2011, an amount that corresponds to about 17 % of the global market for professional investments (excluding alternative investments and private foundations).[1] Of these sustainable investments, just under two-thirds (approximately 6.7 trillion euros) were attributable to Europe, just under 30 % to the United States (2.7 trillion euros), and approximately 4 % to Canada. These three regions together account for some 95 % of the total sustainable investment market (Global Sustainable Investment Alliance 2012). Asia's sustainable investment volume is rather more modest, at US$74 billion, while Australia and New Zealand account for US$180 billion and Africa for US$230 billion.

Since 2002 the European Sustainable and Responsible Investment Forum (Eurosif) has published data on the European market every 2 years. According to their data, between 2011 and 2013 the sustainable investment market in Europe grew by around 3 trillion euros to just under 9.8 trillion euros, an increase of around 45 % (Eurosif 2014). Market growth is extremely dynamic. In only 10 years the market has grown more than 30-fold from around 330 billion euros to over 9.8 trillion euros (Fig. 3).

Thus, at the end of 2013, sustainable investments accounted for approximately half of all investments in the European market as a whole, whereby roughly 97 % of the investment volume was attributable to institutional investors, principally to government pension funds in Norway, Sweden, and France (Eurosif 2014). This preponderance is less pronounced in Germany and Switzerland, where private investors accounted for around half of the investment volume (Eurosif 2014) as can be seen in Fig. 4.

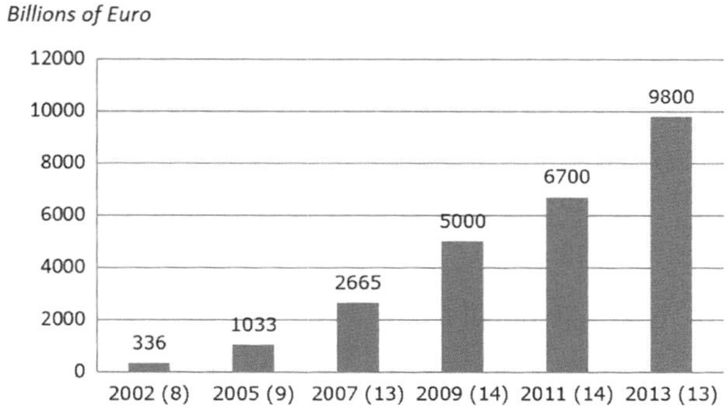

Fig. 3 Estimated volume of sustainable investments in Europe. *Source*: Eurosif (2014)

[1] This estimate is based on market studies by seven regional investment forums (Eurosif, US SIF, SIO, RIAA, SIF Japan, ASrIA, and ASIF), which were consolidated in the "Global Sustainable Investment Review 2012" published by the Global Sustainable Investment Alliance (GSIA). The following investment strategies were considered: (a) negative screening, (b) best-in-class screening, (c) norm-based screening, (d) integration of ESG criteria, (e) impact investing, (f) thematic investments, (g) active shareholder policy.

Billions of Euro	2011	2013	CAGR 2011-2013	Growth 2011-2103
Sustainability Themed	48.0	59,0	10.8%	22.6%
Best-in-Class	238.1	353.6	11.8%	24.9%
Norms-based Screening	2'132.4	3'633.8	30.5%	70.4%
Exclusions	3'584.5	6'854.0	38.3%	91.2%
ESG Integration	3'164.1	5'232.1	28.6%	65.4%
Engagement and Voting	1'762.7	3'276.0	36.3%	85.8%
Impact Investing	8.8	20.3	52.2%	131.6%
EU Industry (EFAMA est.)	13'800.0	16'800.0		21.7%

Fig. 4 Estimated volume of sustainable investment strategies in Europe. Source: Eurosif (2014)

The distribution of different sustainable investment strategies varies widely across Europe. According to Eurosif, impact investing has been the fastest growing strategy since 2009, with an increase of 131 %, followed by exclusions/negative screenings and engagement and voting with growth rates of 91 and 86 % respectively. Between 2011 and 2013 thematic investments grew by 23 %.

In the same period strategies involving active shareholder policy and the integration of ESG criteria have grown by 65 % and 14 % respectively. The predominantly employed investment strategy is negative screening (6.9 trillion euros), followed by integration of ESG criteria (5.2 trillion euros), norms-based screening (3.6 trillion euros) and engagement and voting (3.3 trillion euros). The best-in-class approach) and thematic investments account for just 353 billion euros and 59 billion euros, respectively (Eurosif 2014).

The fact that sustainable investment strategies are significantly gaining in prominence, is important for companies throughout the world. Companies are making large efforts to become more sustainable not least to win the favor of sustainable investors. Today there are few companies listed on the stock market that do not provide information on their sustainability strategy and implementation in order to give interested investors and analysts transparent access to relevant information.

1.3 The Returns and Risks of Sustainable Investments

Sustainability is an analytically well-supported approach, not mere idealistic enthusiasm. Many academic studies prove that sustainable investments achieve returns in line with the market (Mercer 2009). Nevertheless, the critics remain stubbornly convinced that sustainable investing impairs the risk/return profile of a portfolio. This objection should be taken seriously, for both the extensive application of exclusion criteria and a rigorous best-in-class approach considerably restrict the investment universe, which has a negative effect on risk and return expectations. Moreover, the poor performance of certain thematic investments, such as solar energy in earlier days, supports this thesis. Following the renewable energy hype, driven as it was to even greater heights by government-subsidized funding, these

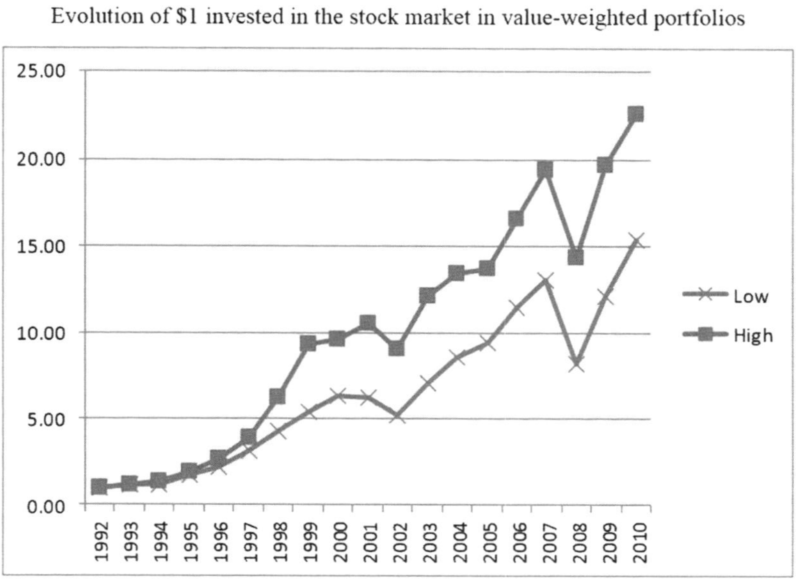

Fig. 5 Performance of high and low sustainably oriented companies. *Source*: Eccles et al. (2011)

securities decreased considerably in value, with some of them even becoming insolvent.

A 2011 academic study performed at Harvard Business School, examined the performance of 180 US companies over 18 years and shows that sustainability actually increases investor returns (Eccles et al. 2011). Between 1993 and 2011, the portfolio of 90 firms that all had implemented a sustainability program with a focus on environmental and social issues (sustainable companies) achieved a share price performance that was almost 50 % higher than the portfolio of the non-sustainable peer group (Fig. 5).

2 Consideration of Sustainability Performance as an Investment Criterion

As indicated, the volume of sustainably invested funds is growing rapidly. Increasing numbers of investors, both private and institutional, are favoring investment strategies that factor in sustainability criteria, be that for reasons of long-term risk/return considerations and or ethical conviction. Leaders of listed companies are reacting to this investor interest and inform them proactively about their sustainability performance, providing sustainability reports and information on the sustainability of their corporate strategy, processes, and results on company websites. The most progressive among them integrate material sustainability information directly in their regular annual reports, thereby underlining the relevance of this information for financial investors. Sustainability analysts in sustainability rating agencies, banks, and asset management firms, analyze this company data, enter into direct dialogue with company management, and then rate the company against its peer group. Financial analysts also make use of this information as part of their financial and economic assessment of a company.

2.1 Sustainability Reporting

Companies generally publish their sustainability reports annually along with the regular annual report. In recent years, more and more companies have started to integrate relevant (material) sustainability information directly in their annual reports (a practice known as integrated reporting). This should make it easier for investors to integrate the relevant ESG criteria into their investment analysis. Ultimately, the method that companies choose to communicate on sustainability issues – whether by preparing a separate sustainability report or by integrating the relevant information into the regular annual report – should not impact the investment decisions of sustainable investors.

As a rule, companies' sustainability reports are modeled on the guidelines and structure set out by the Global Reporting Initiative (GRI). The GRI was set up in the United States in 1997 by the United Nations Environment Programme (UNEP) and the NGO and investor network Ceres. The GRI Guidelines are intended to support companies and organizations in sustainability reporting and thus to offer companies, governments, investors, employees, and interested members of the public a benchmark for making comparable decisions. Reports drawn up in accordance with the GRI guidelines are based on transparency, and their purpose is to provide standardization and comparability. The guidelines are required to provide specific figures and indicators related to the economic, environmental, and social aspects of a company's activities, products, and services to increase comparability. The GRI Guidelines have become an accepted standard throughout the world. Thus, for instance, members of the United Nations Global Compact are advised to submit a sustainability report prepared in accordance with GRI guidelines (UN 2010).

Current GRI guidelines (G4) specifically address the key topics relating to sustainability along the value chain for different industries. Reporting entities are obliged to provide a comprehensive illustration of their value chains and to disclose, in addition to their ownership structure, their relationships to suppliers and service providers, and their environmental and social impact, together with their stakeholders' concerns (GRI 2013).

Sustainability reporting should cover all business processes that comprise part of a company's value chain. The Swiss-based Geberit Group, the leading European manufacturer of sanitary and plumbing systems, provides a good example of sustainability reporting by presenting its sustainability strategy on the basis of its value chain, classified within four areas: procurement logistics, production,

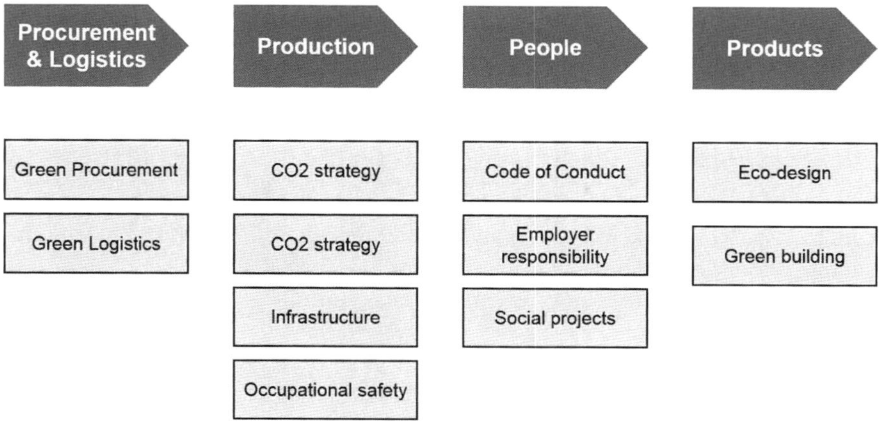

Fig. 6 Geberit AG sustainability modules. *Source*: Geberit (2013)

human resources, and products. To ensure effective monitoring, each module contains clear responsibilities with measurable objectives, the steps necessary to achieve these objectives, and quantified indicators (Geberit 2013) (Fig. 6).

The GRI Guidelines also call for transparency with regard to supply chains. Companies that are committed to complying with sustainability strategy standards usually require their suppliers to sign a code of conduct as well to ensure that suppliers implement environmental standards, socially acceptable working conditions, and fair commercial practices. Many aspects of the code of conduct are based on the principles of the UN Global Compact. Code of conduct requirements are normally imposed on direct suppliers only, and rarely on sub-suppliers, although this is increasingly becoming a requirement of investors.

For companies, these relationships raise the issue of "system boundaries," or the degree to which a company is responsible for maintaining sustainability standards along its entire supply chain. One frequent argument is that a large-scale analysis of the suppliers of suppliers is neither practicable nor feasible. For example, Geberit which currently requires all its suppliers to sign the Geberit Code of Conduct, plans in future to selectively impose this obligation on its second- and third-tier suppliers as well, selectively in particular risk areas as well. This reaction is the group's response to the demands of its various stakeholders.

A relatively new sustainability reporting initiative is coming from the United States. Namely, the Sustainability Accounting Standards Board (SASB 2013) is pressing for the integration of essential ("material") ESG information into a company's regular reporting as part of the annual report. The aim of this initiative is to determine which information relating to environmental, social, or governance issues (ESG issues) is pertinent to a given industry and to ensure that the disclosure of this information in company annual reports becomes a standard requirement. Thus, the financial relevance of ESG issues is emphasized and the assimilation of this information on the part of financial analysts is ensured. The reporting fundamentals, particularly the definition of the key ESG information for different industries, are currently being drawn up with the support of various industry groups.

2.2 Sustainability Analysis and Rating

In contrast to traditional credit or equity analysis, which examines and evaluates the financial information of a company or corporation, sustainability analysis assesses information regarding the environmental, social, and governance issues and performance of a company.

Rating agencies collect data relating to the relevant sustainability criteria; these systems are aligned with the paradigms of sustainability and corporate social responsibility. The fundamentals consist of publicly available information about the business, such as company or sustainability reports, as well as direct written and oral questionnaires (Staub-Bisang 2011). External information is also obtained from databases and from stakeholders such as nongovernmental organizations. An example of this in practice is the Swiss company RepRisk, which assesses the ESG risk exposure of companies and projects (such as reservoir dams or power plants) based on information available on the internet from independent parties (NGO blogs, newspapers). This analysis typically results in a multipage company profile, along with the company's sustainability rating as compared to the control group, a description of the company's activities in various social and environmental fields, and a comparison with its competitors. The purpose of this analysis is to give investors and other interested parties such as customers or employees information about the company's performance in terms of sustainability criteria. Key elements of this process include the personal and financial independence of the sustainability rating agency and of the sustainability analyst, plus the completeness and comparability of the collected data. Portfolio managers with a sustainability mandate use these sustainability analyses and ratings in addition to the financial analyses contained in buy-side or sell-side company case studies by brokerage houses.

There are some 20 established providers of sustainability research in Europe. The market has consolidated significantly in recent years, as different agencies have merged. Because of the evident market potential, large international financial groups such as MSCI have entered this market via various acquisitions in order to profit from the trend whereby investors worldwide include sustainability criteria into their investment processes. Although different rating agencies adopt different approaches, the results are comparable.

The German sustainability rating agency Oekom Research, for example, assesses companies on the basis of roughly 100 sector-specific criteria relating to the major environmental and social issues. Environmental criteria relate to environmental management, products, and services such as eco-efficiency. Social criteria take into account labor and supplier relations, product stewardship, corporate governance, and business ethics. The results are weighted differently depending on the industry. Thus, in the automobile industry, environmental criteria are given a 60 % weighting and social criteria are given a 40 % weighting, whereas in the textile industry the weighting is precisely the opposite.

The leading Swiss sustainability agency Inrate examines the effects a company's activity has on society and, based on this, the degree to which its sustainability policy and procedures have positive effects on the environment and society. This approach considers the entire life cycle of products and services, since it analyzes not only in-house processes, but also the supply chain together with end use by consumers and clients. Inrate's sustainability analyses and ratings cover the four areas of environment, society, personnel (employed by the company and its suppliers), and governance.

Common to all of these approaches is the sustainability analysis of the entire value chain and stakeholder relationships of a company. It is standard practice for the supply chain to be examined as well; however, this is the case mainly within the context of stakeholder engagement and without any great influence on the sustainability assessment of a company. Generally speaking, the sustainability analysis carried out by rating agencies is limited to a company's direct suppliers on the grounds that companies can normally impose a code of conduct only on their direct suppliers and that it is difficult for outsiders to trace the extended supply chain. Some observers would probably argue that the extended supply chain is only relevant for a company's reputation to the extent that it can be traced back and that the company could be accused of "deliberately turning a blind eye" to social and environmental malpractices.

Given the increasing relevance of social media, it makes sense to impose obligations further down the chain on second- and third-tier suppliers and consequently the broadening of the sustainability analysis makes sense given the increasing relevance of social media. Supply-chain risks often only become apparent to financial investors when reputational damage occurs on the demand side (especially in the consumer goods industry) as a result of reports written by NGOs, bloggers, or journalists, followed by a drop in sales. Moreover, the local population in supplier countries as well as society as a whole would benefit from enforcement of greater commitment along the supply chain.

3 The Significance of Sustainability Analysis Results for Investment Decisions

The insights gained from the sustainability analysis will have a greater or lesser influence on the investment process depending on the investment approach adopted by the rating agency or the client – that is, the investor. Investors who adopt a best-in-class approach will form an investment universe made up of companies that are leaders in sustainability. As a result, even companies that are well managed may fail to be included in an investment universe if they do not focus on sustainability aspects or if they fail to communicate their sustainability concerns. Because of the

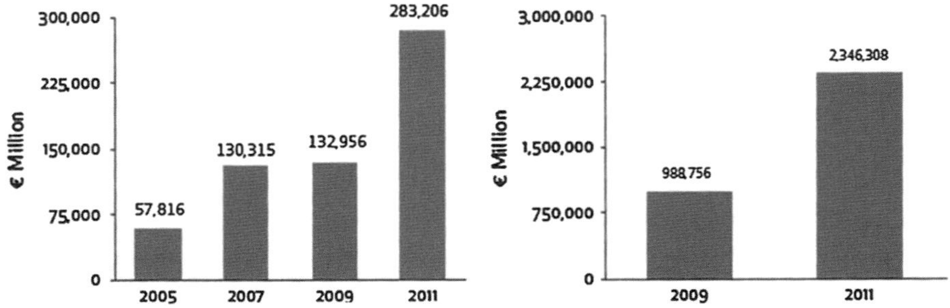

Fig. 7 Investment volumes based on a best-in-class approach and exclusion criteria (Europe). *Source*: Eurosif (2014)

increasing use of exclusion criteria on the part of sustainable investors, companies that do not adhere to the labor or environmental standards of international organizations (ILO, UNO), for example, may likewise find themselves excluded from the investment universe.

Both strategies, the best-in-class approach as well as the application of exclusion criteria, have been rapidly gaining in popularity in recent years, as can be seen from the two charts above for the European market (Fig. 7). The integration of sustainability criteria into the investment process (ESG integration) is increasingly adopted, driven not least by the steadily growing number of signatories to the UN PRI. Many of the world's largest asset owners, such as pension and government funds, have committed themselves to considering sustainability criteria which in their investment processes, in line with the UN PRI. However, the degree to this consideration actually influences their investment decisions in practice varies from investor to investor and is difficult to assess from the outside.

4 Case Studies of Leading Companies

The sustainability risks and challenges vary from industry to industry. For consumer goods companies, the sourcing of raw materials plays an important role. Compared to industrial companies, which operate in a business-to-business (B2B) market, sustainability is generally of greater importance to consumer goods companies, generally, as their business relationships ultimately occur directly with the end consumer. End consumers are very quick to broadcast negative experiences, especially in today's age of social media. Conversely, manufacturing companies tend to emphasize resource efficiency and environmental pollution in their production process. In the raw materials industry, the environmental damages related to the extraction and processing of raw materials commodities as well as the social concerns related to poor working conditions in developing countries are the top priorities. In the financial services industry concerns relate mainly to the environmental and social issues of projects being financed and to the type and content of the services and investment products which are provided to clients.

4.1 Nestlé

With a turnover of over US$104 billion (2012) and more than 339,000 employees, Nestlé is the world's largest food manufacturer. Since its establishment more than 150 years ago, Nestlé has had a strong international focus. Although today about a quarter of its turnover is generated in the United States, its remaining turnover is very broadly diversified geographically, with France, Brazil, and China as its next largest markets, each accounting for approximately 6 % of turnover. Over 42 % of its total turnover is generated in developing countries. Nestlé has production sites in 86 countries and sells its products in over 190 countries. Brands such as Nescafé, Nespresso, KitKat, Maggi, Buitoni, Nesquick, and Nestea enjoy a high level of brand awareness. Moreover, Nestlé is widely diversified across segments, with coffee and other beverages as its largest and most profitable segment, followed by milk and frozen products.

Nestlé's sustainability strategy is underpinned by three fundamental pillars: creating shared value (CSV), sustainability, and compliance. To some extent there is a deliberate overlap of the scope of these three areas. Nestlé understands compliance as the precondition for sustainable action, i.e., adherence to the appropriate legislation and to its own internal guidelines as well as external ones. Nestlé is a member of the UN Global Compact, with which it has aligned its own internal Nestlé Corporate Business Principles. These principles are in turn reflected in the Nestlé Supplier Code, which applies to its 160,000 primary suppliers.

Under the concept of CSV, Nestlé focuses on three key areas, in which the company can create value for society and for the company. In the area of nutrition, Nestlé is working on solutions for societal problems such as under-nutrition and malnutrition that prevails in many parts of the world. Nestlé has, for instance, enriched its products in over 80 countries with various nutritional substances such as iodine, iron, vitamin A, and zinc in order to combat local malnutrition in different countries. Additionally, the company is systematically reducing the sodium, sugar, trans fatty acids, and fat content in its products: in 2011 it made the such adjustments to over 1,200 product recipes. Nestlé is also attempting to reduce water usage in its production processes and to provide its agricultural suppliers with appropriate information about how to save water and improve water quality. In the realm of rural development, Nestlé is committed to creating better living and working conditions for the farmers and communities who supply the company, either directly or through other suppliers. Nestlé buys directly from over 680,000 farmers and indirectly from around 5 million farmers. Overall Nestlé affects the lives of about 25 million human beings. According to World Bank estimates, GDP growth in the agricultural sector is approximately four times more effective than growth in other sectors in the fight against poverty. Against this backdrop, Nestlé is launching programs such as the Nescafé Plan and the Nestlé Cocoa Plan, which will allow the company to buy better quality goods directly from farmers. Importantly, these initiatives offer training programs for farmers with the aim of increasing their productivity and hence their income. Nestlé has also conducted a large-scale initiative involving a systematic audit of its main suppliers using online questionnaires from the Sedex database and by carrying out external audits.

Finally, sustainability relates to the social responsibility of a company like Nestlé. On the one hand, and particularly for a food manufacturer, this means managing its economic operations in accordance with natural resources as well as addressing social challenges. Because of the increasing relevance of social media, companies can no longer conceal information, but must be transparent in every respect. In collaboration with GlobeScan and SustainAbility, Nestlé conducts regular surveys in the area of sustainability so as to determine the relevance of various issues for its stakeholders and for Nestlé itself. Nestlé can then use these analyses as a basis for prioritizing its different endeavors. Nestlé often collaborates with partners such as the Danish Institute for Human Rights, WWF, Greenpeace, Oxfam, or the Fair Labor Association in projects addressing issues like climate change, child labor, or deforestation. Another problem area is food wastage, which, according to Nestlé, is as high as 40 % and takes place mainly at the beginning and end of the supply chain.

Nestlé is the first large food manufacturer to report in line with the GRI A+ standards and tries to keep consumers, its most important stakeholder group, informed about sustainability via social media and other channels. According to Nestlé, two-thirds of all consumers in Europe and North America are interested in sustainability, but see the responsibility for this as lying with the manufacturers and are not prepared to pay more for it. A further 20 % have no interest in sustainability, while 10–15 % specifically shop for sustainable products. Interestingly, products from sustainable companies actually attract greater interest in countries such as China and Brazil than they do in Europe. This is particularly true in places where environmental pollution is all-pervasive.

Nestlé's CEO is responsible for the company's sustainability strategy and is seeking to embed sustainability throughout the whole of its existing structure. Because of the sheer size of Nestlé any improvements or stresses to the supply chain affect hundreds of thousands, if not millions of human beings. Even if controversies still exist in certain areas such as cocoa or water, Nestlé's efforts on so many different levels can be clearly seen and felt by now by many people throughout the world. Nestlé's outstanding financial performance shows that sustainability enhances profitability and creates value for shareholders and stakeholders alike.

4.2 Geberit

The Geberit Group is an international company within the sanitary technology sector, where it is the European market leader. Its products are installed in both new construction and renovations and are distributed largely via the wholesale trade. Its principal markets are Germany, Switzerland, Italy, and the Benelux countries. In 2012 the company achieved a turnover of over 2.1 billion Swiss Francs. Worldwide, Geberit employs over 6,000 people in seven countries.

The Geberit Group made the issue of sustainability a central focus as early as 1990 and introduced environmental managers to its production sites in 1992. The CEO is directly responsible for the area of environment and sustainability and plays a central role in raising the awareness of internal decision-makers. The Group's sustainability reporting is transparent and comprehensive. For the 2012 reporting year, for the first time Geberit used an external stakeholder panel composed of independent professionals to review its sustainability strategy.

Geberit has set itself the goal of integrating its sustainability concept into its business processes across the entire value-creation chain. Its sustainability strategy is organized under four headings: Procurement & Logistics, Production, People, and Products. Objectives and procedures are brought together in what are called modules; the annual sustainability report describes the highlights of each module and the measures planned for the next 2–3 years. On the product side, Geberit's efforts are geared toward an overall reduction of product-related water and energy usage. As the European market leader, the Group has achieved noteworthy results in this area, thanks to its size: from 1998 to 2009, its dual-flush and flush-stop "fleet of cisterns" saved 8.8 billion cubic meters of water, two-and-a-half times the annual water consumption of all households in Germany (Geberit 2012).

Geberit coordinates its efforts to make its products consistently compatible with environmental friendliness, resource conservation, and durability by implementing its eco-design initiative. As part of this initiative, the company holds eco-design workshops with the aim of selecting the most environmentally friendly materials and mechanisms for all of Geberit's new products and product updates; the workshops also aim to introduce these materials and mechanisms at the earliest possible stage of the development process to guarantee a high level of resource efficiency for the production and use of the product. Geberit requires its suppliers to demonstrably comply with its standards of environmentally and socially acceptable production and to sign the company's Code of Conduct. The Code of Conduct addresses compliance with legislation, human rights, health and safety in the workplace, pay and professional development, and environmental protection and integrity. Noncompliance with the Code of conduct results in immediate termination of the business relationship. Since 1991, Geberit has prepared a comprehensive eco-balance sheet for its production facilities as part of its Environmental Management module. Geberit aims to improve its (currency-adjusted) environmental pollution figures per unit turnover by an average of 5 % per annum.

Geberit is perceived as an innovative and sustainable company; in Europe it is the clear leader in the sustainable sanitary-products sector. The company has gained a strong position in the sustainable construction growth market and thus is well positioned for the future. Finally, Geberit demands commitment from its suppliers, such that the effect of their sustainability efforts augments its own. The 8.8 billion liters of water saved and the associated energy savings, together with an appreciably reduced eco-balance, show that Geberit truly lives in accordance with its sustainability beliefs, communicates its results transparently and comprehensively, and thus generates trust and credibility. Geberit's positive margin development coupled with its excellent reputation and market share in the sustainable construction sector demonstrate that sustainability can be compatible with commercial success and can even have a positive effect on profitability and market position. Between 2002 and 2012, Geberit's turnover increased by about 70 %, and the trend in its operating margin was consistently positive, with an increase from 14.6 % to 24.8 %.

Geberit is represented in a total of 13 sustainability indices listed on stock markets, and approximately 15 % of its shareholders are institutional investors with sustainability mandates. Its presence in these indices helps Geberit carve itself a position as a sustainable and innovative company, both vis-à-vis capital markets and the general public. The company is listed in the Dow Jones Sustainability Index, the FTSE4Good Index Series, and other sustainability indices. Since 2008, Geberit has also been a formal member of the UN Global Compact.

4.3 Swiss Re

Swiss Re is the second largest reinsurer in the world (after Munich Re) with collected premiums totaling US$25.4 billion and a profit of US$4.2 billion in 2012. Its business model is regional and is relatively broadly diversified across the different business units. The Americas and the EMEA each account for approximately 40 % of turnover, while just under 20 % of turnover is generated in Asia. Split by business unit, Property & Casualty reinsurance generates 45 % of turnover, followed by Life and Health (reinsurance of life policies and pension funds) at 38 %. The third business unit, Corporate Solutions, offers various services such as a wide range of transfer of risk solutions, for example, for project risks in the energy sector or pension fund risks resulting from increasing longevity. The Admin Re unit buys large blocks of life insurance business from primary insurers and uses them to generate stable cash flows, which generally have low correlation with Swiss Re's other business risks. The premiums collected from the insurance business are managed by the Swiss Re asset management division. With a portfolio totaling more than US$150 billion, Swiss Re is one of the largest institutional investors in Switzerland and, indeed, in Europe.

Sustainable action is embedded as a guiding principle in Swiss Re's Code of Conduct as a guiding principle. The company aims to be a driving force for sustainable development. Swiss Re operates by insuring risks in a sector that is able to make targeted decisions regarding which risks to insure or not to insure on the grounds of sustainability because these risks entail significant reputational risks. Swiss Re's Sustainability Risk Framework addresses this responsibility by examining every transaction in all business areas to determine whether or not it falls under one or more of the Framework's eight policy areas (defence industry; oil and gas, including oil sands; raw material mining; dams; animal testing; forestry and deforestation; nuclear weapons proliferation; human rights violations; and environmental issues in general). If a transaction does fall under one or more of the policy areas, then the Sensitive Business Risks Process must be initiated. This involves internal sustainability experts examining whether there is an infringement of the principles laid down in the Sustainability Risk Framework. On this basis, the committee of experts makes a recommendation either to proceed with the transaction, to proceed with certain conditions attached, or to reject it. In 2011, the sustainability committee examined 158 transactions; it gave the go-ahead to 107 and a conditional go-ahead to 22 but it recommended that 21 be rejected. The recommendations were all followed, and the total commercial value of the cases rejected in 2011 was about US$200 million. The recommendations made were followed in each case. Furthermore, the Sustainability Risk Framework defines certain areas such as antipersonnel mines and nuclear weapons, where transactions are excluded from the outset. The Sustainability Risk Framework exclusion criteria also apply to asset management. In addition, Swiss Re incorporates ESG (environmental, social, and governance) principles in its investment process for all financial assets.

Alongside the aspect of risk, Swiss Re also offers insurance solutions for sustainability challenges. Some of its most important innovations in this sphere are index insurance solutions. In these situations, the insured are not required to claim any specific damage, rather an immediate payment is made as soon as certain values strongly correlated with the actual risk of damage are exceeded. Thus, for example, an insurance payout is automatically made to insured farmers as soon as precipitation in their region exceeds certain threshold values that strongly correlate with the actual risk (of drought or flood). Swiss Re uses satellite images to allow it to offer this insurance in areas where meteorological data is limited. This concept is based on the R4-Initative, which started as a pilot project (HARITA) with 200 farmers in Ethiopia and has since been expanded to include over 13,000 farmers in Senegal and Ethiopia. Those with insurance coverage can pay premiums either with cash or with their own labor in community projects. In cooperation with the International Finance Corporation of the World Bank, these types of index insurance solutions are currently being expanded to a wider market area. In 2011 index insurance solutions were placed with institutional clients such as companies, banks, government institutions, and NGOs in eight African countries.

In addition, Swiss Re is in fact the world leader in the insurance-linked securities market (Car Bonds), instruments that are highly suitable for diversifying what are known as tail risks (that is, very unlikely but nonetheless possible risks such as extreme weather conditions) across a broader risk pool. For investors, the low correlation with other investment classes is particularly attractive from the viewpoint of risk diversification and portfolio protection.

Last but not least, Swiss Re is constantly honing its own activities in relation to environmental friendliness; as early as 2003 it achieved a reduction in its CO_2 emissions of over 54 %. The fact that the Head of Sustainability reports directly to the Chief Risk Officer, which implies that sustainability is regarded as a risk issue, shows that these efforts are taken very seriously at an internal level. This is clear from external awards and rankings: in 2012, for instance, Swiss Re received, for the fifth time in a row, the SAM (Sustainable Asset Management) award as Super Sector Leader in the insurance sector, singling it out as the most sustainable insurance company in the world.

5 Outlook

The importance of sustainable value chains as an investment criterion has grown substantially over the past decade, and signs indicate that this trend is growing. This is especially true for institutional investors who are accountable to their beneficiaries. In most cases these beneficiaries are insured persons who pay into a pool and/or are, as citizens, co-owners of a sovereign wealth fund. In democratic societies, they exert a hard to quantify but ongoing pressure. But they do not want to be entangled either in risky investments or in improper economic or political activities.

This means that companies, especially listed companies, are under pressure from many angles. Their management does not want to be considered short-termist by capital market participants, nor do they want clients and consumers to regard them as unfair business partners.

In this situation, sustainability proves to be a good recipe for a circumspect and rational approach that takes the needs of all stakeholders into account. Moreover, shareholders as well as customers and consumers value this approach by maintaining or increasing their investment and purchasing more products thus making the company more valuable. In essence, the integration of sustainability considerations in the value chain generates added value – both today and in the future.

Bibliography

Eccles, R. G., Ioannou, I., & Serafeim, G. (2011). *The impact of a corporate culture of sustainability on corporate behavior and performance* (Working paper). Boston, MA: Harvard Business School.

Eurosif. (2014). *European SRI study*. http://www.eurosif.org/our-work/research/sri/european-sri-study-2014/. Accessed 20 November 2014.

Geberit. (2012). Annual report 2012.

Geberit. (2013). *Geberit Nachhaltigkeitsstrategie 2012–2013 mit Zielen und Massnahmen*. http://geschaeftsbericht.geberit.com/geberit/annual/2011/gb/German/pdf/SustainabilityStrategy_2012_13_20120215_DE.pdf. Accessed 12 May 2014.

GRI. (2013). *Global reporting initiative*. http://www.globalreporting.org/G4. Accessed 11 June 2013.

Global Sustainable Investment Alliance. (2012). *Sustainable investment review 2012*. http://www.gsi-alliance.org/resources/. Accessed 11 June 2013.

Mercer. (2009). *Shedding light on responsible investments, approaches, returns and impacts*.

PRI. (2013a). *PRI association*. http://www.unpri.org/about-pri/about-pri. Accessed 11 August 2013.

PRI. (2014). *PRI fact sheet*. http://www.unpri.org/news/pri-fact-sheet/. Accessed 12 May 2014.

SASB. (2013). *Accounting for a sustainable future*. http://www.sasb.org. Accessed 11 June 2013.

Staub-Bisang, M. (Ed.). (2011). *Sustainable investing for institutional investors*. Zürich: Verlag Neue Zürcher Zeitung.

Staub-Bisang, M. (2012). *Nachhaltigkeit ist ein pragmatischer Ansatz*. In Portfolio Institutional 10/12.

UN. (2010). *UN global compact*. http://www.unglobalcompact.org/news/50-06-24-2010. Accessed 11 June 2013.

Nestlé: Sustainable Value Chain Management from the Farm to the Fork

John Bee, Peggy Diby, Bineta Mbacké, and Barbara Wettstein

1 Creating Shared Value at Nestlé

1.1 About Nestlé: "Creating Shared Value" as a Guiding Principle

With sales of more than CHF 92 billion in 2013, Nestlé is the world's leading Nutrition, Health and Wellness company. Headquartered in Switzerland, the company employs 333,200 people and has 447 factories worldwide.

Nestlé believes that for a company to be successful over the long term and to create value for shareholders, it must create value for society. This means creating superior, long-term value for shareholders by offering products and services that help people to improve their nutrition, health, and wellness. Henri Nestlé founded the company in 1866 on the success of a life-saving infant cereal. Today, the company aims to enhance lives by offering healthier and tastier food and beverage choices for all stages of life.

To build a business capable of both delivering superior shareholder value and helping people improve their nutrition, health, and wellness, Creating Shared Value (CSV) is the approach Nestlé takes to its business as a whole (see Fig. 1). In addition to nutrition, the company focuses on water and rural development, given their critical importance not only to its business but also to its employees, farmers, suppliers, distributors, and communities where it operates.

J. Bee (✉) • B. Mbacké
Nestlé S.A., Avenue Nestlé 55, 1800 Vevey, Switzerland
e-mail: John.Bee@nestle.com

P. Diby • B. Wettstein
Nestlé Central and West Africa Limited, Plot 11/13 Airport City, Marina Mall, 2nd Floor, Accra, Ghana

M. D'heur (ed.), *Sustainable Value Chain Management*, CSR, Sustainability, Ethics & Governance, DOI 10.1007/978-3-319-12142-0_14

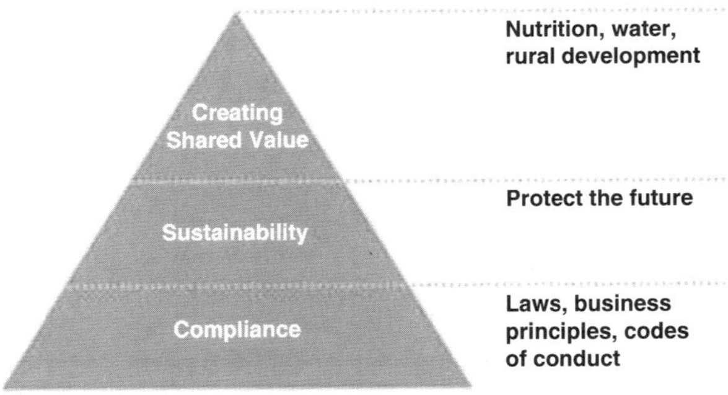

Fig. 1 How creating shared value works at Nestlé (Nestlé in Society 2013)

The practice of Creating Shared Value was born at Nestlé. Harvard Professors Michael Porter and Mark Kramer have described CSV as "creating social and environmental benefit as part of making a company competitive over the long term. Business can help societies progress and all sectors can help business improve and flourish" (Porter and Kramer 2006). They discussed Nestlé's approach to business in "The Nestlé Concept of Corporate Social Responsibility, as implemented in Latin America" (FSG and Nestlé 2006); and found "a company that affects the well-being of millions of lives every day, operating with a long term perspective that goes well beyond the conventional models of corporate responsibility." The Creating Shared Value concept was further articulated in the Harvard Business articles on "Strategy and Society: The Link between Competitive Advantage and Corporate Social Responsibility" (Porter and Kramer 2006); "Creating Shared Value" (Porter and Kramer 2011); and "Innovating for Shared Value" (Pfitzer et al. 2013).

Nestlé continues to actively manage its commitments to environmental, social, and economic sustainability needed for operating its factories, and for the sustainable growth and development of the communities and countries where it is active. This involves substantial training and education of people inside and outside of Nestlé, as well as large investments in technology with lower environmental impact.

Creating Shared Value (CSV) requires compliance with the highest standards of business practice, including international codes and standards, as well as its own Code of Business Conduct, Corporate Business Principles, and Management and Leadership Principles.

As stated by Nestlé Chief Executive Officer, Paul Bulcke, Creating Shared Value is the way Nestlé does business and the way it connects with society at large:

Nestlé is all about quality of life and nutrition – that is what we live for as a company. But the relevance of that is the value that it creates: for consumers, for society and for our business – driving competitive advantage and R&D, being ahead of the curve, and building our brand value. (Nestlé in Society 2013)

1.2 Operationalizing CSV at Nestlé

Within Nestlé's general corporate governance structure, the Chairman, the Chief Executive Officer, and other members of the Executive Board are ultimately responsible for the supervision and management of its role in society and CSV, supported by a number of other governance bodies.

1.2.1 CSV Council

The Nestlé Creating Shared Value Council brings together external experts in corporate strategy, nutrition, water, and rural development to assess its progress and discuss CSV opportunities and challenges.

1.2.2 CSV Alignment Board

A quarterly CSV Alignment Board, chaired by the CEO, has also been established to oversee the strategic implementation of Creating Shared Value across all Nestlé's businesses. It leads the development and evolution of CSV and sustainability objectives and strategies at Group level, while reverting to the Executive Board for input and confirmation. The CEO's presence on the CSV Alignment Board and the CSV Council ensures alignment between both Boards.

CSV strategies are then devised, implemented, and their effectiveness monitored. Some are global in nature (such as the Nescafé Plan and the Nestlé Cocoa Plan) and others are region or country-specific, like the case of the Grains Quality Improvement Project in Central and West Africa which will be analyzed in this article.

Nestlé is always working to add nutritional value at every stage of the food supply chain – before the farm to after the fork: from sourcing ingredients, to processing, manufacturing, and distributing to its consumers. Its work at every stage in the value chain starts with science and adds value to society.

2 Nestlé and Rural Development

Nestlé has a very broad range of activities, which aim at benefiting local communities. For instance, it has partnered with the International Federation of the Red Cross and Red Crescent Societies (IFRC) to provide water and sanitation facilities and promote hygiene in Côte d'Ivoire. More than 100,000 people benefit from this program. In addition, children and communities are benefitting from the 40 schools being built with the support of the World Cocoa Foundation in order to fight against child labor (Nestlé in Society 2013).

In addition to nutrition and water, Nestlé focuses its CSV activities on rural development. Nestlé's well-being and that of farmers is closely tied together, but in many rural communities, a lack of investment in socioeconomic and agricultural infrastructure diminishes the quality and quantity of the raw materials Nestlé depends on. So the company works on capacity building and training farmers in sustainable production methods – both to protect raw materials supply, and to have a positive, long-term impact on local economies and standards of living.

Nestlé interacts with 686,000 farmers through the company's Farmer Connect program (mainly milk and coffee) and other sustainable sourcing programs (Dairy: 411,600 farmers; Coffee: 240,100 farmers; Cocoa: 34,300 farmers). These programs are committed to the local sourcing of raw materials, offering technical assistance and ensuring cooperation to meet the highest sourcing standards (Nestlé in Society 2013).

Other programs in the area of rural development in the Central and West African region include the Nestlé Cocoa Plan and the Nescafé Plan, which aim to ensure a long-term supply of good quality of raw materials while improve the livelihood of farmers. Nestlé aims to support the sustainable development of the rural communities where it sources and manufactures because they are essential to its business. Its contribution to the rural economy also extends to the agricultural support and capacity-building farmer programs that it provides.

Additionally, the Nestlé Healthy Kids Global Programme, which is currently implemented in 68 countries, including Ghana, Nigeria, Cameroon, Côte d'Ivoire, and Senegal, aims to raise nutrition and health knowledge as well as promote physical activity among school-age children. This is based on multi-partnership approaches, including national and local governments, NGOs, nutrition and health institutes, and sport federations. Owing to the fact that one third of children in West Africa are stunted in growth, and that malnutrition is one of the main causes of high mortality rates, this program is very important.

3 Value Creation Starts at the Farm

Nestlé has been present in Central and West Africa since the 1950s. It is currently operating seven factories that produce 80 percent of Nestlé's products sold in the region, including family and infant cereals. A long-term supply of good quality raw materials which are being used to produce Nestlé's goods, such as maize, sorghum, soybean, and millet, is very important in the region as consumer demand for affordable cereal brands like *Golden Morn* (family cereal) and *Cerelac* (infant cereal) is growing.

3.1 The Issue of Mycotoxins

However, sourcing these products from local farmers is a challenge, due to the prevalence of mycotoxins, contaminants that pose significant health threats to humans. Mycotoxins are biological poisons caused by various types of fungi. They secrete these toxins when they are under unfavorable environmental conditions. Depending on the type of fungi, the name differs. Aflatoxin, for example, is secreted by the fungus *Aspergillus* (FAO 2014).

The health implications associated with mycotoxins and aflatoxin contamination are severe, especially for infants and young children. This natural, fungus-based contamination can cause immune suppression, impaired development in children, and liver damage in both humans and animals.

Particularly in Central and West Africa mycotoxins are a serious issue. Here contamination of raw materials can occur in the field, during harvest as well as during storage because of the high level of humidity. While many agricultural raw materials are affected, corn and peanuts are typically the crops with the highest levels of aflatoxin contamination.

The company has therefore set the reduction of high mycotoxin levels in cereals and grains that it sources locally as one of its priorities in the region.

3.2 *Lack of Good Quality Cereal Grains and Legumes*

Before the launch of the Nestlé Grains Quality Improvement Project in 2007 in Ghana and Nigeria, the aflatoxin level of locally grown cereals and grains was high. This meant that approximately 50 % of the grains used in the company's factories (mainly maize, soybean, millet, and sorghum) in these two countries were rejected because they did not meet Nestlé's strict specification (see Fig. 2), leading to significant production line stoppages. This affected the production level and operational efficiency of the factories as well as their ability to meet a growing market demand. It also meant that about 60 % of maize, for instance, needed to be imported.

Not only were farmers unable to sell their produce to Nestlé, aflatoxin contamination also resulted in significant postharvest losses. Up to 30 % of cereal crops in the region were lost to moldiness, insect damages, and aflatoxin contamination. This resulted in high economic losses for local communities.

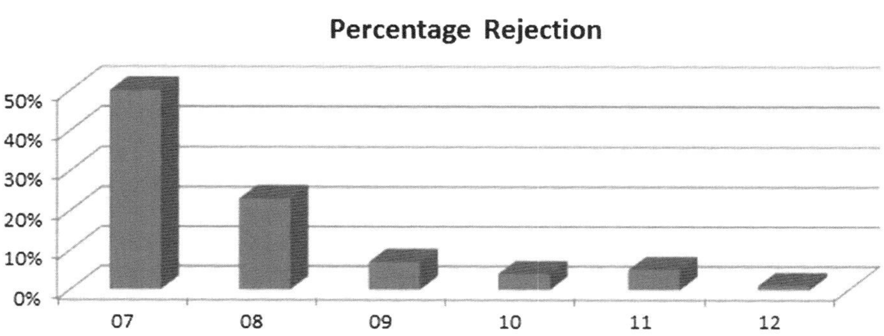

Fig. 2 Rejection rate at factory gate for grain maize, grain sorghum, grain millet, and soybean between 2007 and 2012

In addition, Nestlé was concerned about the hidden health and social implications associated with the consumption of highly contaminated grains. Indeed, local farmers and their families consume a part of their contaminated produce, without being aware of the effect of this on their health. It is estimated that 150 million people in Ghana, Nigeria, and Côte d'Ivoire are exposed to aflatoxin.

3.3 Improving Quality

The Nestlé Grains Quality Improvement Project was developed in 2007 to address these challenges with the following objectives:

- Improve the quality and safety of grains to meet the increasing demands by Nestlé cereal factories in the region.
- Reduce the impact of global food shortages on cereal grain quality and prices.
- Lessen the dependence of Nestlé factories in West Africa on imported raw materials.
- Improve the overall health of rural communities through the consumption of good quality and safe grains.

The project is currently being carried out in Ghana and Nigeria. Through the following activities, it particularly aims at reducing the high levels of mycotoxins in grains and legumes:

- Toxin-reduction activities such as good agricultural and storage practices, developed in cooperation with national extension partners;
- Capacity-building training sessions from Nestlé agronomists;
- Raising awareness of the health implications of contaminated grains among farmers, agricultural extension officials, food companies, retailers, transporters, and wholesalers.

The project provides a platform where beneficiary farmers are assisted to adopt modern agronomic practices to produce high quality raw materials, which can then be purchased by Nestlé for its regional factories.

3.4 Multi-stakeholder Partnership

Recognizing that it could not tackle the problem of mycotoxins alone, Nestlé decided in 2007 to collaborate with the International Institute of Tropical Agriculture (IITA) in a new type of Research-Private Partnership. Together, the two organizations engaged in farmers' capacity building.

The main support adopted to improve the quality of grains has been training and awareness campaigns. The training focuses on the best practices necessary for growing quality grains, harvesting, packaging, transportation, and storage. Training materials, developed with research input from the IITA and the College of Agriculture and Consumer Sciences of the University of Ghana, have been designed to suit various categories of stakeholders. These include pictorial instructions of the 'Do's and Don'ts' for illiterate farmers.

The training module has two core principles, namely:

- Addressing unsafe agricultural practices, such as spreading harvested grains on bare ground.
- Addressing poor storage practices, such as the storage of harvested grains in facilities that expose them to mold growth and insect infestation.

Capacity building courses are not only organized for the farmers and farmer cooperatives, but also for agricultural extension officers. Ghanaian and Nigerian agricultural extension officers are trained by Nestlé agronomists in collaboration with the Northern Rural Growth Programme of the Ministry of Food and Agriculture in Ghana and the Ministry of Agriculture and Rural Development in Nigeria. Trained extension officers in turn train farmers, farm village heads, suppliers, and transporters.

Overall, the farmer training has been a key platform for information exchange between aflatoxin stakeholders and the grain supply chain. The training that farmers receive ensures full integration of optimal aflatoxin management practices into the farming system. It also empowers farmers to make optimal crop management decisions.

3.5 *Results Achieved Since the Start of the Project*

- Over 50,000 farmers, 20 transporters, and 11 suppliers in Ghana and Nigeria have been trained and are now equipped with the skills to reduce mycotoxin levels in cereal and legume grains.
- 60 Agriculture Extension Officers from Ghana's Ministry of Food & Agriculture and the Nigerian Ministry of Agriculture & Rural Development are equipped with new skills in aflatoxin management.
- 150 villages in the region are benefiting from Nestlé's agriculture programs.
- The rejection rate at the factory gate decreased from 50 % to 2 % between 2007 and 2012 (Fig. 2).

This project has created value for farmers and local communities as well as Nestlé. Local communities have improved the safety and quality of the cereals and legumes they produce, leading to improved health outcomes and decreasing health costs associated with the consumption of aflatoxin-contaminated crops. Farmers have also increased their access to markets, therefore improving their revenue opportunities.

To name a concrete example, 40-year-old maize grower and shea harvester Samata Alidu, who lives in Gushie, northern Ghana, is one of 24,000 female farmers who have received training in better agricultural practices through this project. Thanks to the advice she has received from Nestlé agronomists, she has been able to more than double her production. "The impact on the community has been immense," she says. "I can now pay my children's school fees. I can afford to sew them school uniforms. And I even have some extra income."

At the same time, Nestlé factories in the region are now able to buy local grains and cereals that meet Nestlé's specification. Dependence on imported maize and soybeans used in the production of *Golden Morn*, and *Cerelac* has been eliminated with Nestlé now using 100 % locally sourced, good quality grains. This contributes to creating a more efficient production planning as well as a leaner and cost competitive supply chain.

While many agricultural programs in the region are designed to only address food insecurity, that is, ensuring food availability, Nestlé has gone a step further by implementing a multi-prong project that champions food safety, raises awareness of the health implications of mycotoxin-contaminated grains and legumes, and prevents loss of the produce.

The level of resources the company commits to the project is intended to address age-long difficulties in African agricultural value chain, especially in rural communities, seeing this development as a positive precursor to other programs being rolled by the company to promote region-wide actions on nutrition, health, and poverty alleviation.

4 Adding Value and Satisfying Nutritional Requirements

In addition to improving the safety of locally-sourced raw materials, Nestlé is looking to improve their nutritional quality as well as adapting products to the nutritional needs of local consumers. Central and West Africa population suffers from many micronutrient deficiencies on a widespread scale, affecting people of all ages and socioeconomic groups. For example in Nigeria about one third of pre-school children are deficient in vitamin A, while more than three quarters suffer from iron deficiency. Young children and women of child bearing age are particularly vulnerable to iron deficiency because they need higher levels of the mineral for growth and to cover for losses. The consequences of iron deficiency are anemia, impaired mental and physical performance in children, and increased maternal and child mortality.

4.1 Fortification

Known interventions for preventing these deficiencies include supplementation, dietary diversification, and fortification of commonly eaten, widely accepted staple foods. Food fortification is an important strategy to address this issue as it enables broad coverage of target populations with micronutrients. Studies have shown that fortification of commonly eaten foods helps to improve the deficiency status of populations and at the same time does not require changes in dietary habits. Thus, to bridge the micronutrient gap on the continent, Nestlé is adopting various approaches to tackle the deficiencies depending on the severity of the situation at hand.

The Nestlé Research & Development (R&D) Centre Abidjan, Côte d'Ivoire, the first of its kind in Africa, plays a pivotal role in this respect. It focuses on agriculture, raw materials, and traditional African ingredients. It fuses local knowledge with modern bioscience and technology to respond to the specific needs of the region's consumers. This expertise in technology and processing has driven the creation of innovative and affordable products for the benefit of consumers. An essential focus for the R&D Centre has been the development of foods based on nutritious and affordable products that use raw materials, such as maize, soybean, or millet, used in *Golden Morn* and *Cerelac*, grown in the region.

In 2012, Nestlé introduced Vitamin A and iron fortified *Golden Morn,* which is one of the most popular cereal brands in Nigeria with more than 210 million servings consumed every year. According to a micronutrient fortification policy, Nestlé's products need to contain at least 15 % of the recommended daily intake of the specific nutrient per individual serving for a product to be considered fortified.

Good market penetration, frequent consumption, and affordability make cereals good vectors for fortification. By fortifying this product, Nestlé is helping millions of Nigerian families to increase their intake of two essential micronutrients. Furthermore, a consumer engagement campaign highlighting the importance of iron and Vitamin A in cereals and fortification of *Golden Morn* was carried out. Since launching the fortified product in 2012, sales by volume have increased by 56 %.

In order to implement safe, nutritionally relevant and sustainable iron fortification of *Golden Morn*, a number of hurdles had to be addressed. Nestlé's scientists at the R&D Abidjan Centre first evaluated all available iron fortificants in function of product stability and sensory properties, fortificant bioavailability, and cost impact on the final product. Extensive factory trials were run to ensure micronutrient homogeneity in the finished product, and quality assurance methods were specially developed to monitor fortification levels at production sites.

4.2 *Biofortification*

An additional way to increase the micronutrient content of foods and beverages is to improve the quality of the raw materials used. This is why Nestlé launched a biofortification program, based on breeding traditional crops that are naturally rich in micronutrients.

To this end, Nestlé R&D Centre in Abidjan is working with IITA to produce high-yielding, nutritious crops varieties, such as high Provitamin A, minerals (iron, zinc, etc.), and proteins crops; promote their use among farmers; as well as develop new products using biofortified crops. This is done by breeding each crop using conventional genetics methods. This means that for the same crop, varieties which have different characteristics are crossed in order to develop the desired criteria in a new variety. For example, one high yielding cassava or rice variety is bred with another variety with low yield and high vitamin A content to create a new variety combining high yield and improved vitamin A content.

One of the challenges associated with biofortification includes getting high yielding varieties that are biofortified with the desired micronutrients at high and stable level. One of the ways to rapidly increase micronutrients content in crops is to use molecular biology techniques in the breeding process. The R&D Centre Abidjan plans to acquire such techniques in the coming years in order to shorten the time required to develop new varieties.

In 2013, six biofortified products (rice, wheat, maize, sweet potato, cassava, and millet) were in development in Nestlé's R&D Centres around the world, including four in Abidjan (cassava, maize, rice, and millet). The first Nestlé products containing biofortified raw materials are planned to be launched by 2015, as a complement to direct fortification.

5 Case Study/Example: Creating Shared Value Across the Infant Cereals Value Chain

To better serve consumers with lower incomes, Nestlé created its Popularly Positioned Products (PPPs), which bring together the company's nutritional know-how, product quality, as well as extra nutritional value – such as fortification to fight nutritional deficiencies. One example of such product is *Cerelac* in the Central and West Africa Region, which was already mentioned above. PPPs allow consumers to profit from affordable nutrition, health, and wellness benefits, while at the same time bringing economic benefits to all those involved in Nestlé's supply chain, starting with smallholder farmers.

Cerelac is a range of nutrient-dense, easy-to-digest infant cereals that are fortified with micronutrients such as iron, zinc, vitamin A and vitamin C, and probiotics (Bifidus BL). In addition to 400 g tins, the product is made available in 50 g single serving sachets that are sold individually for the equivalent of US$0.40. More than 300,000 sachets of *Cerelac* 50 g are sold every day in Central and West Africa. Each sachet contributes to addressing malnutrition and essential micronutrient deficiencies prevalent in infants and young children in the region.

As already described, childhood malnutrition is a major problem in developing countries, in particular in Central and West Africa. Too often, complementary foods are introduced too soon or too late or are of poor quality. The frequency and amounts of food offered may be less than required for normal child growth, or their consistency or energy density may be inappropriate in relation to the child's needs. In addition, the nutrient content of these foods may be insufficient. Poor feeding practices in the first 2 years of life, such as suboptimal breastfeeding and inappropriate introduction to complementary foods, including food of poor hygienic quality or lacking essential micronutrients, can have an irreversible effect on the physical and cognitive development of a child. Affordable, fortified products, such as Nestlé's PPPs, can be part of the solution to increasing the nutrient value of the diet of young children and giving them a solid foundation for health and wellness throughout their life.

To raise awareness about the importance of good nutrition for infants and young children, Nestlé has built communication initiatives focusing on the importance of nutrition in the first 1,000 days of life, from conception to the second birthday. These emphasize the importance of appropriate infant feeding practices, including breastfeeding and timely introduction of nutritious complementary foods. The objective is to contribute to helping mothers improve their children's diets, by demonstrating for example the importance of iron, zinc, probiotics, and vitamins.

Mothers are taught through these initiatives that their babies' nutritional needs are huge – for their relative weight, they are five times greater than adults – while babies' stomachs are five times smaller. As a nutritionally dense infant cereal, *Cerelac* is especially adapted to fill the nutritional needs of small tummies.

These educational messages are either conveyed to mothers via healthcare professionals, or directly to mothers through TV, radio, and activity-based media campaigns. As an example, the weekly radio program "Nutrition Line" is an instrumental vehicle to educate mothers all across Central and West Africa on healthy nutritional choices, going beyond baby nutrition. Other educational activities are carried out in traditional markets, where mothers have the opportunity to try the product.

Through the Nestlé Nutrition Institute (NNI), the company also provides scientific workshops and continued professional education to healthcare professionals. Themes covered during these events include the importance of breastfeeding and appropriate, timely introduction of complementary foods.

This represents just one link in a chain of benefits, which starts with the local sourcing of raw materials made possible through the Nestlé Grain Quality Improvement Project. 100 % of the maize used in *Cerelac* is sourced locally thanks to this program. In addition, almost 80 % of the millet used in the recently launched *Cerelac Millet* will be sourced locally from Ghanaian farmers in 2014.

The chain of benefits continues in the two factories (Tema in Ghana and Agbara in Nigeria) where *Cerelac* is produced in the region: water consumption and waste are minimized and local manufacturing means creation of jobs for local communities.

Finally, Nestlé's expanding distribution network also ensures availability of *Cerelac* when and where it is needed, at the lowest possible cost and greatest possible efficiency. A growing attention is being given to PPPs, through a dedicated sales force, the Nestlé Sachet Promoters, and specific logistics (e.g., use of tricycles), helping distributors to extend their coverage of Nestlé PPPs. As such, *Cerelac*, in particular in its single sachet format, is available on a widespread scale in small shops and kiosks, easily accessible to local consumers. And this gets to the heart of what a PPP is. It is not just a product, but it is a business model designed from start to finish to provide the best possible products to its customers at the most affordable price, and bringing health and economic benefits from farm to fork.

6 Future Plans

Nestlé has already implemented numerous activities and programs to support the sustainable development of the rural communities where it sources raw materials and manufactures its products, with the objective of increasing farmers' productivity and the quality of their crops. Further measures to improve the nutritional quality of the products and make them affordable to lower-income consumers have also aimed at improving their wealth and well-being, thereby creating benefits across the value chain.

Based on the results achieved with the Nestlé Grains Quality Improvement Project, the company is now planning to roll out similar programs to other African countries where mycotoxin contamination is a problem, such as Zimbabwe and Kenya.

In addition, to reach out and educate more farmers about this issue, innovative communication tools have been developed. For instance, a role-play performance to raise awareness about the consequences related to mycotoxin contamination has been filmed and is being shown at village community centers thanks to a video van that travels across the villages. The training materials are also being translated into local languages.

Moreover, Nestlé is working on further projects to tackle other emerging food safety issues in the supply chains, such as pesticide residues and aluminum contamination, in order to ensure that grains meet the increasing tighter international norms, such as the European Food Safety Standards (EFSA) requirements.

Bibliography

FAO. (2014). *Mycotoxins*. http://www.fao.org/food/food-safety-quality/a-z-index/mycotoxins/en/. Accessed 14 April 2014.

FSG and Nestlé. (2006, March). *The Nestlé concept of corporate social responsibility, as implemented in Latin America*. http://www.issuelab.org/resource/nestle_concept_of_corporate_social_responsibility_as_implemented_in_latin_america. Accessed 23 April 2014.

Nestlé in Society. (2013). *Creating shared value and meeting our commitments 2013*. http://www.nestle.com/asset-library/documents/library/documents/corporate_social_responsibility/nestle-csv-full-report-2013-en.pdf. Accessed 23 April 2014.

Pfitzer, M., Bockstette, V., & Stamp, M. (2013, September). Innovating for shared value. *Harvard Business Review*. Accessed December 3, 2014 from https://hbr.org/2013/09/innovating-for-shared-value

Porter, M. E., & Kramer, M. R. (2006, December). Strategy and society: The link between competitive advantage and corporate social responsibility. *Harvard Business Review*. Accessed December 3, 2014 from https://hbr.org/2006/12/strategy-and-society-the-link-between-competitive-advantage-and-corporate-social-responsibility

Porter, M. E., & Kramer, M. R. (2011, January). Creating shared value. *Harvard Business Review*. Accessed December 3, 2014 from https://hbr.org/2011/01/the-big-idea-creating-shared-value

The Thin Air Factory: The Value Chain Unchained

Julian Borra

1 The Value Chain Unchained: The Revolutionary Nature of Social Networks and Technology

"It was the best of times, and the worst of times" (Dickens 1999).

In framing the conflagration of human suffrage, seismic social shifts, and profound philosophical complexities of the French Revolution as he did in the first line of his great book "A Tale of Two Cities," Charles Dickens sought to find a better way to communicate the human, material, and philosophical lessons inherent in the Revolution; by presenting them as a narrative "whole" made up from the weaving together of two distinct strands to reveal the human truths that lie somewhere between the two.

The resolution demonstrated how the positive and negative natures of each, when calibrated as a complete and integrated whole, ultimately lead humanity's representatives to "do the right thing."

Given the multitude of measures, methodologies, and processes already in existence, why bother seeking yet another supply chain, value chain, sustainability pillar, human resource, or CSR perspective?

The answer? Simplicity. My concern lies in the multitude of exegeses that already exist on how to create and capture more value in the value chain:

- The impacts of sustainability practice;
- How social networks are transforming business;
- The power of "everyday language" to unlock employee action, engagement, and value;
- The role of Human Desire in optimized workforces;

J. Borra (✉)
The Thin Air Factory Ltd, 12 Girdlers Road, London W14 0PU, UK
e-mail: julian@thinairfactory.com

M. D'heur (ed.), *Sustainable Value Chain Management*, CSR, Sustainability, Ethics & Governance, DOI 10.1007/978-3-319-12142-0_15

- Capturing the gains of the value chain;
- The share price case for "Purpose Beyond Profit";
- A compelling case for "Adaptive Governance."

In the world of *act or be acted* upon, I do not believe that an aggregated and simplified human narrative of desire and resilience is just some discretionary strategic luxury.

1.1 The Revolutionary Nature of Social and the "Animal Spirits" of the Value Chain

Given the intra-connectedness and brutal transparency the social networks bring; revealing wildly varying thresholds of living standard and liberty; blurring cultural and geographic orders and borders; firing populist fervor and social contagion; and driving leaps and shifts in both the fabric and systems of society and industry; to waste any energy whatsoever trying to either completely ignore or apply controls to any degree of evolution is an increasingly futile pursuit. The irrepressible massing social organism will simply work its way around the side and through the gaps, however tiny, random, or unrelated they may seem.

This fluid shifting mass of local and global tensions, turbulences, influences, and effects, and the volatility of human desires that underwrite them is a phenomena that I have chosen to identify as a form of **Animal Spirits**[1] (*after* Keynes 1997). They are demonstrating an increasing ability to disturb, undo, elevate, illuminate, and improve the structure, fabric, and nature of value chains both vertically and horizontally.

The old blocks of business (focus: Reputation) and brand (focus: Advocacy), its operationalized self, its value chain, HR, and stakeholder constituency have become atomized: sliced and diced and disaggregated by the impact of technology and social networks (see Fig. 1).

The compounding, truncating, and reintegrating nature of this action is what shapes the future model: transforming the company into a unified evolving intelligent intuitive entity; one imbued with far greater adaptive abilities and therefore far more resilient in nature.

[1] **Animal spirits** is the term John Maynard Keynes used in his 1936 book *The General Theory of Employment, Interest and Money* to describe the instincts, proclivities, and emotions that ostensibly influence and guide human behavior, and which can be measured in terms of, for example, consumer confidence. It has since been argued that trust is also included in or produced by "animal spirits."

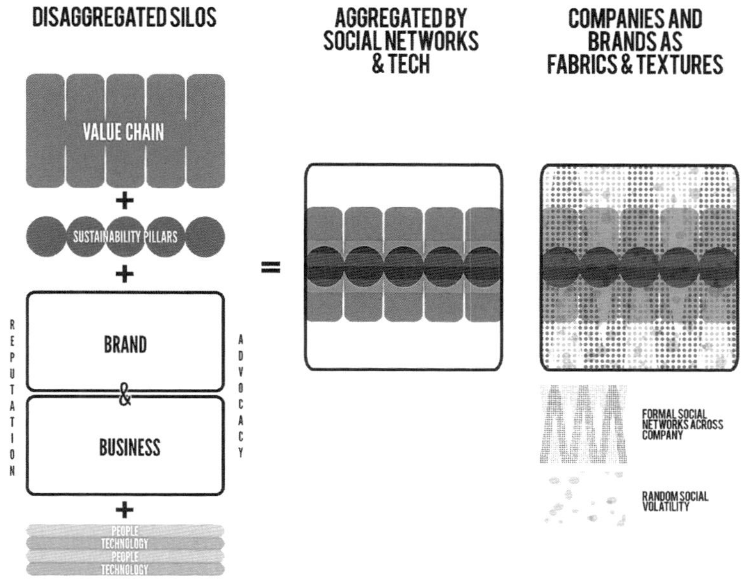

Fig. 1 The old silos of brand and business aggregated and atomized by the social networks and technology

1.2 *The Value Chain Unchained: A Journey Between Two Lines*

Using the Dickens metaphor of the *Two Cities*, I see the two cities that will influence the transformation of value chains as those of "**Shared Resilience**" and "**Mutual Desire**," as shaped by the tensions and dynamics of the two Ages in which we find ourselves: that of Transparency (and its effect on our adaptive capabilities) and Austerity (and its effect on our models of desire):

- I shall use them to both explore and humanize the complexities of the value chain. This has a number of purposes: To illuminate the truly "social" human fabric and nature of companies even in the deep dark, highly engineered, modeled, and controlled corners of their supply and value chains
- To illustrate the inevitability and immutability of the human condition at work in companies – the flux and flow of human nature as it acts against the machine – in the age of the social network
- To better capture in more compelling and human terms the value created within a company especially through the pursuit and application of more sustainable material, systemic and human strategies, processes, and behaviors
- To demonstrate that the most important red thread that runs from the top of a brand to the bottom of a business's operational self – and from one end of any value chain and the supply chain it envelops to the other – is the flux of human nature and the storytelling that it can generate

- To protest that the "animal spirits" of the value chain world – the politics and capricious nature of human desire – must be more formally accounted for as they are both the valve on the value chain pressure cooker and a powerful source of resilience

1.3 Shared Resilience: Transforming the Philosophical Nature of Brands

Shared Resilience and Mutual Desire are the pillars of a business's most compelling communicated and socialized self. Shared Resilience intends to provide a more holistic and representative context in which to model a company's efficient response to an ever changing environment.

Reaching beyond the definition of Resilience already common to smart and adaptive value chain modeling, which focuses purely on "inside-out" mechanisms that allow for the incorporation of event readiness – an systemic trait that allows the company to adapt to risk that affects its capacities enabling the company to provide an efficient response, often rendering the value chain capable of recovering to its original state or even better post the disruptive event.

Shared resilience is resilience that is also predicated on that company's ability to map, measure, model, and incorporate for "outside-in" influencers and dynamics on the performance and adaptive nature of the value chain that are not held within and subject to the controlled environment of the value chain but which have increasing influence over its ability to perform – most particularly:

- The increasing volatilities and impact of social networks and the fluid and interrelated nature of individual and collective concepts of desire across the whole value chain stakeholder group,
- The impact of broader social upgrading driven by those value chains on both their immediate employees and partners and the broader economies in which they exist.

1.4 The Revolutionary Nature of Social Creation and Social Activism

'Simplify. Simplify.' (Thoreau 2007)
'Just one 'simplify' would have sufficed. (Emerson – in response)

That the thriving knowledge economies of our postindustrial age amply demonstrate the transition of the world's leading and emerging economies from a solid to a more liquid concept of modernity is undoubted. What is revealing is the speed of passage and scale of impact of that transition on the nature, structure, and management of power and responsibility across civic, corporate, and domestic dimensions.

We are shifting *from a time* where to be a modern and thriving industrial society meant upholding and securing that society predicated on a model *of Solidity and Order punctuated by intermittent Chaos; to a time* where to be a postmodern and postindustrial society capable of capturing added value in the upper knowledge tiers demands that one can accommodate a society that thrives on a model *of fluidity and Chaos punctuated by intermittent order*.

The fixed communities of the old enjoyed a solidity of identity sheltering within the protection of a highly structured social contract negotiated between the elite, the representatives of the populous, and those governing them, their security, and their prosperity.

These communities are evolving, deconstructing and in doing so shifting in exponential numbers into peg communities. Peg communities enjoy a volatility of identities sheltering under the protection of a fluid social contract – a contract increasingly renegotiated directly with both the elite and those representing them and subsequently rewritten directly by the people through the facilitation and collective voice of the social networks.

The peg communities of this volatile fluid mass thrive and multiply in the new universe provided by the social networks and the 1.7 billion users that constellate themselves within them.

The relevance of this for value chains comes in the exponential multiplication and amplification of *human nature* the social networks generate on, around and within them: its flux, its morals, values, concepts of purpose and righteousness: its instincts, irrationality, behaviors, allegiances, randomness, and emotional charges – and ultimately not only how they impact in every corner of domestic life and how they shape cultural and social norms – but also in regard to how they might predict best engagement and deployment of labor and the professional classes inside the value chain to greatest effect.

These are for me the value chain equivalents of what Keynes called the *animal spirits* in economic modeling. They are the immutable, extant, and dynamic embodiment of the human condition and how it plays out its concepts of security, risk, fulfillment, and survival in the everyday: and ultimately their role in shaping human DESIRE.

In his prescient book "The Third Wave," Toffler sets out the psychological volatility and shifts the new age brings:

> "The link between communications and character is complex but unbreakable. We cannot transform our media of communication and expect to remain unchanged as people. A revolution in the media must mean a revolution in the psyche" (Toffler 1981).

The social media revolution has generated two dynamic influencers on the *psyche* of a company: those of *social creation and social activism*.

These influencers are found to be at work in the governance, materiality, and systemic nature of a twenty-first century company in some shape or form. The democratized nature and immediacy of the social networks reach and scale give them an ability to fundamentally alter and influence what were once highly controlled and mostly invisible models, structures, and strategies of a company.

Social Creation is the generative outpouring of collaborative communities and the sharing economy. Massed communities of shared interest, purpose, and desire are swapping intelligence, ideas, and data to create new ways of thinking and doing for better. The social networks have fueled the rise and delivered the communications tool of choice for the social entrepreneur, enabling social purpose and mission based nonprofit/profit businesses to expand their altruistic networks and embed themselves into physical and virtual communities for action and support.

In the pursuit of creating a more agile problem-solving, shift-shaping, people-powered, solution-centric culture, large corporates have begun to draw the social entrepreneurs and their particular way of shaping teams around a task to invigorate the corporate system from the inside out and the ground up – to drive improved performance and perspectives into the governance and the systemic operational nature of their businesses.

From Tech and network hacks to the Maker Movement and the open source and co-creation cultures originated on multiple campuses, these highly socialized collectives are proof that random groupings of people clustered around one pure focus can impact on both the inbound logistics and operational pillars of the value chain. The paradigm-shifting technology of 3D printers – a Maker Movement innovation – is currently rewriting the old rules of manufacture and short-run production, democratizing the process and fracturing the old temporal, material, and capital frameworks of product design and manufacture.

The shared economy/recycle/reuse culture being championed by the likes of Yerdle, the shared economy website in the San Francisco Bay Area takes the "yard sale" and the human cultural truth of *swapsies and sharesies* to a whole new social level. The dynamic real time insights, data, and intelligence of the Shared Economy are going to eventually impact on the inbound and outbound logistic dimensions of the value chain. As these shared economy sites begin to develop scale and mass, they are reaching out to the brands most likely to reoccur in their domain. In doing so they are potentially recreating the concept of "Cradle to Cradle" planning; elevating it to a sort of "Cradle To Cradle Plus" – where brands factor the secondary, reusable sharable nature and traits of what they produce as well as their by-product potentials into their formal value chain planning. This is a direct impact on value chain modeling driven by social culture – life cycles shaped and rendered in social networks as opposed to lab based academia. Add to this the cultural shifts created inside the value chain with the proliferation of co-creation and open source sites, inspired not only by the leading-edge hackers and makers but also by the transformation of global businesses like GE through their *Ecomagination* initiative and social creation is plainly upon us.

Social Activism centers on how social networks allow the populous to organize themselves to voice and action against organizations and companies that are felt by the fluid, volatile mass to be against the collective good, acting irresponsibly or abdicating their role in maintaining the world we share.

Having (accidentally or otherwise) been seen to embrace everything from abusing workers, tax evasion, poisoning rivers, battering orangutans, and bloating children to funding sweatshops running on child labor, the companies and corporations who have arrogantly dismissed the power of the social networks have come to rue the day. The social media networks and the communities that thrive within them have used the tools at their disposal and the nature and critical mass of collectivism and VOICE to take on things as diverse as the Bonus Culture at ING to creating a new people powered market economy in Energy Supply price in South West England, driven by a *Facebook* group of homes and businesses; focusing their anger at energy company prices hikes and monopolies through direct collective action.

In their own way the proliferation of causal and philanthropic platforms that enable the end user and engaged advocates to connect directly with socially responsible actions and initiatives by the companies, and share, collaborate, and invest in those initiatives has created an expectation and culture of shared action to mutual benefit – the ability for the human to "act" on their emotions and in line with their value and belief systems either for or against a company, institution, or organization in whole or in part.

1.5 Immutable 'Animal Spirits' and Their Impact on Value Chains

Not only have our ***animal spirits*** proven that they are not going away any time soon: they have proven themselves capable of un-writing everything from a brand's reputation to the social contract of a First World economy. The natural shifts, chaos randomness, and turbulence inherent in their collective nature, values, and actions wild-firing across the social landscape are reshaping governance. Their celebration of creation and activism are beginning to overlap, impact, and inspire in ways that will fundamentally change the way companies model and manage their value chains. As these social (and by their nature cultural) dynamics are drawn to an increasing degree from the purely social world into the systemic and cultural fabrics of the commercial and industrial worlds, the social nature of them, the human collective and individual instincts, proclivities, and emotions inherent in their generation and written through the fabrics and contexts they weave become a variable innate within those worlds – and therefore become part of their dynamic DNA.

The animal spirit is what compels us to scrutinize the concepts of purpose and responsibility that are innate in those value chains and deciding whether they are acceptable, authentic, and ultimately fit for purpose in the eyes of the societies and cultures in which the value chains sit and seek to maintain themselves.

Now the walls of the factories and the boardroom have been rendered in glass – and the social contract is being rewritten through the all-seeing eye of smart phone cameras, YouTube, Twitter, and Facebook.

The ability of those social networks to provide WISDOM – high speed, low cost organization and communication designed to inform inspire and ignite action across sometimes expensive, polycentric networks – diverse or widely spread clusters and groups of varying strengths of identity, and WITNESS – recording those events, diarizing them, integrating and corroborating them across other sources and networks to optimize action and impact – has given everyone the clear idea that they can act against any entity of any size.

Social Networks are forcing a re-imagination of value chains by driving the intangible/irrationals of wisdom of the crowds (co-creation, open source innovation methodologies, and bald-faced pressure-group politicking), concepts of social conscience, morality, and responsibility into the old dark and highly controlled corners of value chain optimization – especially those previously out of sight and subject to no-one. This is creating new pressures on companies. It is testing their capacity to be adaptive both across their value chain models – especially in regard to their marketing, sales, and services pillars where animal spirits are most likely to be at play.

The social impact of technology, how it is adopted and utilized by vertical economies and the society and communities both inside and outside the company will become more and more integrated with the Human Resource function because of the massive social impact technology will have in the coming decades. This is about far more than just matching existing tech maturities of the labor force with the operational and logistical tech requirements of the value chain and how that impacts on the ability of the business to optimize its tech applications.

In a recent leader piece in the Economist Magazine – titled "Coming To An Office Near You" – the author points to the enormous impact of incoming technology on societies: "...where income gaps will widen, causing huge social dislocation, and perhaps even changing politics" (Economist 2014). The social influence and impact of the technology embedded in any given value chain is not restricted to the factory floor, the R&D department, or secretarial pool. The same article points to there being a very large gap and lag between when new technologies start driving down employment numbers, creating social dislocation and inequality in their wake and the time when it finally upgrades the economy as a whole, elevating everyone to a higher standard of living and prosperity.

The anger that rising inequality creates will be played out using the most powerful collective tool known to the people – the social networks on which they gather and share and through which they have already begun to act. This is an animal spirit that value chains must be resilient too. It impacts directly on their ability to function optimally because a company's value chain is inextricably linked to its social responsibility to every stakeholder group.

This volatility will be exacerbated by the fact that governments are themselves failing to create contingencies to deal with the scale and duration of social turbulence new invasive technologies will bring to every shape and form of workforce.

2 Shared Resilience: The Age of Social Transparency and the Elevation of Resilience

I wish to explore the dimensions and influencers most closely related to what I see as the role of human desire, and the drivers of social dynamics and concepts of prosperity on the modeling efficiencies, economies, and performance of value chains over time and across cultures.

Happy companies have become the state du-jour. Psychotherapists of every method are sweeping through the old industrial world seeding self-assertion and self-realization programs to both serve the employee and the employer. The effects of these employee initiatives are in turn being measured as proof of the company's adaptive and human-centric modeling for success. Client companies are using consultants to seek explore and set out the most impactful compound indices of engagement, satisfaction, and resilience in their workforces and management structures and comparing them against financial performance metrics to ascertain the best dynamic tension point and balance between productivity and sociocultural disposition of the business to succeed.

The advisors and architects of value chains and supply management are adopting tools and behaviors from the social media landscape to transform themselves, and by doing so building the levers and pulleys of the social world into their systemic selves – and thereby creating a foundation at least for an amplified degree of resilience based upon laws of similarity. For example, the TUI "Spreading Smiles" model uses storytelling in shareable portable formats and the power of social networks (both internally and externally) to distribute the assets.

Companies are now wholly subject to the new social laws of immediacy, scale, and action brought about by the conflation of smart phone technology, nanosecond connectivity and share, and the reach and volume of social networks. The enforced value chain transparency they bring and its attendant turbulence are forcing new forms of adaptive governance to be applied horizontally along the value chains of substantial businesses across almost every sector.

The increasing perturbations of uncontrollable social impacts and influences are unchaining the old fixed value chain models in favor of more adaptive and resilient models: The intention of hybrid value chains is to create collaborative entrepreneurship at global and country levels by aligning the needs and resourcefulness of both social and corporate actors. They specifically "set out" to identify complimentary competencies, shared benefits, and new more qualitative networks of growth and competition because it exists for a different reason. This is about choosing to optimize actors from two sides to create a greater unified whole. It is a methodology that reaches into traditional companies who are by their nature no social entrepreneurs and illuminates a greater, more socially attenuated model for success.

Equally, disaggregated and fractural value chain models point to the need to engage and align the needs and resources of multiple social and corporate actors to offset the perturbations by not having done so. They are all examples of adaptive thinking – where the brutal transparency and blunt social tool of collective activism offered by the new social world has acted upon and compelled a model to adapt.

The company and its value chain is therefore better able to allow itself to flex within a desired state to absorb both existing and unexpected turbulences, because it has chosen to act before it is acted upon – and by doing so becomes more resilient. In the world of global business we tend to be surprised by the degree to which some of the most visible corporate actors have adapted, especially those previously regarded as inherently and stubbornly toxic in their operational and cultural nature.

In doing so, some high-profile multinational companies have developed a greater resilience to absorb and metabolize future volatility and weather the social storms, hugely improving their innate adaptive abilities for the medium and long term.

We will also notice that these actors have to a greater degree tried to fathom their storytelling around a shared resilience – a "what's good for you, is good for us, is good for all of us" thread. And we find the power of social being used to deliver dual source benefits of communication, connectivity, co-creativity, and action in many brands – with platforms being created to link together the end user and the source market to mutual benefit.

TUI AG, for example, the global travel business, has wrapped its inventory of CSR and sustainability in a feel-good initiative called Spreading Smiles – a brand storytelling platform initially designed to share in-destination, in-company, and in-market CSR and sustainability storytelling amongst colleagues, partners, and suppliers. Ultimately it is designed to spread outwards to the customer, to drive preference and add value in the consideration and transactional space.

TUI's ethos of enlightened corporate responsibility of how a company might act for good with a degree of authenticity, and in doing so take "do good" from a hypothetical narrative to a living action inside and outside the business aligns closely with that cited in the Centre for Strategic and International Studies "7 Revolutions" rolling research project:

> "A well-run business that applies its vast resources expertise and management talent to problems that it understands and in which it has a stake can have a greater impact on social good than any other institution or philanthropic organization" (CSIS 2014).

TUI, like McDonalds, Starbucks, and many other brand operations cannot ignore the social noise around the downside of their industries, whether that be around the carbon footprint of planes in the sky, the provenance, dietary quality, and environmental impact of low-cost meat-based fast foods, or the balance and fair nature of the direct and indirect trading relationships that exist between global corporations and small growers of crops like coffee beans. These are sources and measures of integrity upon which animal spirits act quickly and fiercely in the social networks if a company falls short or foul of their responsibilities or commitments and is found wanting.

Take the example of micro to mass crossover in Global Activism featured in the paper "Communicating Global Activism" by Lance Bennet from the University of Washington (Bennet 2003). It began with an email from Jonah Peretti to NIKE. Peretti, a culture jammer took exception to a Nike site promising greater consumer freedom through the addition of personal sloganeering on its shoes. Peretti submitted an order with the word "sweatshop" as the preferred slogan on the Nikes. A round of amusing exchanges around Nikes refusal to put this word on his shoes and their breaking their own contract of customer freedom. The emails escalated with Peretti sharing the exchanges with friends. The Nike sweatshop story eventually reached a potential audience of anything from several hundred thousand to several million. Animal Spirits have a very loud roar.

It makes supreme sense to fathom and scope how to use the behaviors, tools, and platforms of the very social culture that might act against you to equal and opposite purpose – to weave a more robust flexible and resilient culture and fabric for your value chain. We already find evidence of social media tools being deployed as the driver of resilience in category management practice in Procure-to-Pay (P2P) value chain management.

In a recent piece by *Andrew Bartolini*, called *"Business Or Pleasure – How social Networking will reshape the the P2P value chain"* – Bartolini inspires businesses to seize on the enormous power and systemic benefits of social media concepts like Twitter, wikis, and Facebook to reshape their businesses for greater relevance and resilience (Bartolini 2011).

To truly reflect the new truths of the socially amplified world in developing more resilient value chains, we need to create a "shared concept of resilience": one where a company's strategy of resilience not only matches and compliments or adds value to their stakeholder's concept of their own resilience – but also generates a social culture around the mutual benefit of escalating resilience.

It is only then that you begin to set out the shared language of shared resilience – and shape the foundation stone of mutual DESIRE.

3 MUTUAL DESIRE the Age of Austerity and the Transformation of Desire

3.1 The Transformation of DESIRE

The model or measure by which we can to a reasonable degree plot the "animal spirits" in and around a value chain – in such a way as to increase shared resilience and to engender mutual desire – is providing the foundation to operate more resilient value chains going forward. Human nature's role as an influencer on every corner of the value chain and the role of social networks and technologies in enabling and increasing the spread and scale of that influence is in desperate need of further exploration.

The social networks have transformed the nature of human desire by increasing, the number of variations of it, the speed at which it can flux and transform, the structures and collective dynamics of its communities, and the ability to organize and act upon it.

It is in that space that the relative nature of particular groups in society and their ability to agitate and activate toward what they find most desirous and the specifically human dimensions of the value chain become clearer. Everyday human desire, that of the employee, the factory manager, the corporate executive, the grower, or the trucker, and their concepts of what a thriving and prosperous life looks like or should look like, through the collectivism of the social networks, become influencers and effecters of the value chain and its ability to operate optimally.

As we become more and more aware, better informed and more prone to action, the impacts of how we consume and what is done in our name become integral to the human condition and how we shape, project, and act on our desires. More importantly, the responsibility of companies in delivering concepts of a thriving life – fair labor and decent communities, respect for environment, support of the disadvantaged in the communities of labor – become part of what is a desirous life. This does not stop at a job's ability to pay for the right beer brand and make of cotton-buds.

In the age of labor rights and pressure groups this makes fluctuating concepts of local prosperity of direct concern to a value chain model. If the social networks, through influence, information, or comparison, can shift a worker's concept of prosperity and respectful thriving and in turn change their opinion of how a company delivers against those concepts, that is a source of perturbation that needs to be managed – because the company's ability to control secure and sustain the performance and commitment of a workforce at a foreseeable and manageable cost is crucial to the resilience of its value chain. Concepts of prosperity and a worker's right to thrive within them are certainly becoming more volatile.

Material austerity and the instability of financial markets have crushed the mirage of the old American Dream of prosperity – of infinite growth and of unfettered ever-increasing production and consumption. The social networks and their communities of shared action and interest bring with them a relentless flow of real-time reminders that we exist in a finite material world defined by diminishing resource and degrading environment and expanding human need.

Social, Material, and Environmental pressures are demanding that we reimagine the way we consume. This is not just seen as a strand of living. How we consume is a signifier of our concept of prosperity and what it means to thrive. To create a new and more enduring aspiration and sustainable lifestyle to pursue we must reimagine prosperity. We simply cannot afford to continue with the old world model of prosperity.

It is not just individuals: but communities, cultures, and governments who are realizing that we have to change. We have to transform our desires to make room for a more enduring aspiration. A company's ability to improve its resilience – by becoming more adaptable to the turbulence of resource scarcities, natural disasters, financial collapse, volatile markets, and monopoly market players is only one half of the challenge.

We see communities, tribes, and cultures seeking to reimagine their own concept of prosperity, seeking to find a more culturally attenuated, local, and more achievable concept of what it means to thrive. Companies must find a way to better absorb the shifting nature of concepts of prosperity and the fluctuating effect of the human desire for a prosperous life and how that impacts on them through the constituency of their value chain stakeholders without changing the desired state in which they exist.

One of the ways in which they could do that, is by using storytelling and communications to make the sustainable nature of their value chain more relevant and valuable and therefore more desirable to their customer – and in turn reaffirming their role as an elemental part of a prosper strategy.

For it to succeed, we need to reframe the storytelling of sustainable truths, actions processes, and behaviors to underwrite a more positive form of desirable consumption. But we must also more formally account for the circular nature of how desire acts upon the value chain from the outside in.

The new social world has turned the previously impervious, culturally fixed visions of prosperity and one-way social dynamics into a porous, open, two-way trans-national "share and compare" culture. This means that the concepts of desire and prosperity that any value chain exists within, are not just those of the fixed indigenous population. The new and more porous nature of our joined up world allows outside concepts of prosperity, need, fairness, and freedom of voice to influence the workers "inside the cultural gate."

The human truth of the shifting concepts of prosperity we now live within is imminent in many economies today. On their return from émigré labor markets like the United Kingdom, Polish workers were found to desire a lifestyle quite different to those of their stay-at-home cousins. The returning workers concepts of prosperity had upgraded, as had the concepts of social construct they were prepared to work and exist within (Orenstein 2014).

The most recent recalibrations of what constitutes a thriving, successful, and more enduring lifestyle are to be found in nations trying to throw off the shackle of the old imported American version and reassert their own culturally attenuated version of prosperity. More and more nations are beginning to seek a more authentic and meaningful imagining or reimagining of prosperity – one that is truer to their local social financial and material means and the more home-spun strategies for smarter, lighter living – the politics of thrift and smarter leaner consumption – as seen through a very local filter of long standing cultural wisdoms, vernaculars, and truths. This influence will in turn shape a new more locally or culturally attenuated concept of prosperity.

If the value chain embraces as a truth the concept of "social upgrading" not only as an unavoidable and inextricable fact of life – that social perturbations only multiply – but also as a proven dynamic outcome of best in class value chain management (see Gereffi and Fernandez-Stark 2011).

If the vertical economy in which the value chain or particular part of the value chain exists presents the opportunity for effecting social movement, the value chain needs to be able to absorb the cultural impacts of that social movement on its

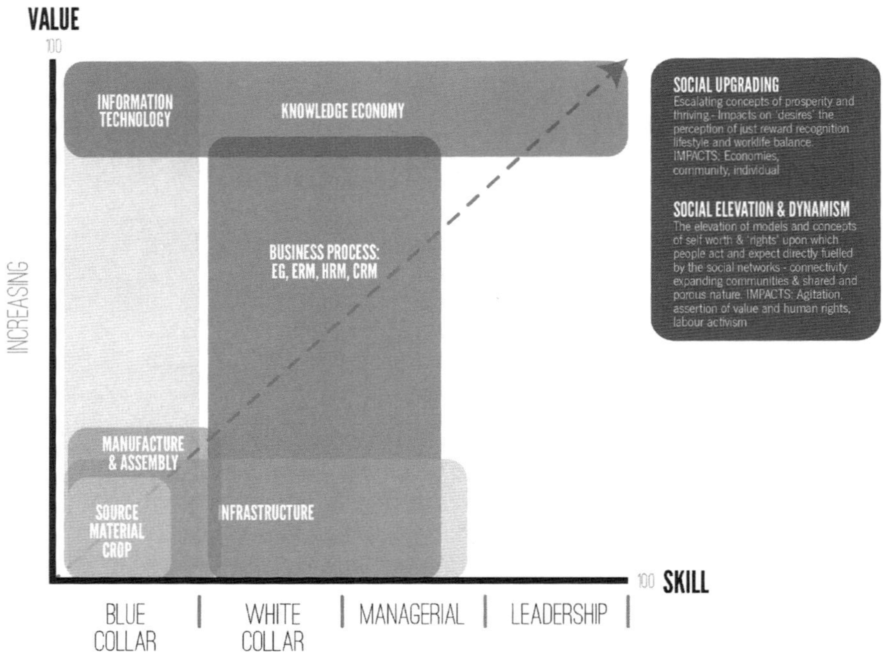

Fig. 2 Social upgrading along the dimensions of skills and value (Thin Air Factory 2014)

Human Resource support activity. As we move up the value axis (Fig. 2), we see the emergence of a greater sense and degree of burgeoning identity, as well as the structural and financial utility of the thriving desirable life that the individual worker seeks.

As we move up the skill/value axis, we will find a relative increase in the influence and impact of human desire – individual, collective, and sociocultural – on that value chain's ability to function both optimally and sustainably. This in turn creates a new perturbation – that of sustaining the scale of labor at the right value level to sustain and secure that link in the value chain.

A recent example of the direct impact social networks and the wider online media can potentially bring to Value Chains on cultures and communities climbing up the skills/value axis centered around anecdotal evidence of the shift in Labor expectation in both India and China but most pointedly China, in regard to the emerging middle class (of some 800 million people) and the evolution in what they see as constituting a thriving life and prosperous living and how that impacts on the value chains they work within.

As their Quality of Life improves in regard to income, leisure time, and their right to a more vocal opinion on the ethics and responsibilities of the private sector in which they work, the emerging middle classes grow in confidence and both assert and vocalize their position and their own and their communities rights to improved treatment. They are pressing for greater labor rights, higher wages, and better recognition and rewards. This puts direct pressure on the value chains of the businesses they are working in – most pointedly on their ability to sustain the price advantage of manufacture to remain dynamic and growing, regardless of whether they are feeding an export or domestic marketplace.

These fluid and shifting perceptions and expectations of social upgrading – the right to a better life as delivered directly through the company someone works for, especially if that person is part of a highly lucrative and immensely profitable global value chain – are being stoked by the increasing influence and transparency of social media and online media both within and without – even in a heavily controlled media state like China.

Everyone from Wikipedia to Global Post bloggers and academics in Georgetown University point to how the labor-rights issues around employee welfare and child labor at Foxconn, Apple's component partner company in China, have affected the nature and reputation of Apple. The reporting, originating in a Daily Mail article and subsequently wild-fired across the networks, resonated with both Chinese and Western workers, who also believed the employees deserved better of the company in whose Value Chain they existed.

This had a direct impact on the Reputation of Apple both in the liberal early adopter creative community and broader heartland of the brand.

The exposure also seems to have pressed a far greater and more adaptive form of governance into being at Apple to absorb the turbulence created by the social and traditional media backlash of the worker's rights abuses and the increased and intense scrutiny under which the company now operates.

Some examples of the kind of reporting and opinion still going on in late 2013.

In the Global Post

> In early 2012, The New York Times published a Pulitzer Prize winning series of reports that exposed poor working conditions in factories that made Apple products. The reports focused on i-Pad manufacturing in China, describing harsh labor conditions, safety concerns and difficult work environments. Living conditions, overtime demands and other egregious workplace red flags, like underage workers and falsified records, were documented as well. The report did a lot to bruise the reputation of Apple. The tech giant is now taking steps to improve its track record.

In a white paper on Corporate Social Responsibility:

> "By increasing transparency and subjecting Apple and its suppliers' practices to more public scrutiny, Apple and its suppliers will be forced to improve working conditions for factory workers or face negative media attention and possibly even legal actions" (Myers 2013).

And finally on the broader note of Social Upgrading:

> 'Middle class expectations in emerging and developing countries are rising and evolving as their countries' economic situations improve, following Hirschman's "tunnel effect". They are no longer satisfied...' 'The rising expectations of the expanding middle class in developing countries......will these middle classes be agents of change?' (Pezzini 2012).

This is why I see a circular nature to the concept of social upgrading – where it features not just as an outcome driven by the presence of a global or local value chain at work in any given vertical context – I see social upgrading and the subsequent shifts in human desire it creates in both collectives and individuals as becoming a primary impact on and of primary concern to the inbound dimensions of that value chain.

Social upgrading in a vertical economy, is a particular sociocultural outcome of the existence of a value chain and the inextricable and influencing nature of human desire in value chain modeling (Gereffi and Fernandez-Stark 2011). Social upgrading cannot be viewed as a fixed or smooth state: it is volatile and open to value chain models in direct relation to each other; not some dislocated and abstracted set of effecters or entities. The commissioners and modelers of value chains would be wise to heed the dynamics of global social activism encapsulated in Castells entreat in The Network Society 1997 to "grasp the transformations of space, society, and **identity** that are associated with digital communications networks" (cited in Bartolini 2005).

Value chains must accept that by their provision of employment to larger groups and communities, they are inextricably linked to creating a stable and achievable model of prosperity within the vertical economies in which they exist. This role in the fabric of prosperity and its provision makes them vulnerable to any flux and turbulence that may run through society in relation to that model of prosperity and people's degrees of satisfaction with the private and civic actors involved in underwriting it – especially given the social tools that are now at their disposal. Global activism enabled by the social networks delivers concepts such as that of scale shift – as identified by McAdam et al., centering on the diffusion of protest networks and the accompanying transformation of identities – must be factored into the systemic heart of the value chain (McAdam et al. 2001).

3.2 The Role of Adaptive Leadership in Sustaining Value Chains and Telling Better Stories of Shared Resilience

Resilient storytelling is storytelling that can both inspire every stakeholder to the company) and challenge excuses for inaction or evasion. Resilient storytelling is rooted in a substantial narrative that cannot be set aside or dismissed as thin, inappropriate, inauthentic or irrelevant; or ultimately dismissed as 'gloss': a faddy, fluffy, short-sighted myth generated in an over-populated brand workshop somewhere. By its very nature resilient storytelling must be both resilient in itself and inspire greater resilience in its telling. But it must be fit for purpose and be forged from a whole picture of the company – not just its individual functions and layers.

To create storytelling that can absorb the turbulence and flux of the ever changing, ever evolving world a company seeks to thrive in, that storytelling must embrace the four corners of the company; from the top of its brand to the bottom of its business value chain, and from one end of its value chain to the other.

Resilient storytelling creates a clear sense of unified purpose beyond profit, a clear central tenet of adaptive governance to shape, manage, and distribute the mutual endeavor that purpose demands and the shared benefits it offers. It ultimately frames the integrity of every relationship the company generates and engages itself in; and every piece of communication the company produces in undertaking and maintaining those relationships. Resilient storytelling is storytelling that can relentlessly inspire and drive advocacy in every stakeholder.

Resilient storytelling is one that reconciles and reframes the most compelling, differentiated, and most valuable points of systemic, operational, and material resilience (the sustainability and CSR aspects and traits in the company) to the greatest number of shareholders with meaning and effect. It is the voice of the company's mutual desire for shared resilience: a reflection of the strongest and most compelling points of shared resilience, and a primary source of increasing resilience in itself. Resilient Storytelling is completely informed by every dimension of the company – and only in its simplest and most human telling can it take the smartest, most enduring, and most innovative ideas from *thinking to doing*.

If we are to reimagine a more enduring model of prosperity beyond the toxic one we have now, we must transform desire – and we can only do that by reconciling the desires of the people who are demanding with that of the people supplying. And we can only do that by contemplating, identifying, and communicating the points of **shared resilience** in such a way as to elevate the state of **mutual desire** across every stakeholder: where the desires and best interests of owner, employee, supplier, partner, customer, and the three degrees of immediate social separation around them are met equally.

Resilient storytelling is storytelling that does not suffer the stones of in-authenticity; or that which could be thought of as trite, or lacking in character and meaning. It is storytelling that draws together the two cities of **mutual desire** – the singular most compelling model of desire that drives every important stakeholder group in the value chain – and **shared resilience** – the actions, materials, and behaviors of authentic and integral resilience that bind every stakeholder of the company around a singular differentiated purpose. A company's ability to engineer an increasing resilience both in its preexisting models of value creation and capture, and through its systemic operational structure and nature will be shaped and defined by its ability to adapt to both increasing social influence and impact and transformative desire across every stakeholder group.

Moreover, the dynamic nature of that resilience will also rely on its ability to create compelling storytelling around the two crucial drivers of that resilience. Its ability not only to absorb turbulence and stay true to itself systemically and operationally, but to be able to communicate the nature of that resilience with value and relevance to every stakeholder in its universe in such a way as to qualify the company's ability to fulfill the desires of those stakeholders (regardless of whether they be customer, supplier investor, or employee) will be critical to the success of the company in the newly transparent and explicitly desirous world.

Simply put, this is about a company's ability to clearly communicate in singular and differentiated terms

- First, its *model of shared resilience*: its purpose, integrity capability, and commitment to stay in the business it is in, delivering the value it does without the need to drastically transform the nature of that business, in the face of turbulence presented by the increasing material scarcity and relentless social scrutiny of the world it exists within.
- Second, it is to communicate its *map of mutual desire*: across every dimension of its value chain – those of its workers, support, and supply partners – the desires of its managers and leaders – in such a way as to identify a mutuality of desire to bolster and fuel the systemic resilience of the company.

To that point even with the greatest intention and rigoros systems in place it is still possible for the resilience and the adaptive nature of the company to fall short or fail, predicated on a traditional flaw: that of the leadership viewing the company's ability to adapt as being the role of everyone else other than them. But I would venture that equally, if the desire is not there in the leadership to manage and secure the company's particular and singular model of shared resilience and relentlessly communicate the value of it in a language common to all, the company will falter regardless of the integrity of its value chain modeling.

The extreme degrees of transparency, the shifting nature of concepts of prosperity, and the pressures of material scarcity, are demanding that the governance and leadership structures and behaviors managing these value chains also take on a new and more highly *adaptive* nature. The need for greater resilience drives an increasing demand for a more resilient and adaptive form of leadership – the increasing need for leaders offering wider perspectives and the lateral skills of a less siloed and more enlightened style of leadership – are directly driven by their need to absorb, process, and act upon a wide ranging set of drivers and influencers at once. This is to manage the increasing pressure to secure the purpose and the integrity of a company beyond pure profit and set out its responsibility to all of its stakeholders (especially those most vulnerable to economic, environmental, and cultural shocks). To manage the nuances and impacts social networks are driving into the nature of doing business demands a more diplomatic and empathetic form of leadership response. Besides this, it is to manage the increasing pressures of material scarcity and increasing demand on a growth-centric business and the exponential ascent of the Renaissance product, service, and business: those which require their executive to both plan and act with exceptional adeptness across multidimensional, multi-skill, multi-capability, multi-market processes, stakeholders, environments, and theaters of action). Companies are managing themselves across ever more complex value chains, shifting market dynamics, proliferating partnerships, and highly sophisticated distribution models and networks. The new model leader needs to manage the increasing pressures of material scarcity, the voracious market demand for a wholly growth-centric business; and the exponential ascent (and some would say tyranny) of developing Renaissance products and services (predicated on the '*have it all; does it all*' multi-functional, multi-capability, multiple benefit philosophy 'de jour' of the *phone camera computer music diary entertainment system wellbeing watch bracelet*). This new world requires the Executive to both plan and act with exceptional adeptness across multiple dimensions, disciplines, skill-sets, source markets, processes, stakeholders, environments, and theaters of action.

An increasing need for understanding and reconciling the natures and needs of public relations, corporate affairs, marketing communication, human resources, internal communications, and R&D in one efficient and economical piece of integrated storytelling requires leadership with an audacious amount of social and systemic empathy.

This form of leadership is deemed better placed to deliver more resilient companies because they are attitudinally better disposed to unlocking the old solid structures, while embracing dynamic evolving technologies and methodologies. They are also far more likely to pursue untapped opportunities delivered by the social networks, previously deemed by the old Executive guard as an anathema or just some passing fad.

They are seen as critical to the quality and degree of resilience the company demonstrates in this new fluid turbulent circumstance, because they demonstrate the kind of attributes and attitudes as set out in CSIS's *7 Revolutions*, where the effective leader of the future is described as "*...someone who will jettison vertical integration information hoarding and dogma in favor of optimization, recalibration, and negotiation.*" (CSIS 2014).

But I would go further. I would say that one of their primary roles and objectives lies not only in their ability to use the audacious breadth and depth of vision and knowledge to see the whole picture but to explicitly and relentlessly state the goal of **mutual desire** to secure all that is good in the value chain to drive exceptional resilience.

Ironically one of the greatest factors in a company's ability to remain resilient lies in its ability to communicate its most resilient traits in such a way as to create a sense of shared desire amongst every stakeholder in its value chain, to secure, manage, and maintain those traits. The specific leadership objective here is to create a *task force* of **mutual desire**. To do so they will need to deconstruct the siloes of marketing communications, Human Resources, Corporate Affairs, and Sustainability, directing those various actors in no uncertain terms to work in unison, to find the single, most compelling and human storytelling. This narrative must shape a clear concept of **mutual desire** based upon the authentic traits and behaviors of **shared resilience**. In doing so the singular narrative effectively creates a value chain reaction, shifting the company from behaving reactively as a fixed defined organization to behaving proactively, as a fluid evolving social organism, with an innate reflex capable of acting before one is acted upon.

4 Escalating DESIRE

The word **desire** best illustrates what I see both as a universal descriptor of the human condition and of the human social product or outcome of high performing value chains. This is in particular regard to how the value chain outcome of social upgrading becomes recycled in a circular manner to affect the inbound nature, dynamics, and support activities of the same value chain over time (a more circular and holistic view of a model currently seen in more linear and modal terms.). This is predicated on the belief that as working conditions and status and choice improve, the rational nature of a job – its function as a generator of income, stability, and security for the individual or family unit – becomes shaded to increasing degree by the emotional nature of a career or profession – its role as an elevator and illuminator of the rare and differentiated individual.

I venture the idea that the outcomes of that transition and recalibration creates new pressures and tensions on the inbound aspects of a value chain, especially through its supporting activities – in particular in regard to Human Resources (remuneration, training, retention of talent) and the influence, adoption, and impact of technology in the workplace (professional/domestic overlap of technology and broader social impacts of technology on labor).

To that end **desire** for me denotes: a sense of emotional compulsion or magnetic attraction toward a thing, a feeling, an experience, state, or structure that embraces and fulfills both a functional/rational and spiritual/emotional purpose for the person accommodating and fueling the compulsion. It illuminates an aspirational duality – the profundity of human contradiction – by being both about the identity and the utility dimensions of why people pursue or engage in anything.

In this particular instance – that of value chain optimization – there are primary and secondary aspects to this **desire**. The primary aspect is that of Labor and Performance. The secondary one is of market and demand. But both are effectively formed around concepts of compatibility and mutuality. In regard to the work aspect, a compatibility of **desire** is crucial to sustained performance and engagement of the workforce. This boils down to how the particular calibration of their personal model of utility/identity "fits" with the desires (the vision, ambitions, values, traits, and rewards) of the company they work for; the presence of a shared or **mutual desire**; how the collective nature of them fit a wider ecosystem and furthermore what makes the product of this collective toil attractive and saleable on a sustained basis. This requires us to factor in the desires of the other most important stakeholder – the customer. The compatibility or fit between the desires driving the inside of the company and those driving others to purchase the product of those desires.

For example in product design terms you will find this duality of **desire** – this exceptional combination and measured calibration of the emotional and the functional – in what Italians refer to as "*La Bella Figura*" – the perfect blending of perfect form and function – harnessing visual aesthetic and exceptional performance in one perfect "it."

In *Yvon Chouinard's* biography *Let my people Go Surfing*, the founder of the trailblazing Clothing and Accessories brand Patagonia refers to a basic design truth, which I believe shows in quite brutal relief the calibrating nature and tensions of need and want; and how concepts of identity and utility fuse in one act of *production* (Chouinard 2006). He refers to the traditional Irish women of the coastal communities: who knit heavy gauge fishing jumpers from exceptionably durable and weather resistant wools. They decorate these fishing jumpers with highly individual stitching to set them apart. This is done as matter of craft, pride, and individuality: but it is also to enable their husbands and sons to be identified if they become lost at sea and subsequently washed ashore. It is the variance of the relationship between these structural elements of **desire** – utility and identity – that need measuring and understanding to a far greater degree in increasingly active and chaotic social world.

4.1 Desire as a Narrative Lever

I use the word **desire** for very specific reasons throughout this text. There may perhaps be a far better word to put in its place, but I use it for now for the three following narrative reasons: It is a device by which we can compel ourselves to continuously view, review, and reappraise value chains as models subject to and inextricably effected and affected by the fluid, reoccurring nature of human flux and flaw – the global value chain architects' version of the economist's (Keynesian) "Animal Spirits." This human turbulence, recently unfettered by the rise of technology and social networks, can create unexpected, irrational, and anomalous pressures, impacts, schisms and shifts in the otherwise perfectly engineered and orderly theories, systems, and models of the value chain.

It is a "feel" word – it forces us to compliment the engineered, analytical, and scientific nature of global value chains by elevating and illuminating the human insights and revelations that may come from the application of a more intuitive interrogation process. It also potentially sets two polarities between which we might begin to measure "animal spirits" in every stakeholder across the value chain: those of identity and utility. To that end, I would seek to create a compound index of measurements to map and measure each stakeholder across these two dimensions and thereby assess the degree of social flux – degrees of disparity and commonality – in the particular value chain.

This focuses directly on the concept of value as a desirable and enduring outcome in every link of the value chain.

4.2 The Identity and Utility Elements of Desire

Again, in Yvon Chouinard's Autobiography: *Let My People Go Surfing*, in the section covering philosophies, he touches on a study by *Dr. Thomas M Power* at the *University Of Montana* that showed that only 10–15 % of the money Americans spend on goods and services is necessary for survival (our base utility need). So a minimum of 80 % of their expenditure is to an increasing degree concerned with identity and social upgrading at an individual level.

If we were to undertake the same research in every society in which value chains exist and what is more, focus particularly on the role and impact of social networks and technology in helping define identity beyond utility, we might find that there are variations sufficient enough to demand or warrant specific actions on behalf of the company to secure a greater more resilient link at that point where these influence the value chain most directly.

Studies like the one referenced to above might suggest that in using the axis of utility and identity we would create a far more robust understanding of how the animal spirits of desire and their particular balance of identity and utility might impact on the engineered dimensions of a value chain. To add depth to what might otherwise be a rather flat piece of research, and given our objective of shared resilience, there is also the question of whether the form of utility or identity is ultimately of a positive or negative nature. Current models of desire (in regard to how we consume products and services) have a broadly negative effect on the environment, depleting natural material resources, degrading communities, creating unsustainable levels of consumption and models of aspiration that ultimately are not built to be enduring. Multiply these impacts by an ever increasing global population hosting an ever increasing global middle class, impacted upon by environmental flux and catastrophe; and this is the stuff every risk mitigation strategy in every value chain has to deal with (Fig. 3).

For this reason I would venture that the second axis should be one of positive and negative, taken in the context of the value chain and the economy in which it exists. This is not to say that we should seek for them to be either one or the other. The human truth again lies somewhere between despotic totalitarianism and the highly controlled concept of collectivism.

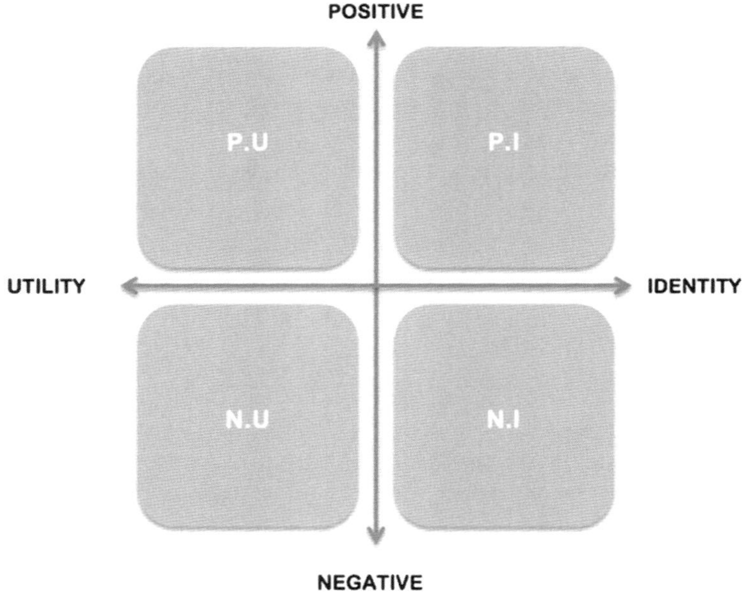

Fig. 3 The four dimensions of measurable desire (Thin Air Factory 2014) – a proprietary approach ©thinairfactoryltd2014

4.3 Mutual Desire and Shared Resilience

Having used human desire and the measurement of it across the stakeholder group to assess the social nature, traits, and behavior I believe that the relationship between that social nature and behavior and the **shared resilience** of the value chain needs to be assessed. The positive and negative readings on personal concepts of identity or utility, i.e., to what degree does the increasing realization of them deliver negative or positive impact, in the context of that person's life and the wider social communal commercial and environmental world around them. This enables us to view the resilience of the social dimensions of the value chain at every link stage. It will also enable us to better see points of tension – of similarity or dissonance – between the resilient nature of the company operationally, materially, and philosophically and those who serve it (either directly or indirectly at every stage in its value chain). A clearer understanding of the nature of **mutual desire** across the value chain will better enable us to define the most successful tenor and specification of **shared resilience** because it will be crafted against a clear understanding of which particular aspects of the company's improving and resilient nature are most compelling and motivating to every stakeholder – which are mutually compelling as part of a collective mindset.

5 Conclusion

There are three beneficial outcomes to our mining and modeling this deeper understanding and reading of the "animal spirits" of **desire**. The first is the one we originally set out. In undertaking this exploration and mapping of the "animal spirits" of **desire** across the value chain, we become better placed to recognize the mutually aligned and reoccurring models of **desire** that best suit the company (and can therefore inform far more efficient and impactful HR management strategies in regards to recruitment, training, and talent retention).

Second, by using the threads of emotional similarity and shared cognition, we are better able to identify and elevate the points of resilience (what some would call the Sustainability and CSR strategies, methodologies, and actions of the company) that are most compelling and inspiring to every stakeholder. This in turn creates a more cohesive and aligned community of shared belief and action which in turn helps to create a more resilient company: through engagement and heightened performance and commitment to a state of **mutual desire** and **shared resilience**. Third, if we secure the relationship between **mutual desire** and **shared resilience** in such a way as to be able to use the insights, dynamics, and sense of collective self as leverage to positive effect – to help drive added value into the social, systemic nature of the company to create real differentiation, focusing innovation and R&D, and inspiring employees – it will in itself have become a positive factor in improving the resilience of the company by fusing the ambitions and horizons of the systemic and social dimensions of that company.

Bibliography

Bartolini, S. (2005). *Restructuring Europe: Centre formation, system building and political structuring between the nation-state and the European Union*. Oxford: Oxford University Press. http://hdl.handle.net/1814/23882. ISBN 9780199286430.

Bartolini, A. (2011). *Business or pleasure – How social networking will reshape the P2P value chain*. http://www.basware.com/sites/default/files/upload/ardent_partners_-_business_or_pleasure_-_how_social_networking_will_reshape_p2p.pdf?p2pdl=NL_31. Accessed April 30, 2014.

Bennet, L. (2003). Communicating global activism. *Information, Communication and Society, 6*(2), 143–168.

Chouinard, Y. (2006). *Let my people go surfing*. New York: Penguin.

CSIS. (2014). *Seven revolutions*. http://csis.org/program/seven-revolutions. Accessed April 30, 2014.

Dickens, C. (1999). *A tale of two cities (reprint)*. New York: Dover.

Economist. (2014). *Coming to an office near you*. http://www.economist.com/news/leaders/21594298-effect-todays-technology-tomorrows-jobs-will-be-immenseand-no-country-ready. Accessed April 30, 2014.

Gereffi, G. & Fernandez-Stark, K. (2011). *Global value chain analysis: A primer*. http://www.cggc.duke.edu/pdfs/2011-05-31_GVC_analysis_a_primer.pdf. Accessed April 30, 2014.

Keynes, J. M. (1997). *The general theory of employment, interest, and money (reprint)*. Amherst, NY: Prometheus Books.

McAdam, D., Tarrow, S., & Tilly, C. (2001). *Dynamics of contention*. Cambridge: Cambridge University Press.

Myers, C. (2013). *Corporate social responsibility in the consumer electronics industry: A case study of apple Inc.*, Georgetown University. http://lwp.georgetown.edu/wp-content/uploads/Connor-Myers.pdf

Orenstein, M. A. (2014, January/February). Poland – From tragedy to triumph. *Foreign Affairs*.

Pezzini, M. (2012). *An emerging middle class*. OECD Observer http://www.oecdobserver.org/news/fullstory.php/aid/3681/An_emerging_middle_class.html. Accessed May 28, 2014.

Thoreau, H.D. (2007). Walden (reprint). Minneapolis: Filiquarian.

Toffler, A. (1981). *Future shock: The third wave*. New York: Bantam Book.

BASF: Measurability – A Prerequisite of Shared Value Creation in Agriculture

Markus Frank, Katharina Fischer, and Dirk Voeste

1 Sustainability in Food Production

A focus on sustainability in the food value chain has become a basic prerequisite for suppliers and consumers. There is no other industry where so many product and production characteristics are marketed as "sustainable" as in food production. Yet, how sustainability is manifested varies widely in the different value chains of the food sector. This means, for example, that there are different assumptions for fresh goods on the one hand and processed fruit and vegetables on the other. In an environment with many sustainability assessments, the quest for widely accepted indicator sets and their measurability is therefore unavoidable. The successful development and marketing of sustainable solutions is the core of the BASF company strategy and is therefore also a theme for the Agricultural Solutions business segment. With AgBalance™, BASF has attempted to facilitate a realistic and precise depiction of the food value chain. Measuring sustainability can be a central key to steady improvements toward sustainable agriculture. It is therefore an essential requirement that it succeeds in translating results from complicated life-cycle analyses into farmers' everyday reality and to derive specific recommendations for action from this.

To what can the trend toward the introduction of higher sustainability standards be attributed? Although some observers believe that the later stages of the value chain drive the issue strongly (Giovannucci and Ponte 2005), other authors consider the active role to lie with end consumers and their increasingly troubled trust in the food industry (cf. Spaargen and Oosterveer 2010; Willmroth 2011), and others lean

M. Frank • K. Fischer (✉)
BASF SE, 67117 Limburgerhof, Germany
e-mail: markus.frank@basf.com; katharina.fischer@basf.com

D. Voeste
BASF SE, Carl-Bosch-Str. 38, 67056 Ludwigshafen, Germany
e-mail: dirk.voeste@basf.com

M. D'heur (ed.), *Sustainable Value Chain Management*, CSR, Sustainability, Ethics & Governance, DOI 10.1007/978-3-319-12142-0_16

to the side of the food corporations and supermarket chains (cf. Traill 2006). This development certainly imposes increasing demands on all levels of the chain – from the supply of agricultural resources to plant and animal production to food processing and marketing.

Global mega-trends will bring increasing challenges with them. A dramatically increasing global population of approx. nine billion people by 2050 (UNFPA 2012, p. 9) and a rapidly growing middle class in developing countries such as China and India with a demand for high-quality and nutritious food, with an additional increased appetite for meat and dairy produce (cf. Oosterveer and Sonnenfeld 2012; Gerosa and Skoet 2012, p. 30), mean further pressure on agricultural production, based on the poor energy conversion in the production of meat. In light of this, sustainable agriculture and a sustainable supply chain are imperative. One stage in the chain is not sufficient on its own; we need a joint effort across all stages.

2 Shared Value Creation as the Key to a More Sustainable Food Chain

Numerous existing sustainability programs often have their roots in Corporate Social Responsibility (CSR) activities, which are not integrated into the core business of the company (Porter and Kramer 2011). However, companies such as Unilever and Nestlé have recognized that these types of initiatives not only represent an investment in their brand and reputation, but they can also help to effectively resolve of one of their primary problem areas: supply chain security. Lawrence and Burch (2007) and Frank (2010) report that food producers and supermarkets increasingly rely on cooperation with suppliers and long-term contacts for a secure supply of high-quality products, particularly in product categories with high potential for differentiation. These contacts may include multiple layers of the value chain concerned, as they strive for considerably closer cooperation with the producers (Nestlé 2013).

Coffee represents the first industry where the step toward shared value creation in the sense of the "shared value" concept defined by Porter and Kramer has been taken as a result of CSR programs. For example, in 2004 Starbucks launched the C.A.F.E (Coffee and Farmer Equity) project which set out sustainability criteria for the whole coffee supply chain. Producers receive a bonus depending on the sustainability value of their production method in the areas of economy, environment, and society (MacDonald 2007). Since then, Starbucks has also measured the influence of the C.A.F.E practices on biodiversity (Starbucks 2013a, b, c). Today, the company sources most of its coffee – around 200 million pounds per year – from plantations adhering to C.A.F.E. practices. In addition to this, Starbucks has, in cooperation with various nongovernmental organizations, introduced adjusted guidelines in accordance with the C.A.F.E. standards for further ingredients in their products, such as cocoa and tea (Starbucks 2013a). Shared value creation via sustainable processes in the whole supply chain has become an integral part of Starbucks' brand promise (MacDonald 2007). There are also examples of shared value creation across the whole supply chain outside of product categories such as luxury foods, which by their very nature have great potential for differentiation and accordingly boast a promise of value.

2.1 Example 1: Ben & Jerry's

For example, the Ben & Jerry's brand launched the "sustainable milk" initiative in Europe in 2003 with their Dutch partner Cono Kaasmakers. In order to introduce higher sustainability standards for suppliers, a program called "Caring Dairy" was launched which aimed for binding standards for various sustainability factors and continuous improvements (van Calker et al. 2005). Besides animal welfare and husbandry, the focus of the initiative was primarily on various economic key figures and environmental parameters. Farmers, who produce in accordance with higher sustainability standards, receive a premium price of EUR 0.50 per 100 kg milk. Cono Kaasmakers pays an additional bonus for providing pastures for cows to graze on.

2.2 Example 2: Unilever

Five large Brazilian tomato growers (total production: 440,000 tons per year) in the state of Goiás have been cooperating with Unilever Brazil for 10 years to establish sustainable production technology. Unilever's technical consultants visit the agricultural operations on a weekly basis and advise the managers on issues such as conservational tillage, erosion protection, irrigation, organic plant protection, and waste prevention. In doing so, the tomato growers benefit from lower production costs and an increased yield.

The logic of this shared value creation includes an attractive business model for all stages of the value chain – from the farmer to the manufacturer to the supermarket. According to Hamprecht et al. (2005), this business model is most lacking in global value chains for soft commodities such as cereals, oilseeds, etc. It is not automatically possible to differentiate this. Accordingly, there are also problems for companies such as Unilever in implementing "sustainable sourcing" in these soft commodities (Unilever 2013). All the initiatives described pose the question, which areas of sustainability – environment, economy, and society – and which sustainability indicators form the basis of shared value creation. Additionally, there is a need to check the sustainability performance in terms of the selected indicators. This verification requires sustainability to be measurable. Complete transparency concerning the scope and premises of the various measurement methods is essential, in order to create transferable and practice-orientated results.

3 Assessing Sustainability

Measurability requires indicators. These reflect different aspects of sustainability in different supply chains. For example, "fair trade" is a typical sustainability indicator in the coffee and tea supply chain and is also seen in certain tropical fruits such as pineapples, bananas, and mangoes. Conversely, in soft commodities such as cereals or rapeseed, the carbon footprint is more often considered an important factor. Furthermore, "organic" or "regional" production and consumption is often simplistically presented as equal to sustainability (Ernst & Young AG 2007). Although this apparent equality is easily understandable for the consumers, it often disregards the conflicts of interest inherent in sustainability. These include, for example, a potential overestimation of the influence of transport and logistics as well as the underestimation of storage in refrigerated warehouses, etc. (de Haes and de Snoo 1997). This therefore almost inevitably raises the issue of widely accepted sets of indicators and their measurability.

An important initiative for the development of a broader set of initiatives is represented by the Keystone Initiative in the USA and its "market" program. In "Field to Market," a broad group of stakeholders – seed companies, agricultural producers, processors, trade, and nongovernmental organizations – joined together in 2006 to establish sustainability standards in the American agricultural supply chains, particularly those of soya, maize, and cotton, and to make them measurable. The aim of "Field to Market" is to identify relevant sustainability criteria for the relevant supply chains, which acknowledge the potential role of diverse technologies (e.g., genetically modified seed stocks) and support farmers in aligning themselves with these criteria (cf. The Keystone Center 2013; Constance 2010; Field to Market 2013a, b). The aim here is to roll out widespread general minimum standards and thereby raise the baseline to some degree. The priority is heightening the sensitivity of farmers to central sustainability issues and thereby facilitating their implementation in their businesses. The "Fieldprint Calculator" is a central tool of the "Keystone" initiative. Farmers can use it as an aid to check their overall sustainability in terms of energy, soil, and water consumption, and use and their effect on climate. In doing so, the "Fieldprint Calculator" compares the method and the performance of individual farmers with the average in their region and in their state. Access to the "Fieldprint Calculator" is free and it is very easy to use (Field to Market, Fieldprint Calculator 2013). The aim is to increase the awareness of farmers on the issue of sustainable production by comparison with a benchmark.

Whilst "Field to Market" was predominantly driven by American farmers' associations, the RISE initiative and the SAI (Sustainable Agriculture Initiative) platform began with food companies, first and foremost with Nestlé. RISE (Response-Inducing Sustainability Evaluation) is an indicator-based method for assessing the sustainability of an agricultural business. RISE has been developed since 2000 by the Bern University of Applied Sciences with strong support from Nestlé. An updated version, RISE 2.0, was released in 2011. RISE aims to increase the sustainability of agricultural production by "circulating and rooting sustainable philosophy and practice" (HAFL Bern University of Applied Sciences, School of Agricultural, Forest and Food Sciences 2012).

The target group for RISE covers all stakeholders in agriculture, society, and economy who share this vision. Users of this method are most notably employees in agricultural consultancy, training, and development coordination. RISE was and is being developed gradually through joint projects with partners including Nestlé, the Research Institute of Organic Agriculture (FiBL), Syngenta, the Federal Office for Agriculture (BLW), and the German International Cooperation (Geselllschaft für Internationale Zusammenarbeit, GIZ) in the form of cooperative projects, commissioned work, and training. Users also include universities in Germany and abroad.

The model covers 12 indicators from the areas of economy, environment, and society. The indicators are energy, water, soil, biodiversity, emissions, pesticides, waste, cash flow, profit, investments, local economy, and the social situation of the business (Häni et al. 2003). Like "Field to Market," RISE is not a control method or certification. Users commit themselves to complying with a code of conduct, where all collected and processed data is treated with strict confidentiality. The RISE method has already been implemented in over 1,200 companies in 36 countries. The types of businesses analyzed include dairy, vegetable and arable farms, mixed farms, coffee, cocoa, and tea plantations, small African operations, and nomadic herdsmen (HAFL Bern University of Applied Sciences, School of Agricultural, Forest and Food Sciences 2012). While the strength of RISE lies in its ability to create a truly detailed analysis of an agricultural business, one weakness is the high costs involved in data collection and the work carried out.

In order to be able to evaluate suppliers more quickly and economically, Nestlé, Unilever, and Danone joined forces in 2002 to launch the SAI platform. SAI relies heavily on simplifying and focusing the sustainability indicators to enable broader coverage of agricultural suppliers (Hamprecht et al. 2005; SAI Platform 2010). Today the SAI platform has over 40 members solely from the food industry. In 2009, the SAI platform issued the first set of sustainability indicators, covering issues such as energy consumption, humus balance, and greenhouse gas emissions.

In addition to these initiatives driven primarily by commercial companies, there are further approaches which deal with certain aspects of sustainable agriculture. This includes, for example, INDIGO (Bockstaller et al. 1997), KSNL, KTBL (Association for Technology and Structures in Agriculture (KTBL) 2008), REPRO (Küstermann et al. 2008) and on a national and international level, IRENA (Indicator Reporting on the Integration of Environmental Concerns into Agriculture Policy System) (European Environment Agency 2005), and the OECD approach (OECD 1993, 2001). These indicator systems generally focus on the practical applications for agricultural operations. In doing so, these methods do not make use of the so-called life-cycle analysis, where the whole footprint of a product or a process can be mapped from the "cradle," i.e., the harvesting of the raw materials, potentially right to the "grave," namely the consumption of the food and the disposal of the packaging. As a result, they cannot create a realistic image of the share that various stages add to the value chain and tend to overestimate the influence of individual factors on the whole footprint (de Haes and de Snoo

1997). Furthermore, these methods do not logically address all three dimensions of sustainability. This is a task which BASF faced when developing the "AgBalance" in 2011: A sustainability method which guarantees the breadth and flexibility of a life-cycle analysis whilst also involving an acceptable balance between data requirements and scientific depth.

4 AgBalance™ as a Contribution Toward a More Sustainable Value Chain

The successful development and marketing of sustainable solutions is the core of the BASF company strategy and is therefore also a central theme for the Agricultural Solutions business segment. A fact-based sustainability analysis is indispensable to be able to achieve this goal, in order to critically assess BASF technologies and solution approaches as well as to support producers in their positioning toward their customers.

With AgBalance™, BASF has attempted to facilitate a realistic and precise depiction of the food value chain, in order to be able to derive specific recommendations for action from this. Against the backdrop of almost 20 years of experience in life-cycle analysis, this scientific approach was also selected for observing the agricultural chain. The result is a sustainability method which includes ecological, economic, and social aspects in its analysis and which can – with simple adaptations to the local circumstances – be used around the world (Frank et al. 2012). Depending on the issue, it is equally possible to observe a certain technology or production system as it is an entire value chain. In doing so, these methods make use of the so-called life-cycle analysis, where the whole footprint of a product or a process can be mapped from the "cradle to the grave." As agriculture is one of the world's most fully globalized markets, AgBalance must be broad enough to cover the most important sustainability concept of all regions, while also promoting a practical application adapted to local conditions (Frank 2011). The reality of agriculture is depicted in AgBalance using 69 sustainability indicators. The project team at BASF developed the indicators in cooperation with academics, nongovernmental organizations, political and consumer organizations in the EU, the US, and Brazil and gave them appropriate weighting factors. That last step is necessary to be able to derive an overall statement from a multitude of indicators and to therefore offer simple representations on a scientific basis.

In order to ensure the quality of sustainability profiles from the outset, a broad range of stakeholder groups have participated in developing and weighting these criteria. This is and will continue to be an irreplaceable part of the development. Amongst others, the representative consumer surveys on the significance of individual factors in the society of multiple countries (including Germany, France, Great Britain, USA) influenced the weighting.

Using official statistics and established scientific data sources (e.g., the IUCN biodiversity indicators, FAO statistics, etc.) as well as field studies, the 69 sustainability indicators are calculated and their results expressed in a relative form, which brings light to the specific differences between two production systems. The indicators are divided into the following 16 categories within AgBalance™:

Environment
- Biodiversity
- Soil
- Land Use
- Energy Consumption
- Resource Consumption
- Water Use
- Emissions
- Ecotoxicity Potential

Economy
- Variable Costs
- Fixed Costs
- Macroeconomy

Society
- Farmer
- Consumer
- Local Community
- International Community
- Future Generations

AgBalance™ therefore delivers results that enable farmers, the food industry, politicians, and society to objectively evaluate processes in terms of their sustainability profile. In doing so, a vast amount of information on individual factors can be ascertained in addition to overall statements on the sustainability of agricultural practices (e.g., ploughing). AgBalance was finalized in mid 2011. In September 2011, the methodology was given independent assurance by the global test providers TÜV SÜD, DNV Business Assurance, and NSF International.

AgBalance™ can be used to map an individual farm or the whole agricultural sector in one region, for example. The focus can either be on the agricultural production system alone or on the processes that have established themselves downstream in the value chain, such as logistics or processing.

A corresponding case study with the holding company SLC Agricola in Brazil involved an internal benchmarking of two large farms, each with over 10,000 ha, to identify the central sustainability drivers for their crop rotation consisting of soya, maize, and cotton and to derive follow-up opportunities for their continuous improvement from this (Frank et al. 2012). An average cultivated hectare for each of the two farms, Panorama (Bahia state) and Planalto (Mato Grosso do Sul state) were compared on the basis of the operation data from the 2009/2010 season. The indicators from all three sustainability dimensions – environment, economy, and society – were investigated using a holistic approach over a section of the life cycle that starts with the raw materials used in production (the "cradle" of the process, for example, phosphorus extraction or oil production) and ends with the delivery of the harvested goods at the nearest port.

The analysis shows that the Planalto farm is substantially more sustainable than the Panorama farm, which is largely due to better results in the economy and environment categories (Fig. 1). With the launch of the AgBalance method in 2011, BASF also introduced a demonstration tool for the visualization of the results (Fig. 2).

Fig. 1 Relative sustainability index of the two farms Panorama and Planalto. Planalto achieved a 40 % better result

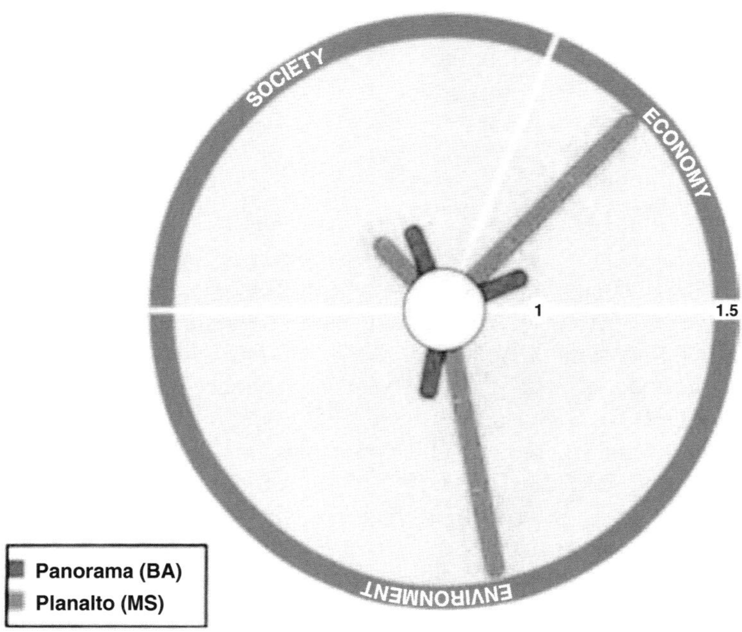

Fig. 2 Representation of the sustainability index in terms of the three dimensions of sustainability. The length of the bar indicates higher sustainability. Each time the worse alternative is normalized to the value 1

The logic of AgBalance now allows further, detailed analyses of the individual indicator systems that originally provided these results. Figure 3 shows the comparison of the farms investigated on the level of the individual indicator categories in the three dimensions of sustainability.

Whilst Panorama achieved better results in two social indicator categories, namely in the "Further education" (in the category "Future Generations" and the working conditions in the upstream chain (in the category "International Community") indicators, Planalto proves to be the benchmark in all other categories. The most important drivers in terms of economy turned out to be an improved cost situation and an increased profit (in the category "macroeconomy"). Both were partly due to a more efficient use of resources and a better logistical starting point (in contrast to Panorama, Planalto had a rail link). In terms of the environment, the most important driver turned out to be a slight in-balance for nitrogen and above all, phosphorous in the soil. This increased use of fertilizer is reflected in the higher energy and resource consumptions by Panorama, which is due to the production and transportation of the fertilizer. In respect to the emissions from production and the field, the increased use of fertilizer plays a central role in Panorama. Furthermore, the "Soil index" category reflects the unbalanced nutrient content in the soil. A second important factor in the Environment category was the pesticide regime. Despite the low volume, with not more than 2 % of the total amount being from pesticides, the use of organophosphates as soil insecticides in the production of cotton leads to poorer results in the "Ecotoxicity" category.

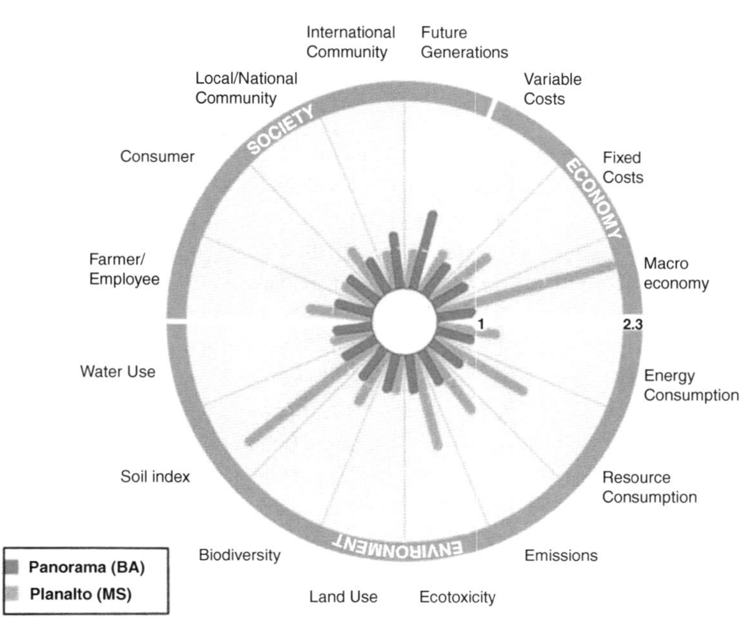

Fig. 3 Representation of the sustainability index in terms of the individual indicator categories in AgBalance

According to initial calculations, the optimization of the fertilization regime in Panorama could lead to savings of almost 15 million kWh of energy (this corresponds to the energy use of roughly 2,000 households in Brazil) in addition to substantial cost savings. The CO_2 equivalents saved using AgBalance™ amount to almost 8,000 tons per year. These results, together with the additional findings on pesticides, can serve as the starting point for a continuous improvement program at SLC Agricola. With its knowledge base, BASF supports a suitable product portfolio throughout the whole life cycle and works toward creating common solutions toward greater sustainability.

5 Sustainability of Shared Value Creation in Agriculture: Where Is the Trend Heading?

Measuring sustainability can be a central key to steady improvements toward more sustainable agriculture. This should have been made clear by the examples given above. The essential prerequisite for this is its success in translating results from complicated life-cycle analyses into farmers' everyday reality for farmers and to derive specific recommendations for action from this.

In light of this, the central requirement of the measurement systems destined for implementation into agriculture, such as "Field to Market," SAI or AgBalance, is to maintain the correct balance between requirements regarding data and scientific depth on the one hand and practical relevance on the other. Acceptance by farmers ultimately decides the success or failure of such systems, with the aim of establishing a basis for shared value creation. Furthermore, there is a demand to be able to map the whole value chain.

However, it has been shown numerous times that agriculture can have a large share of the whole sustainability profile of various value chains in the food and animal feed industries. At the same time, logistics, transport, processing, and consumption can play a substantial role in this. In relation to the entire value chain, SET (Sustainability-Eco-Efficiency-Transparency), the BASF initiative for applied sustainability, advises clients from the food and animal feed industry in Europe, North and South America, and Asia. SET applies sustainability analyses and strategies to specific client products and makes product sustainability measurable. From resource use to consumption by consumers, SET covers optimization potential along the entire value chain. Progress becomes transparent and communicable. This approach should help BASF clients to differentiate products which make a larger contribution to sustainable development from the competition. In 2012, for example, BASF analyzed the CO_2 balance for veal and beef products with the client Westfleisch, supporting them in improving the sustainability of their meat production along the whole value chain. With the combination of AgBalance and SET, BASF is the only service provider in a position to completely and holistically map a value chain in the food sector.

However, the methodical limits must not be disregarded. Each method is only as good as the quality of the input data. Careful quality assurance and control is essential. With the broad and consistent inclusion of stakeholders in the creation of the studies – from further stages of the chain, associations, nongovernmental organizations, and academics to regulators and politics – the relevance of the results depends on the sustainability measurements for shared value creation along the supply chain. Moreover, not all aspects can be logically integrated into a life-cycle analysis. This particularly includes ethical and some social aspects such as the protection of indigenous rights or safeguarding against child labor. These aspects cannot be treated as one conflicting goal among many, given that they are unacceptable or even illegal practices. These topics should be evaluated separately and assessed by relevant stakeholders and – if possible – a solution or certain counter measures should be introduced. Finally, the topic of sustainability in agriculture has an increasingly strong regional character, despite global supply chains. Implementation in the agricultural business takes place solely on a local level. Both the measurement method and the subsequent implementation plan must therefore be flexible enough to facilitate local solutions.

In light of the various measurement methods and principles that have been partially presented above, creating transparency for the common principles and indicators as well as the differences in the various systems must be of particular interest to the whole food value chain. Only this will allow a clear relationship to be found between the results from different methods and allow the measurement results to really be used in a complementary manner.

BASF took the first step in this direction in March 2013 with the international symposium "Perspectives for Agriculture – Progress through Sustainability Assessment." Upon this foundation, BASF will create transparency in a future initiative with stakeholders from the food industry, farmers and also include representatives of relevant interest groups. This will be an additional important step in ensuring that sustainable thinking is more strongly rooted in agriculture, an issue which must be addressed immediately in view of the urgent global problems.

Bibliography

Bockstaller, C., Girardin, P., & Van der Werf, H. (1997). Use of agro-ecological indicators for the evaluation. *European Journal of Agronomy, 7*, 261–270.

Constance, D. H. (2010). Sustainable agriculture in the United States: A critical examination of a contested process. *Sustainability, 2*(1), 48–72.

de Haes, H. A. U., & de Snoo, G. R. (1997). The agro-production chain: Environmental management in the agricultural production-consumption chain. *The International Journal of Life Cycle Assessment, 2*(1), 33–38.

Ernst & Young AG. (2007). *LOHAS. Lifestyle of health and sustainability*. Ernst & Young AG.

European Environment Agency. (2005). *Agriculture and environment in EU-15: The IRENA indicator report*. Kopenhagen: European Environment Agency.

Field to Market. (2013a). *Field to market*. http://www.fieldtomarket.org/about-us/. Accessed 23 April 2013.

Field to Market. (2013b). *Fieldprint calculator*. http://www.fieldtomarket.org/fieldprint-calculator/. Accessed 23 April 2013.

Frank, M. (2010). *BASF's expansion into biopesticides: Opportunities and challenges. A case study*. MBA Dissertation, School of Management, University of Surrey.

Frank, M. (2011). AgBalance: A clearer view of agricultural sustainability. In von B. Hallier (Ed.), *Von der Krise zur Kompetenz* (pp. 140–141). Bonn: Orgainvent GmbH.

Frank, M., Schöneboom, J., Saling, P., & Gipmans, M. (2012). Holistic sustainability assessment of winter oilseed rape production using the AgBalance method – An example of 'sustainable intensification'? *Proceedings of the 8th International Conference on Life Cycle Assessment in the Agri-Food Sector (LCA Food 2012)* (pp. 62–67). Saint Malo: INRA, Rennes.

Gerosa, S., & Skoet, J. (2012). *Milk availability: Trends in production and demand and medium term outlook*. ESA Working Paper 12-01, Agricultural Development Economics Division, Food and Agriculture Organization of the United Nations.

Giovannucci, D., & Ponte, S. (2005). Standards as a new from of social contract? Sustainability initiatives in the coffee industry. *Food Policy, 30*, 284–301. HAFL Berner Fachhochschule.

Hochschule für Agrar-, Forst- und Lebensmittelwissenschaften. (2012, September 14). *RISE (Response-Inducing Sustainability Evaluation) version 2.0: Maßnahmenorientierte Nachhaltigkeitsanalyse der Agrarproduktion auf Betriebsebene*. Zollikofen.

Hamprecht, J., Corsten, D., Noll, M., & Meier, E. (2005). Controlling the sustainability of food supply chains. *Supply Chain Management: An International Journal, 10*(1), 7–10.

Häni, F., Braga, F., Stämpfli, A., Keller, T., Fischer, M., & Porsche, H. (2003). RISE, a tool for holistic sustainability assessment at the farm level. *International Food and Agribusiness Management Review, 6*(4), 78–90.

Kuratorium für Technik und Bauwesen in der Landwirtschaft e. V. (KTBL). (2008). *Kriteriensystem nachhaltige Landwirtschaft (KSNL)*. Darmstadt: KTBL.

Küstermann, B., Kainz, M., & Hülsbergen, K. J. (2008). Modeling carbon cycles and estimation of greenhouse gas emissions from organic and conventional farming systems. *Renewable Agriculture and Food Systems, 23*(1), 38–52.

Lawrence, G., & Burch, D. (2007). Understanding supermarkets and agri-food supply chains. In D. von Burch & G. Lawrence (Eds.), *Supermarkets and agri-food supply chains. Transformation in the production and consumption of foods* (pp. 1–28). Cheltenham: Edward Elgar Publishing Limited.

MacDonald, K. (2007). Globalising justice within coffee supply chains? Fair trade, starbucks and the transformation of supply chain governance. *Third World Quarterly, 28*(4), 793–812.

Nestlé. (2013). *Nestlé in society: Creating shared value and meeting our commitments*. http://www.nestle.com/csv. Accessed 23 April 2013.

OECD. (2001). *Environmental indicators for agriculture. Methods and results* (Bd. 3). Paris: OECD.

OECD. (1993). *OECD core set of indicators for environmental performance reviews*. Environment Monographs 83. Paris: OECD.

Oosterveer, P., & Sonnenfeld, D. (2012). *Food, globalization and sustainability*. Milton Park: Earthscan.

Porter, M. E., & Kramer, M. R. (2011, Januar–Februar). The big idea: Creating shared value: Rethinking capitalism. *Harvard Business Review*, 1–17.

SAI Platform. (2010). *SAI platform: Who we are*. http://www.saiplatform.org/about-us/who-weare-2. Accessed 23 April 2013.

Smith, B. G. (2008). Developing sustainable food supply chains. *Philosophical Transactions of the Royal Society, Biological Sciences, 363*, 849–861.

Spaargen, G., & Oosterveer, P. (2010). Citizen-consumers as agents of change in globalizing modernity: The case of sustainable consumption. *Sustainability, 2*(7), 1887–1908.

Starbucks Corporation. (2013a). *Kaffee aus nachhaltigem Anbau*. http://www.starbucks.de/responsibility/sourcing/coffee. Accessed 23 April 2013.

Starbucks Corporation. (2013b). *Kakao aus nachhaltigem Anbau*. http://www.starbucks.de/responsibility/sourcing/cocoa. Accessed 23 April 2013.

Starbucks Corporation. (2013c). *Tee aus nachhaltigem Anbau*. http://www.starbucks.de/responsibility/sourcing/tea. Accessed 23 April 2013.

The Keystone Center. (2013). *Field to market: The keystone alliance for sustainable agriculture*. https://www.keystone.org/policy-initiatives-center-for-science-a-public-policy/environment/fieldto-market.html. Accessed 23 April 2013.

Traill, W. B. (2006). The rapid rise of supermarkets? *Development Policy Review, 24*(2), 163–174.

UNFPA. (2012). *Weltbevölkerungsbericht 2012*. Hannover: UNFPA, Bevölkerungsfonds der Vereinten Nationen.

Unilever. (2013). *Nachhaltige Beschaffung*. http://www.unilever.de/sustainable-living/sustainablesourcing/. Accessed 23 April 2013.

van Calker, K., Hooch Antink, R., Beldman, A., & Mauser, A. (2005). Caring dairy: A sustainable dairy farming initiative in Europe. *Proceedings of the 15th Congress of Developing Entrepreneurship Abilities to Feed the World in a Sustainable Way* (pp. 81–88). Campinas: IFMA.

Vermeir, I., & Verbeke, W. (2006). Sustainable food consumption: Exploring the consumer 'attitude-behavioral intention' gap. *Journal of Agricultural and Environmental Ethics, 19*, 169–194.

Willmroth, J. (2011, Oktober 25). *Lebensmittelindustrie: Kampf gegen die Vertrauenskrise*. http://www.wiwo.de/unternehmen/handel/lebensmittelindustrie-kampf-gegen-die-vertrauenskrise/5724346.html. Accessed 23 April 2012.

Telling the Backstory: Transparency in Global Value Chains

Georg Lahme and Volker Klenk

1 The Role of Transparency in Global Value Chain Management

Implementing sustainability principles into existing supply chains is a considerable endeavor. However, from day one, it can be used as an asset in corporate and brand positioning, to build trust, and to protect against reputational damage. This requires the careful planning of communication strategies – and ideally, communication departments are actively involved in the planning of these strategies – to enable dialogue with stakeholders to ensure strong, long-term relationships.

When companies put sustainability at the center of their positioning, they begin an important yet challenging journey. Companies such as Apple, Adidas, C&A, Nike, and KiK have recently undertaken this long-term process of trying to change and manage an international supply chain. One particular challenge in these processes is handling the increased pressure from various stakeholder groups demanding more transparency.

Implementing sustainable production standards, quality management processes, and fair working conditions in a textile factory in Bangladesh, for example, is a highly challenging management task. To supervise adherence to those standards and resolve any drawbacks, as well as to make them transparent and credible for external audiences, is the correspondent and yet challenging task for a communication department.

Early sustainability communication paradigms assumed that companies could only position themselves as sustainable in an active and credible way once implementation had been completed. That no longer holds true. External stakeholders expect companies to actively show their commitment to establish sustainability in

G. Lahme (✉) • V. Klenk
Klenk & Hoursch AG, Walther-von-Cronberg-Platz 2, 60594 Frankfurt am Main, Germany
e-mail: georg.lahme@klenkhoursch.de

M. D'heur (ed.), *Sustainable Value Chain Management*, CSR, Sustainability, Ethics & Governance, DOI 10.1007/978-3-319-12142-0_17

all processes of the value chain and to engage in a dialogue about sustainability hot spots and ongoing achievements in critical areas. This has to be an open dialogue, a long-term and problem-oriented conversation. Companies that willingly show areas of weakness, accept criticism, and credibly demonstrate that they are prepared to work on the elimination of problems, build trust. Those that cover up their weaknesses, behave non-transparently and are not willing to work on the solutions to the problems they cause, will be confronted with trust crises.

Implementing sustainability within broad and global product portfolios that are heterogeneous and demand-driven while undergoing fast innovation cycles, is a great challenge. Communication specialists who support such processes have to withhold the reflex to promise too much too fast. Communication and marketing departments think in campaigns, messages, and labels that are aimed directly at the heart and mind of the customer. When it comes to sustainability goals, these are often communicated too quickly – and are overly positive about timelines – with the effect that corporate reality cannot catch up to the sustainability messaging and positioning.

Companies that set out to create transparency about their sustainability objectives and initiatives, soon come to the realization that in all dimensions of sustainability, the journey remains the reward. The weaknesses of supply chains are regularly attacked by angry consumers, critical stakeholders, and investigative media reports. These outbreaks mainly affect big corporations, branded companies, and their suppliers. It also affects those which have set out with significant investments to credibly implement sustainability. For communications, that means to never communicate objectives and wishful thinking as facts, especially regarding sustainability. Instead, always stay grounded and explicitly illustrate the journey that is still ahead. Attempts to practice "green-washing" are counterproductive. Companies become transparent and permeable through well-organized stakeholders, digital media, and the countless external interfaces that every organization has.

2 Demands of Customers and Consumers on the Visibility of Global Value Chains

2.1 Transparency and Accountability

Consumers are increasingly demanding more transparency and accountability from companies regarding environmentally and socially acceptable actions. 82 % of Germans would like to see organizations become more transparent and 80 % even support stricter legal regulations to that effect. These are the results of our transparency survey from 2012 (Klenk and Hoursch 2012). "By now, customers perceive transparency as a basic right," say journalists Christiane Langrock-Kögel

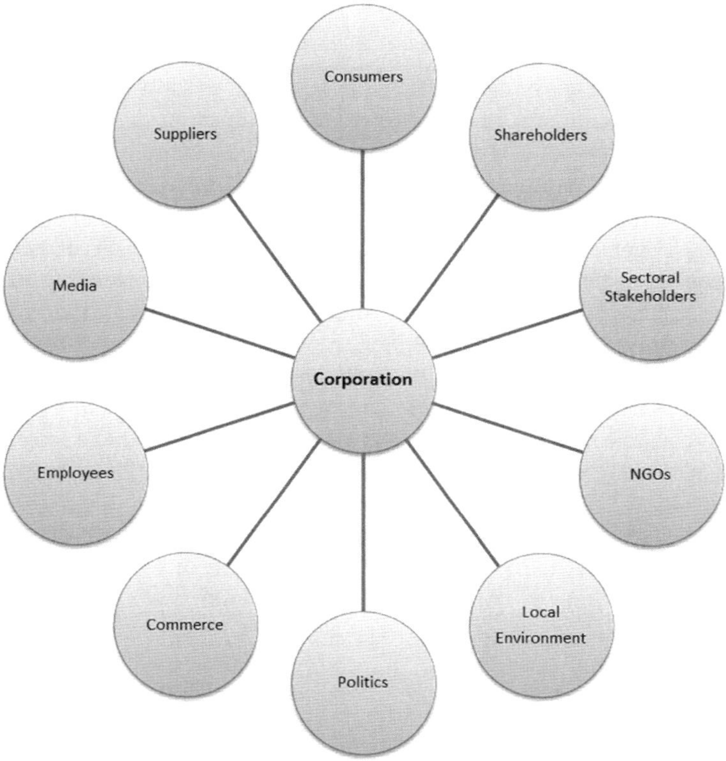

Fig. 1 A company's stakeholders: Different groups of interests articulate their demands toward companies and organizations in a more professional manner. *Source*: Own research

and Marc Winkelmann. "They are no longer compliant receivers, but aware of their own power" (Langrock-Kögel and Winkelmann 2012).

Single critical stakeholders – who used to be on their own in the past – form coalitions and networks via Internet in order to critically analyze and assess the global supply chains of companies (see Fig. 1).

If a company offends the transparency demands of stakeholders, this lack of transparency is often made public along with a call for change; pressure can also be upped with purchase boycotts. Corporations live in glasshouses nowadays. Every corporate action is public. Any company which does not meet the expectations of stakeholders runs the risk of accusation – with negative effects for reputation, turnover, sales, and share price (Klenk and Hanke 2009).

2.2 Rising Consumer Awareness of Supply Chains

In the past few years, consumer awareness in regard to the supply chain of a finished product has been on a constant rise. As various studies show, this is especially true for industries like foods, textiles, and electronics. More and more customers want to know about the origins of the clothes they wear, where the vegetables they consume were grown and under what conditions. Also, they ask for information about how much money workers get and where the "rare earths" used for manufacturing smartphones are sourced. Such questions cannot always be thoroughly answered. However, stakeholder demands for transparency and information within all areas of global value chains are already reality today.

People are increasingly aware of the influence that big corporations have when it comes to the working and living conditions found in their supply chains. A growing number of consumers expect corporations to use their power to make sure suppliers adhere to environmental standards and fair working conditions. If corporations fail to do this, a loss of image may occur, as has happened to some of the largest global corporations. Sporting goods manufacturers such as Nike and Adidas have been severely criticized by non governmental organizations for tolerating child labor; Apple had to account for outrageous working conditions among suppliers like Foxconn; online-retailer Amazon has been attacked for their treatment of contract workers.

3 Approaches to Build Visibility and Transparency

Corporations have to open up and realign their corporate actions in order to assert themselves on the market

Regarding the implementation of transparency, three types of companies can be distinguished:

1. Deniers who only carry out their duties in terms of reporting
2. Followers who consider transparency as important and who are relatively transparent in some sections
3. Confident corporations that have already incorporated transparency as part of their business strategy and equity-story, and thus have entered into an open dialogue with their stakeholders.

Most listed companies belong in the "followers" category. By now almost every DAX30 company or Fortune500 corporation has implemented triple bottom-line reporting and accordingly established transparency within all economic, social, and ecological aspects of their corporate actions.

The tendency toward more transparency, and especially within global value chains, is not going to reverse course. Corporations must act. No brand manufacturer that wants to be successful can avoid examining the topic of (voluntary) transparency. Peter Eigen, founder of Transparency International, states, "The future of business culture and added value lies in an active, voluntary openness, since topics such as trust and transparency are not going to disappear from stakeholders' agendas" (Eigen 2009).

Companies need to find out what their environment wants to know about the origin of their products and the environmental and societal impacts of their products. Also, they must proactively enter into a dialogue with stakeholders including critical NGOs and develop a transparency strategy. They need to learn how to reduce risks und to utilize positioning messages.

Some corporations have already adjusted to these new realities – the British retailer Tesco, for example, asks its suppliers for comprehensive accountability in terms of risk management, supply chain management, and their ecological footprint. Unsatisfactory standards and non-transparency lead to dropping products from their shelves. However, many companies still ignore the development toward more corporate transparency.

Transparency has to be credibly incorporated into the business and sustainability strategy

When companies want to become more transparent and sustainable, those target dimensions have to be strongly linked with changes on a strategic and operational level. Transparency is not just a topic for communications. Transparency has to be deeply integrated into the business strategy and culture. Hence, it is not communicating about transparency, but rather, it is about being transparent. Regarding the transparency process, communication departments help drive the process forward.

4 Operationalizing Voluntary Transparency to Reduce Risks and Enhance Company Value

In order to establish the requirements for targeted, transparent, and credible sustainability communications, four basic steps must be taken:

Step 1: Assess, Evaluate and Balance Opportunities and Risks

Opportunities and risks have to be assessed, evaluated, and balanced. Questions have to be clarified, such as: What are direct competitors doing and what are their results? Which information should not leave the company under any circumstances? Furthermore, analyses of potential competitive advantages through more transparency and risk scenarios with increasing pressure on the company through politics, NGOs, consumers, employees, or others have to be carried out. On such a basis, advice can be given in terms of how transparency can increase advantages or at least how it may be used to mitigate damage.

Companies that work together with a number of suppliers and sub-suppliers primarily have to uncover potentially critical elements within the supply chain and then effectively supervise those on a continuous basis. Only through such mechanisms they can establish the conditions for credible and transparent communications.

If it is still possible today that organic eggs are introduced into the market without actually being organic, how difficult will it be to keep track of international diversified supply chains and the actions of suppliers and sub-suppliers in emerging countries? Can this be achieved by internal management? Are certification processes and regular audits sufficient? Should local organizations be included that can comprehensibly review whether working conditions and environmental standards comply within the sourcing of raw materials and manufacturing? The answers to these questions are diverse. Some form industry initiatives, while others work together with local stakeholder organizations. The fashion retail company C&A developed a code of conduct for dealing with suppliers in 1996. The code regulates supplier relationships, working conditions, and environmental aspects, always in compliance with the general principle of "fair and honest acting". In 1996, a new entity was formed, the Service Organization for Compliance Audit Management (SOCAM), to control the supply chain as well as the compliance with the code of conduct. 17 years later, during the first months of 2013, C&A came under pressure regarding security standards and working conditions among its suppliers in Asia (Hecker 2012). This shows that the establishment of sustainability together with efficient control mechanisms along the supply chain is an essential requirement for a real and stable transparency strategy, which moreover fosters corporate reputation.

Step 2: Stakeholder Analysis
An analysis of stakeholders forms the basis for the development of transparency strategies. Stakeholder surveys not only provide insights into the sustainability performance of businesses, but also the expectations of different stakeholder groups regarding active and passive transparency. All key stakeholders of all fields of action have to be considered.

It is also crucial to evaluate how important stakeholders consider transparency regarding the products or services of corporations. Customers strongly distinguish between sectors and product categories when it comes to transparency requirements. They place emphasis especially on transparency if it is about money, health, or nutrition. Accordingly, transparency is of major importance in sectors like insurance and financial services, pharmaceuticals, and food. But also other industries may be threatened with communication crises if substantial deficits become public.

Step 3: Development of a Transparency Strategy
A transparency strategy should be based on the stakeholder analysis. This strategy defines what kind of information about topics and processes of the supply chain

the company delivers to which stakeholders. Also, it defines which stakeholders should become dialogue partners. The dimension of transparency should always be kept in mind within the context of sustainability processes and corresponding analyses.

Step 4: Communications Planning
The planning of short- and medium-term communication activities can be carried out only after the strategic foundations have been set. One challenge for communications management is to define their own tasks in measurable quantities that correspond to the control mechanisms of the organization. To put it concretely, if a company uses a balanced scorecard as a management system, the communications department has to also manage measurable command values such as trust, reputation, and company value via key figures. The quantitative and qualitative results of the evaluation have to be regularly played back to senior management and also kept in mind within the planning process.

5 An Appropriate Communication Strategy Is Crucial

In order to establish corporate transparency and to strategically market it as a reputation asset, an appropriate communication strategy is needed.

Implementing sustainability within the value chain is an important asset for corporations that has been insufficiently accompanied by communication activities in the past. It is essential to close the gap between sustainability management and sustainability communications. Communications have to support sustainability actions earlier (during its implementation), continuously (it should not be limited to reporting), and more effectively (sustainability communications often lack substantiality). For this purpose, sustainability representatives firstly have to acknowledge the added value of well-conceived sustainability communications internally and externally. Secondly, communicators have to better understand sustainability and to communicate it more credibly. A good sustainability report is always only half the battle. A credible and constructive dialogue with stakeholders depends on transparent and long-term sustainability communications that are closely interlocked with the management system.

The question, how transparently cross-cutting issues like sustainability, corporate governance, and compliance are handled in detail, should be discussed early on within the respective strategy processes and implementation of the management systems. The requirement for a transparent dialogue is the continuous feedback between communications and the management system. This should be understood and practiced as an ongoing process in which top management takes decisions of direction, while the role of the corporate communications department is to be responsible for all internal and external communication activities. This is reasonable since hardly any other corporate function has such a full understanding of all stakeholders of the business, their expectations, and prospective problems.

Internal and external communications are ideally closely linked. First of all, communications must be introversive. One task to begin with is to initiate topics such as sustainability and transparency within the company because along the way of a dialogue-oriented corporate strategy, internal provisions have to be eliminated. Questions that always used to be off-limits suddenly become public issues. For example, the transparency requirements of GRI are subject of controversial discussions among family businesses that never reported on key figures or profit distribution to the owners before. Regarding manufacturing businesses it is about potential critical topics along the supply chain that have long been avoided within external communications strategies but that cannot be excluded if corporations want to position themselves actively and credibly. Within the food industry, marketing-driven companies for which sales were the number one priority in the past, start to actively communicate about nutrition and lifestyles – topics which were largely ignored a few years ago. Such development requires extensive changes within corporate thinking and culture. Corporate leaders and middle management have to dismiss privileges concerning questions about governance and compliance. Another difficult issue is how much information will be made available for competitors in sustainability reports. In order to prepare companies for transparent communication activities, senior management and employees have to be convinced of the validity of the new strategy. Hence, they should be informed early on about the reasons for comprehensive change toward responsibility, sustainability, and transparency. Even more, they should be actively involved in such processes.

As soon as the milestones of the strategy are fixed and an acceptable overall picture has been set, which can be credibly communicated externally, an active dialogue with external stakeholders can be initiated. It is not about presenting a "ready solution," but about starting a transparent problem-solving process that responds to concrete stakeholder expectations. It should be openly communicated which business units need optimizing and the schedule for the improvements.

Many businesses believe that they do meet the external demands of transparency regarding sustainability and value chains by publishing sustainability reports once a year, or communicating good news about environmental initiatives or sustainable products to the press. Reports are doubtless an important step within sustainability communications – during the process of reporting, the sustainability strategy and topics become systematized for communication activities. Yet active sustainability communications far exceed the process of reporting. It is about repeatedly communicating positive developments within sustainability management along the supply chain for the purpose of actively positioning the corporation, both internally and externally. And again, to actively enter into a dialogue regarding problem areas in order to minimize reputation risks.

The path to more corporate transparency and therefore to more consumer trust can be illustrated in the form of a step model (see Fig. 2).

In the first step (Tell you), corporations gradually make more information available to stakeholders, also regarding other critical topics. The majority of

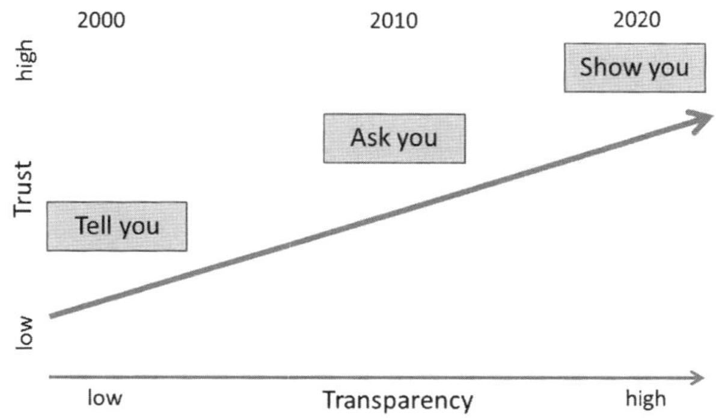

Fig. 2 Step model: Voluntary Transparency based on stakeholder participation. *Source*: Klenk and Hanke (2009)

corporations has already gone beyond this step. As mentioned before, there are hardly any large companies that have not implemented triple bottom-line reporting and accordingly established transparency about economic, social, or ecological aspects of their corporate actions. At this functional level, there has been a change towards a communication that depicts the company's position within society. Since this step of development is still unsatisfactory for some stakeholders and their pressure will not decline, more and more companies choose to establish a diversified stakeholder dialogue. This is the second step (Ask you), when stakeholders participate within the process of transparency creation because trust is only built through such initiatives. In the third step of development (Show you), corporations provide opportunities to convince stakeholders of the corporate reality and to control the statements of the business, be it in person, on location, or communicated via media.

Many medium-sized companies are starting to take the first step, "Tell you." Due to increasing information requirements of clients, they have started to report on sustainability and to prepare active communications along the value chain. These companies can easily be guided by global standards from the Global Reporting Initiative (GRI). According to a survey of accounting firm KPMG, already 80 % out of 250 of the world's biggest companies apply this standard. 28 Dax30 corporations report on their sustainability performance and almost all of them follow the GRI standards. Also, 68 % out of Germany's 100 biggest corporations report as per GRI. Many companies have reservations regarding the transparency requirements of GRI, and especially traditional, owner-managed companies. But the GRI provides useful guidance also for these businesses since introducing obligatory standards for sustainability reporting is currently under debate. This will certainly result in legislation amendments with regard to transparency obligations within the next years. There are alternative standards such as the CSR norm ISO 26000, e.g., that help to underline corporate responsibility that has been strategically and systematically implemented without overcharging the willingness of those companies to provide information.

6 Why Voluntary Corporate Transparency Pays Off

Transparency is an effective tool to create trust and to further strengthen the reputation of a brand or a corporation (see Fig. 3). In turn, a good reputation boosts the strategic positioning of a business within its competitive environment and indirectly affects turnover and profits.

Important: This applies to voluntary corporate transparency only. Companies which become transparent due to public pressure from external stakeholders will not experience benefits regarding trust or reputation. Only voluntary transparency is rewarding in terms of active reputation management.

One example to illustrate this is the German drugstore-chain operator dm. The company has been focused on transparency for quite some time now – both in- and outwards. Already defined within the employee guidelines is that all employees should be enabled "to recognize the extent and structure of their company . . .." This system of "internal glass structures" renders the business extremely powerful since it facilitates organizational learning effects and a dynamic development (Baier 2011a). dm also appears to be transparent to its customers. None of its competitors allows such an in-depth insight into its price policy such as the trade chain. In 1994, dm announced a permanent price guarantee. In 1998, a guarantee for a uniform basic price policy was introduced. Both policies allowed customers to easily understand the corporate pricing (Baier 2011b). Also, it was one of the first companies in its sector to make use of social media, where it appeared extremely open and conversational. dm already has 1.7 million Facebook fans whose questions about the company and their products are promptly answered and accessible. As our transparency survey shows, consumers acknowledge such transparent corporate culture: dm was nominated as Germany's second most transparent company. The favorable reputation has been reflected for years in the economic success of the brand. The drug store is one of the fastest growing retailers. While the turnover in 2005/2006 was 3.67 billion euro, in 2012/2013 it climbed to over 7.5 billion euros (DM 2013). Whether and how strongly transparency expectations of stakeholders

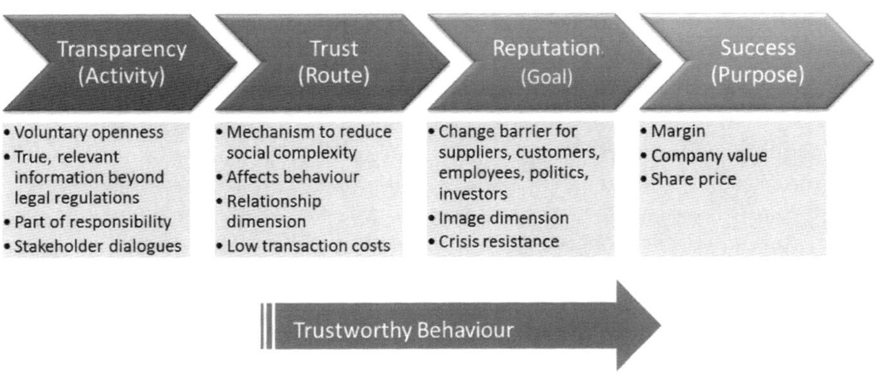

Fig. 3 Causal chain for transparency as a factor of success. *Source*: Klenk and Hanke (2009)

are reflected in sales was analyzed in our transparency survey in 2012. The result: Quality and price are still the most important determining factors for purchase decisions. But transparency has become a buying criterion. Already 22 % of Germans consider transparency an important purchasing criterion. For at least 56 % of respondents, transparency occasionally influences buying decisions. 44 % are even willing to pay more for a product of a transparent corporation. Hence, transparency is a sales factor. A 2011 online survey of Aegis Media Group underlines that also sustainability influences the purchasing decisions of consumers. 5,500 consumers between the age of 15 and 64 were asked about the role of sustainable criteria on purchase decisions. Already every second purchaser pays attention to the origin of products when buying agricultural products such as eggs, fruits, or vegetables. The product origin is even more important than price in this context. The more processed the food is, the less relevant origin becomes. For highly processed foods, price remains the most relevant purchase criterion. But even here the factors of transparency and sustainability have gained importance and have an effect on buying decisions (Aegis Media Resolutions 2011).

7 In Practice: The Challenges of Transparency and Sustainability Communications Along the Value Chain

7.1 The Rewe Case

German retailer Rewe has been practicing sustainability activities as well as sustainability communications for years. There was a comprehensive strategic reorientation, clear standards for its own markets and suppliers, stakeholder dialogues, a clear positioning of the corporate brand, the introduction of organic and fair trade products, private labels, and regional segments (Rewe 2013a). In short, Rewe's corporate and point of sale communication focuses on organic and sustainability messages. Rewe has received multiple awards for best practice with good reason. With its reorientation, Rewe made a clear difference, and not only through the optimization of its own processes and related sustainability figures. The company has become an agent of change for many consumers, for the businesses in the value chain, and additionally, it has become a benchmark for competitors.

Comprehensive sustainability communications have to be carefully counterbalanced given a product range of over ten thousand products. By no means do all consumers consider sustainability aspects while shopping. There is fierce, cutthroat competition in the industry. Sales and turnover objectives and fast innovation cycles determine the business. It is obvious that there are products on the shelves that have nothing to do with organic, fairness, or sustainability. And not all suppliers are willing to or can reorganize their products and processes. There have to be gaps within the system given such a big number of suppliers and diversified value chains.

A report by the German public television (ARD) about the Rewe sustainability label "Pro Planet" in early 2013 explains what communication challenges the positioning of sustainability means for large businesses. If public sustainability promises are broken, there is a loss of trust.

Since 2009 Rewe has been labeling their home brand products that have "positive ecological or social qualities" with the so-called "Pro Planet"-label. On the label website the company states it seeks to improve the conditions of harvesters as "part of the cultivation" (Rewe 2013b).

Journalists of the German TV show "Markencheck" (brand check) traveled to Spain to examine whether Rewe has kept this promise. After visiting manufacturers on-site they concluded that this was not the case. They judged it be "insufficient fairness." The company had promised more on their website than they could actually deliver. Following the program, numerous media reports took up the topic and highly criticized the business. Rewe rejected the accusations of the journalists and declared them incomplete and misleading. Damage to Rewe's image could no longer be prevented. Edeka, a competitor without a sustainability label, had also been analyzed within the report but was spared from criticism.

Rewe can be seen as an example for many companies that want to approach stakeholder expectations regarding transparency and responsibility but face challenges due to the diversity of their value chains. Often management and communication professionals stand between the conflicting priorities of their desire to position themselves as clearly sustainable in order to meet stakeholder expectations and their organizational environment that does not enable them to live up to those standards (Zerfaß and Müller 2012). If corporations start marketing their activities and emphasize positive actions and ignore negative ones, they run the risk that the image portrayed by communications cannot keep up with corporate reality. Today, such gaps are easily revealed by the critical public and distributed online, resulting in loss of trust or reputation crises. Brand expert Torben Hansen comments: "If it is about Corporate Social Responsibility (CSR) [...], people have zero tolerance! [...] Those who promise more than they can actually deliver, will also not be trusted on other subjects. And trust will be withdrawn. CSR communications [are] definitely not image or sales promotion as such [...], but permanent trust building activities, for which the formula in case of doubt rather is "20 % smaller than life": To deliver more than promised. Otherwise the well-intentioned activity turns into a boomerang (Hansen 2013)." Consequently, those responsible for internal processes and communication management have to team up so that there will be no communicative gaps that destroy the trust between stakeholders and businesses.

7.2 The Example of PUMA

In 2011, PUMA was the first multi national company to present an Environmental Profit & Loss Account (E P&L). The manufacturer of sports equipment announced publicly that their corporate acting along the value chain had caused environmental damages of 145 million euro in 2010.

By revealing the impact of water consumption and CO_2 emissions in financial values, the corporation set new standards of ecological corporate reporting. The company sought a dialogue with governments, environmental organizations, and other manufacturers of sports equipment to establish a business model that does not work against, but in line with nature. PUMA positioned itself as a company that consciously deals with the effects it has on the environment. PUMA put massive pressure on its competitors – especially by declaring that Asian suppliers that also worked for Nike and Adidas had caused most of the environmental damage. A number of positive media reports followed the publication of these environmental records. The Financial Times Deutschland wrote for example: "PUMA's neat idea" (FTD 2011).

Primarily powered by the personal engagement of manager Jochen Zeitz, PUMA was the clear impetus for more discussion about sustainability and transparency in Germany by revealing the negative costs of its corporate actions. Such attempts of businesses that do not wait for the development of new standards, yet actively promote sustainability and transparency, are important drivers for the public dialogue about the further development of reporting standards and legal obligations. Prof. Jochen Pampel, Head of Sustainability at KPMG, states: "Managing transparency also means to realistically display the actual state of affairs. This is what the public appreciates. It is more convincing to reveal existing gaps and development potentials and to communicate planned measures than superficially claiming that everything is okay" (Pampel 2012).

Yet Puma has made sustainability not only a core topic within agenda setting and corporate communications. Sustainability is part of the company culture and many parts of the value chain. In 2010, PUMA and their major suppliers (from China, Vietnam and Cambodia) agreed on preparing sustainability reports as of 2011. The suppliers involved, which manufacture about two thirds of all PUMA products, are trained by GRI-approved partners (RNE 2010). Throughout the reports, sustainability issues that would not have attracted attention through auditing, were unmasked. In 2011, PUMA converted its packaging. Also, PUMA published, together with other manufacturers like Adidas, C&A, and Nike, a plan of action that should help to stop harmful chemicals from getting into the environment by 2020.

In early 2013, the company presented a collection of clothes which have a completely closed product life cycle. According to the company, shoes, bags, and accessories of the InCycle product line are either entirely biodegradable or recyclable. Additionally, those are 100 % "Cradle-to-Cradle" certified. A shoe from that collection, e.g., has been manufactured from a mix of organic cotton and linen plus biodegradable plastics. Hence, it can be entirely composted after being used. A tracksuit top has been remanufactured from polyesters that originate from used PET bottles. After its use it can again be converted to polyester granulates that can then be used as subsidiary raw material for manufacturing other products. In order to support the recycling process, PUMA globally installed recycling boxes in their stores, into which consumers could put used clothes (PUMA 2013).

With the new collection, the manufacturer reacted to the results of the E P&L of 2010. Those had revealed that nearly 60 % of environmental impacts are due to the production of raw materials such as leather, cotton, or gums. One of the corresponding objectives is to increase the amount of products that are made from sustainable materials.

8 Conclusion and Prospect

Companies that seek to optimize their value chains and to establish transparency can follow five essential recommendations:

- Be as transparent as possible: Transparency drivers force businesses to act transparently anyway.
- Be proactive: Seek dialogue before a dialogue is forced by external pressure.
- Foster relationships: Maintain dialogues even in relatively calm periods.
- Integrate employees: These are the most important stakeholders, advocates, and ambassadors.
- Be prepared for all scenarios: Develop risk and crisis scenarios in case negotiations and dialogues fail.

In the future, successful companies will be those that succeed in implementing effective sustainability management measures along their supply chain. Reputation gain is the cherry on the cake of every comprehensive sustainability process. Those who practice transparency and have strategically grounded sustainability communications can also better capitalize on sustainability both internally and externally. However, successful sustainability communication activities do not start by only communicating initiatives or writing sustainability reports. That is only part of the process. Along with that, a company needs to moderate critical dialogues, create clarity within complex issues, and create excitement for innovation.

Future communicators should not only treat transparency, sustainability, and CSR as communications issues – they also need to get down to the nitty-gritty and push crucial innovations and long-term change processes along the value chains. They are the fresh impetus to colleagues and for the consumers and end users of the products. It is important that these communicators work together with all functions within the company and are a part of the strategic process. This implies the setup of competencies both content-wise as well as the willingness to work with experts from corporate finance, manufacturing, sourcing, supply chain management, human resources, and environmental and sustainability management – communicators must be interconnected on a multidisciplinary basis.

Bibliography

Aegis Media Resolutions. (2011). Nachhaltigkeit – große Worte oder große Taten? http://www.aemediaresolutions.de/sites/default/files/Nachhaltigkeit_Text_September2011_0.pdf. Accessed 11 June 2013.

Baier, A. (2011a). http://www.transparenz.net/dm-drogeriemarkt-innovative-transparenz-auf-vielen-kanalen/

Baier, A. (2011b). http://www.transparenz.net/dm-drogeriemarkt-mutige-preis-transparenz/

DM. (2013). Presse. http://www.dm.de/de_homepage/presse/. Accessed 11 Juni 2013.

Eigen, P. (2009). "Transparenz und Erfolg", in Erfolgsfaktor Transparenz. http://www.transparenz.net/?p=1039. Accessed 11 June 2013.

FTD. (2011). Pumas saubere Idee, financial time Deutschland (16.05.2011). http://www.ftd.de/unternehmen/industrie/:oekobilanz-pumas-saubere-idee/60053184.html. Accessed 11 June 2013.

Hansen, T. (2013). Gastbeitrag ARD-Markencheck: Lieber mehr halten als versprechen! http://www.handelsblatt.com/unternehmen/handel-dienstleister/gastbeitrag-ard-markencheck-lieber-mehr-halten-als-versprechen/7616942.html. Accessed 11 June 2013.

Hecker, J. (2012). "C&A, Transparenz in der Lieferkette", in Erfolgsfaktor Transparenz. http://www.transparenz.net/?p=4705. Accessed 11 June 2013.

Klenk, V., & Hanke, D. (2009). *Corporate transparency. Wie Unternehmen im Glashaus-Zeitalter Wettbewerbsvorteile erzielen*. Frankfurt am Main: Frankfurter Allgemeine Buch.

Klenk & Hoursch AG/Innofact AG. (2012). Transparenzstudie. (unpublished).

Langrock-Kögel, C., & Winkelmann, M. (2012). Piraten an Bord. *enorm, 2*, 20.

Pampel, J. (2012). Die interessierte Öffentlichkeit honoriert Transparenz. In *Erfolgsfaktor Transparenz*. Accessed June 11, 2013 from http://www.transparenz.net/?p=5200

PUMA. (2013). *Sustainability*. Accessed June 11, 2013 from http://about.puma.com/sustainability/

REWE Group. (2013a). *Nachhaltigkeit*. Accessed June 11, 2013 from http://www.rewe-group.com/nachhaltigkeit/

REWE Group. (2013b). *Was ist Pro Planet?* Accessed June 11, 2013 from http://www.proplanet-label.com/

RNE. (2010). *Sportartikel-Hersteller verlangen Nachhaltigkeitsberichte von Lieferanten. Rat für nachhaltige Entwicklung* (09.06.2010). Accessed June 11, 2013 from http://www.nachhaltigkeitsrat.de/news-nachhaltigkeit/2010/2010-06-10/sportartikel-hersteller-verlangen-nachhaltigkeitsberichte-von-lieferanten/

Zerfaß, A., & Müller, M. C. (2012). CSR-Kommunikation in Deutschland. Empirische Studie zu Rahmenbedingungen und Vorgehensweisen in deutschen Unternehmen. *PR Magazin, 43*(11).

Infineon: Integrated Supply Chain Architecture to Support Sustainability

Kurt Gruber, Christian Pophal, and Hans Ehm

1 Corporate Social Responsibility at Infineon

The term "Sustainability," which is closely connected to Corporate Social Responsibility (CSR), covers the economic, ecologic, and social aspects of our business approach at Infineon. These three traits were already included when the term was defined as part of the United Nations Report "Our Common Future" (United Nations 1987) and the principles described in this report (often called the "Brundtland" report) remain important for Infineon until today.

With the initiation of the United Nations "Global Compact" by Kofi Annan in 1999, these three principles of economic, ecological, and social responsibilities were turned into visible and tangible business principles. The UN Global Compact is a strategic policy initiative for businesses that are committed to aligning their operations and strategies with ten universally accepted business principles in the areas of human rights, labor, environment, and anticorruption.

By doing so, businesses, as a primary driver of globalization, can help to ensure that markets, commerce, technology, and finance advance in ways that benefit economies and societies alike (UN Global Compact, 2014). In 2011, the European Commission recognized the role of sustainable business practices by redefining Corporate Social Responsibility (CSR) as "the responsibility of enterprises for their impact on society" (European Commission 2011).

In order to act sustainably, Infineon considers economic, environmental, as well as social aspects simultaneously and offers a product portfolio that is focused on energy efficiency, mobility, and security. Infineon takes responsibility for its actions based on six pillars that build the foundation of Infineon's CSR strategy (Fig. 1). By doing so, Infineon contributes to sustainability by focusing internally

K. Gruber • C. Pophal • H. Ehm (✉)
Infineon Technologies AG, Am Campeon 1-12, 85579 Munich (Neubiberg), Germany
e-mail: hans.ehm@infineon.com

M. D'heur (ed.), *Sustainable Value Chain Management*, CSR, Sustainability, Ethics & Governance, DOI 10.1007/978-3-319-12142-0_18

Fig. 1 Supply chain management is one of the six pillars of corporate social responsibility at Infineon (2012)

(in the Infineon operations) and externally (through sustainable applications and end products) and shows its commitment to create a sustainable society.

This article examines the contribution of the Infineon supply chain (and its associated people, processes, and systems) to support sustainable business practices and to ensure timely delivery of sustainable products to global customers. After introducing Infineon's supply chain, which is rooted in the highly complex semiconductor industry, additional challenges and potential solutions will be examined. Finally, sustainable value creation will be illustrated using the example of the Infineon carbon footprint. It will be demonstrated how the supply chain contributes to a lower CO_2 emission with every chip produced through process control, an end-to-end approach, flexibility, and a high standard of education.

There are two reasons why an integrated, fast, and flexible supply chain creates sustainable value: Firstly, as a result of globalization supply chains became more important for ecological sustainability. Secondly, the holistic approach of Infineon's supply chain makes it possible to maintain an effective and efficient supply chain which supports sustainability.

2 Sustainable and Integrated Supply Chain Management

2.1 Volatility Meets Efficiency: A Supply Chain Perspective

The transistor was invented in 1947 by the American scientists Shockley, Bardeen, and Brattain, who won the Nobel Prize in Physics for this invention in 1956, and by Welker and Mataré, two German physicists, who presented almost simultaneously a transistor based on the semiconductor material Germanium (Bosch 1994).

Ever since these groundbreaking inventions, semiconductor manufacturing has become the most innovative production of all – developing at the pace of Moore's Law, that stipulates that the performance of integrated circuits doubles approximately every 3 years (Moore was one of the founders of Intel) (Moore 1998).

Moore's Law is the semiconductor law and it implies in simple words that more features, intelligence, and control options are available on the same space – i.e., on a silicon disc, which is called wafer. Moore basically says that this speed of innovation is applied beyond memory or processor chips to other semiconductor areas like sensor chips or power chips or even to systems containing chips. Translated into environmental sustainability this equates to a reduced amount of CO_2 emissions for each additional feature now able to fit in a given area of the silicon disc.

Furthermore, the semiconductor industry is characterized by short product life cycles, high market volatility, mainly because semiconductors are produced at the end of a long global supply chain. The chain starts with a customer who buys, for example, a car from an OEM (Original Equipment Manufacturer). The OEM orders from a Tier 1 supplier (for example, an automotive supplier), who then orders from a Tier 2 supplier (for example, a lighting supplier) and eventually the Tier 2 supplier buys from a semiconductor manufacturer. Through the bullwhip effect, which describes the amplification of demand fluctuations along the supply chain (Lee et al. 1997), the semiconductor industry suffers from extreme volatility (Fig. 2). As a result the semiconductor market varies ten times stronger than the global gross domestic product (GDP).

To be successful in this capital-intensive and innovative semiconductor industry, an integrated and flexible global supply chain architecture is required. This includes integrated processes, electronic data processing, as well as highly qualified and experienced staff. Infineon's supply chain network has evolved from the production

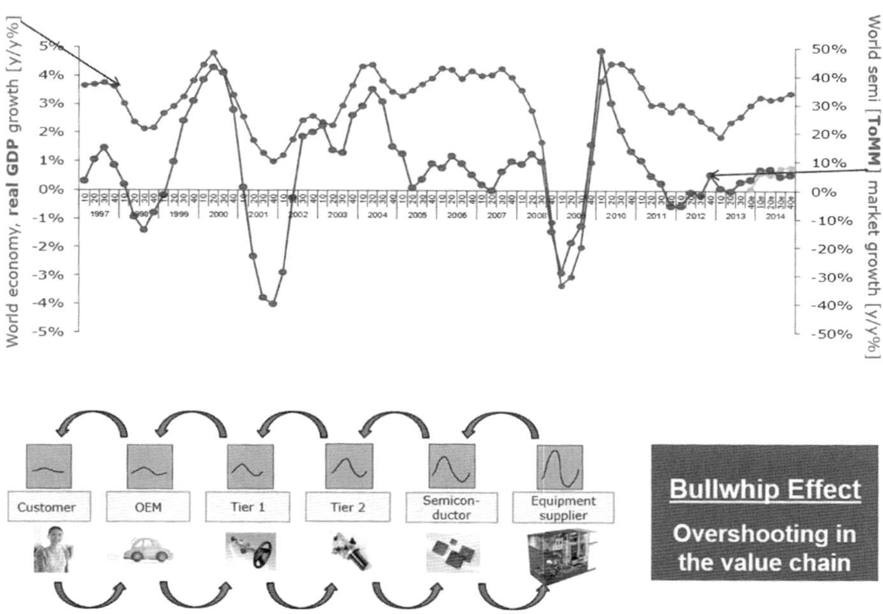

Fig. 2 A central challenge in the semiconductor supply chain is the bullwhip effect (Infineon 2013; Lee et al. 1997)

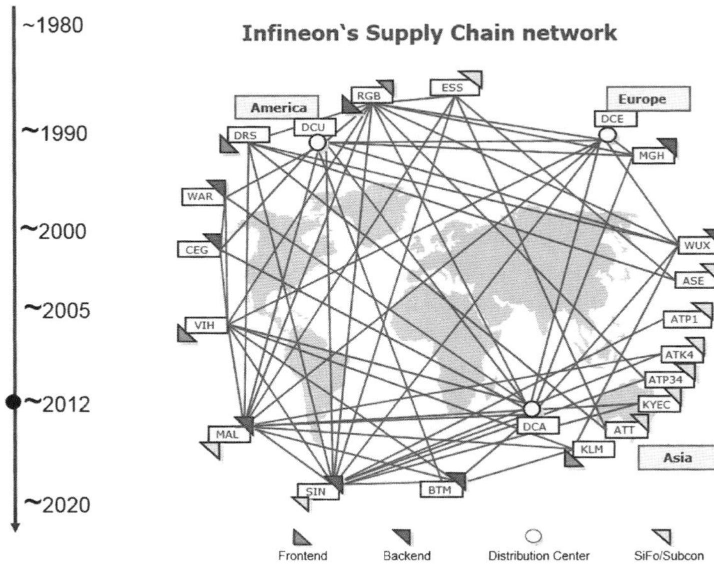

Fig. 3 The Infineon supply chain today is a global network (Infineon 2013)

of almost all products at one single location, to a complex, global network sharing resources with external partners like silicon foundries, subcontractors, research institutes, and universities (Fig. 3).

2.2 A Supply Chain Architecture for Flexibility and Environmental Sustainability

Infineon (at that time called Siemens semiconductors) started out as a single site in Munich in the 1960s. It evolved into several separate supply chains, usually with the Frontend Production (where the silicon wafers get produced) in Europe and the Backend Production (where the wafers are embedded in the chip package) in South East Asia in the 1990s. The level of interaction between these chains was initially low. Later on these individual chains became highly interlinked, which enabled them to benefit from each other. The 2000s saw the establishment of distribution centers close to the global markets. At about the same time and – continuing until today – Si Foundries (this is the domain specific name for silicon foundry subcontractors for wafer production) for Frontend production and subcontractors for the Backend production were integrated in the Infineon Supply Chain as outsourced partners. Today, the Infineon Supply Chain is a flexible, global network, as shown in Fig. 3.

As well as these physical extensions and integrations, Infineon reorganized its internal supply chain operations in 2003, following the Supply Chain Operations Reference (SCOR) model of the APICS Supply Chain Counci (ASCC 2014), which has been in use at Infineon since 1998. The ASCC is a global, independent, nonprofit organization supported by approximately 500 member companies from various industries and academia. The process reference model, developed by the ASCC, is the recognized standard for processes in the supply chain area. The SCOR model is an end-to-end supply chain model, covering not only the supply chain of a single company, but the entire chain from the suppliers' suppliers to the customers' customers. The five main processes of the SCOR model: plan, source, make, deliver, and return are applied in Infineon's supply chain (Fig. 4).

The global supply chain is a key driver for economic sustainability, which ensures delivery reliability and factory utilization in a volatile environment through the provision of high production flexibility: Initially, the focus within the supply chain was almost exclusively on cost reduction, but this has changed significantly in recent years. The supply chain is no longer limited to pure logistics of deliveries and storage, but covers sample production and manufacturing processes throughout the entire life cycle of products, including visibility of the return process. Today sustainability is one of the drivers of innovation, which can be measured in the reduction of CO_2 consumption per cm^2 wafer. A wafer is a silicon disc where the integrated chips are produced, a process which is done in the Frontend production. Once the Frontend process is finished the chips are separated, connections are made to from inside the chip to the pins on the package, and finally chips are encapsulated and tested. This process in the Backend production is called assembly and test.

A key process of an integrated semiconductor supply chain is the PLAN (Planning) process. This process is of special importance for an economically sustainable chain. Innovations which enable more functions with the same amount of CO_2 usage, can benefit from global resource sharing – this must be planned well ahead, to be able to realize it later in the process. Sustainability within the PLAN process implies for Infineon to deal with capital-intensive global production, complex and long process chain, short product life cycles, and high market volatility. The PLAN process considers all available capacities beyond its own production sites, including those from subcontractors and silicon foundries, which are then balanced and used at

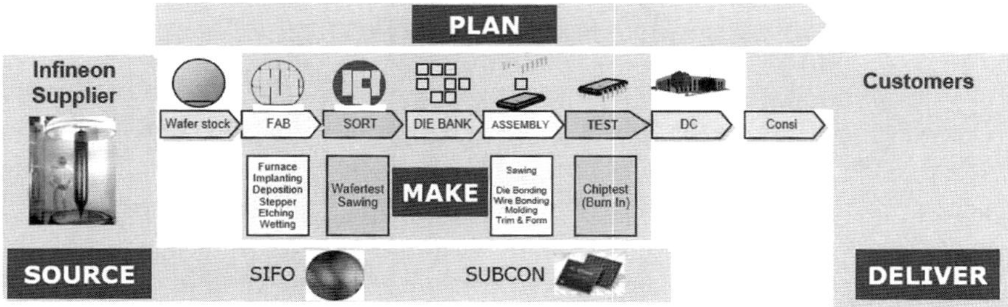

Fig. 4 Infineon supply chain processes according to the SCOR-Model (Infineon 2013)

an optimum level (i.e., economically and ecologically sustainable). The ecological impact of the supply chain is the reduction of CO_2 emissions per chip, more precisely per circuit function, in the production. This goes in line with the capacity utilization as it will be shown in the following paragraphs.

The central process in the semiconductor industry is the "MAKE" process, i.e., the production process. This process covers not only Infineon, but also partners. It covers the silicon wafer fabrication process (FAB), the wafer testing, resp. sorting (SORT), the assembly (ASSY), and finally, the finished component test (TEST). Semiconductor production is a highly complex process. Thus from the wafer start to the finished component it can take 5 months or more. If production is not performed by Infineon alone, the production partners are called Silicon foundries (SIFo), if they perform FAB (Fabrication) and SORT (Sorting) process steps. If they perform ASSY (Assembly) and/or TEST, they are called subcontractors (SUBCON). Figure 4 shows on the left side a pure silicon bar which represents the SOURCE process, in the middle the PLAN process, the MAKE process on the right hand side, leading into the final step of the DELIVER process.

Mastering the processes minimizes the impact of the volatility on the supply chain, resulting in an optimized and flexible utilization of globally distributed capacities. By taking into account the low CO_2 emission in transport compared to the CO_2 consumption during production, an efficient and sustainable utilization of the production sites comes with a great potential for reduction of CO_2 emission per chip (or per cm^2 of wafers). The contribution of the supply chain is highly significant, as it enables global utilization of capacities and innovations.

The semiconductor supply chain is complex. By optimizing the supply chain and increasing flexibility, the impact of volatility can be minimized. The SCOR model helps organize and measure/optimize the supply chain. By using SCOR and improving flexibility, all resources can be fully utilized thus reducing CO_2 per function.

Infineon's supply chain approach has been awarded with several prizes: The award for education of the European Supply Chain Council 2010, the Global Supply Chain Award of the Supply Chain Council in 2010/2011, the German Supply Chain Award in 2012, the SCM Logistics & Manufacturing Excellence Awards from Singapore in 2013, and the European Supply Chain Excellence Award in the Category Automotive, Aerospace and Industry in 2014.

3 Sustainable Products: The Infineon Carbon Footprint

Infineon's main contribution toward sustainability is the development of energy efficient products. This is a key factor in our ambition to save energy and to actively address climate change. In order to calculate the ecologic footprint of our products, we developed a model based on international industry standards, considering the accumulated CO_2 emissions during production (light grey area in Fig. 5), but also the CO_2 savings, which are being made during the use phase of our products (dark grey area in Fig. 5) (Reinhard and Liebi 2011). Recovery of materials through recycling and the related CO_2 savings are relatively minor for those tiny products.

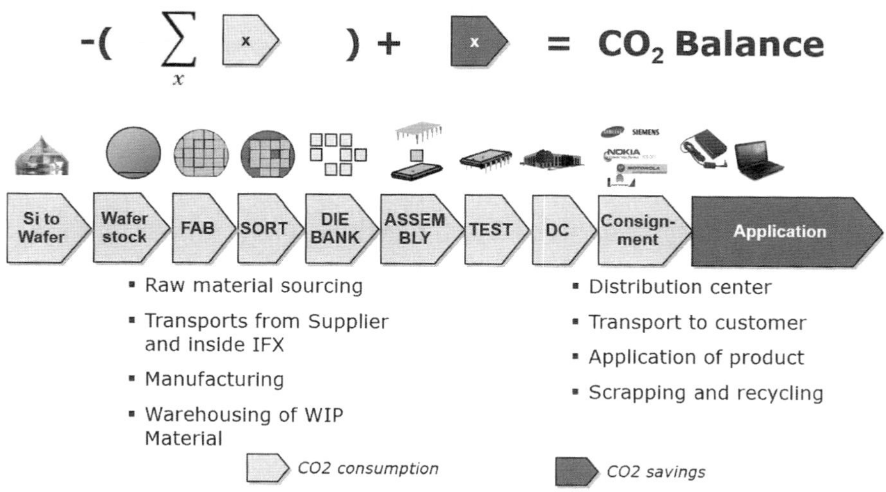

Fig. 5 CO_2 balance determination based on the simplified life cycle assessment (Infineon 2012)

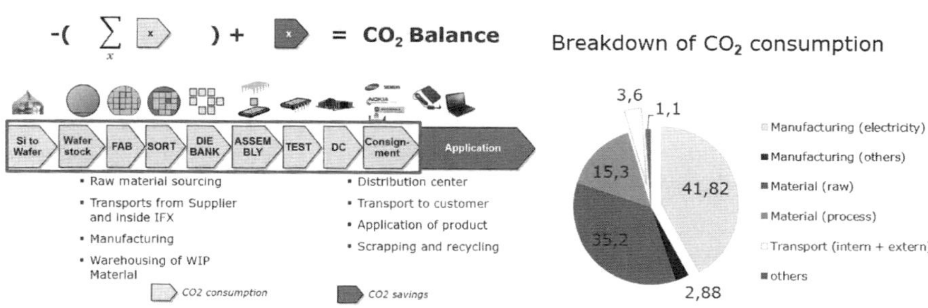

Fig. 6 CO_2 Consumption of Infineon divided by the polluters (Infineon 2012)

For this reason we omitted to consider this positive impact of recycling to Infineon's CO_2 balance. Overall, we consider the CO_2 record of our products according to a Life Cycle Assessment, which is also known as eco balance – based on ISO 14040 (FIS 2013).

At Infineon we address both sides of our CO_2 record: our own production and our contribution to the creation of a sustainable society through our products and solutions. The breakdown of our CO_2 emissions in production, as shown in Fig. 6, shows that the largest part of our emissions is produced by the materials we use.

It may seem surprising that the share of transport is a very small part of the CO_2 footprint, especially because almost every chip is shipped around the world via air freight to make use of the best manufacturing conditions in the frontend production, backend production, or for special processes, before it actually arrives at the customer. Nevertheless, the very low weight of semiconductor products limits the environmental impact of air freight. Thus the highest CO_2 savings can be reached by reducing the required electricity. This is reached by fully utilizing the capacity, via generating and using flexibility in the plan process and by reducing CO_2 through the raw and process material usage in production.

Despite the fact that benchmarking has shown that our production processes and material usage are amongst the most efficient in the semiconductor industry, we continuously improve our production processes to reduce CO_2 and other emissions. For Infineon, responsibility and sustainability is more than just acting according to legal requirements. The IMPRES initiative (Infineon Integrated Management Program for Environment, Safety & Health) enables us to connect our responsibility to humans and nature with economic success. It also includes our commitment for efficient resource management to protect the environment and enable ecological innovations. Through IMPRES Infineon was able to implement key aspects of sustainability in the organization and has had remarkable success in doing so. When compared to the global average of surveys conducted by the global semiconductor organization WSC (World Semiconductor Council), the Infineon frontend production locations in Europe consumed in 2011 50 % less power and 70 % less water per square centimeter of processed wafer surface. The amount of power that Infineon has saved, as a result of increased energy efficiency between 2002 and 2013 amounts to 2.3 terawatt hours (1 terawatt hour is 1 billion kilowatt hours) in Europe alone. This is the equivalent of the annual consumption of a large European city with 2 million inhabitants.

Also at the Infineon headquarters in Munich energy saving is an important matter. The buildings are connected to the district heating, which extracts thermal water at a temperature of 122 °C from approximately 3.5 km depth (Geotherme Unterhaching 2014). This allows Infineon to have access to an almost CO_2 free electricity and heating source. Additionally, in 1998 and in accordance with the Kyoto Protocol Infineon committed to reduce its PFC gas emissions (calculated in the equivalent of CO_2) to the value of 1995 minus 10 % by 2010. Due to increased efficiency, for example, through implementation of remote plasma cleaning technology, substituting alternative process gases, and through the use of exhaust air cleaning concepts, this target was already achieved in 2007. Infineon has also voluntarily started a halogen free initiative, which regulates the use of chlorine and bromine in flame retardants.

The systematic reductions of emissions in production, as well as the large variety of programs and initiatives that exist to protect the environment and the societal sphere, which extend further than the fulfillment of legal requirements, show the importance of CSR and environmental sustainability. Infineon is constantly striving for improvements, which are not implemented out of a sense of duty, but as an element of our daily business activities.

To calculate our contribution to the creation of a sustainable society through our products and solutions, Infineon uses own methods but also cooperates with universities. For example, calculating the CO_2 balance during the use phase of our products was developed in collaboration with the Swiss Federal Institute of Technology (ETH) Zurich as part of the master thesis "Future Trends and Potential of the CO_2 Balance in the Semiconductor Industry" (Reinhard and Liebi 2011). First, a set

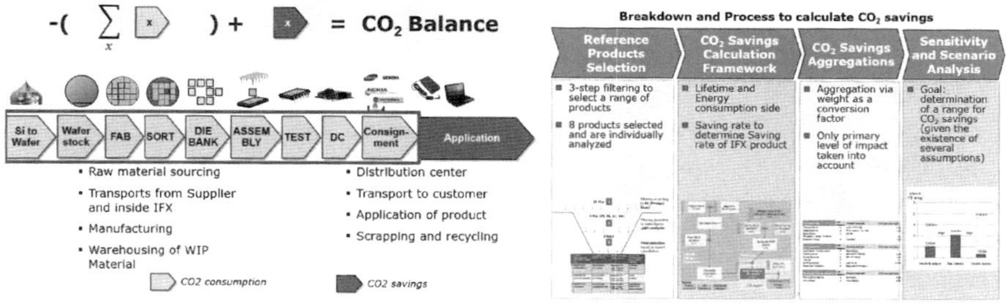

Fig. 7 Calculation of the CO_2 balance for the use phase of the product (Infineon 2013)

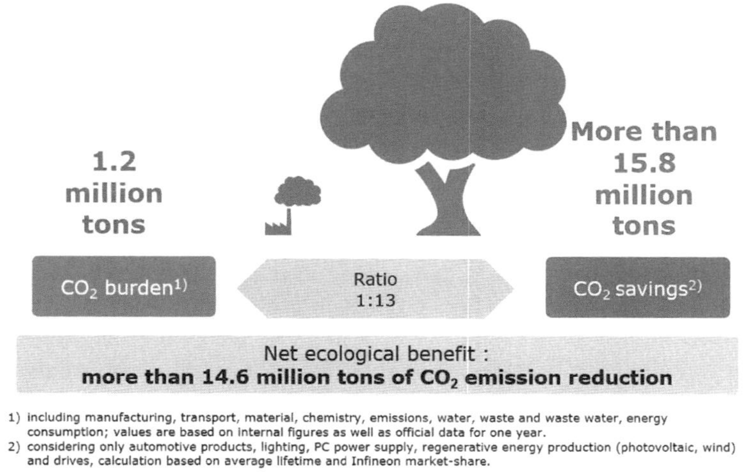

Fig. 8 The Infineon CO_2 balance (Infineon 2013, p. 89)

of reference products, representing Infineon's products was defined. For these reference products (such as a chip for applying engine torque, which makes sure that energy is only used when it is necessary), the CO_2 savings over an application's lifetime were determined as well as the chips' impact on these saving. The results of the savings of reference products were aggregated. To verify the reliability of the statement, a sensitivity analysis was used (as shown in Fig. 7).

The results show that our own carbon footprint of about 1.2 million tons CO_2 needs to be seen in comparison to our annual saving of 15.8 million tons (in 2012), which was made possible by our products. Consequently, Infineon's products save an average of over ten times as much CO_2 during utilization than they consume during fabrication. Through its activities, Infineon generates a positive net contribution to the environment and society (Fig. 8).

4 Conclusion and Outlook

The supply chain is a key contributor of significant value to business today, not only as a key component of business itself but also in the field of sustainability. In the past, sustainability was regarded mainly as a topic which targeted production. Meanwhile, the supply chain has great impact on sustainable business through creating flexibility and optimizing planning. Consistent processes, the best IT tools, and well-trained employees are able to master the processes. Tools and communication provide efficiency and flexibility, which leads to an optimal use of the global capacity; despite the volatility inherent in the semiconductor and electronic supply chain. This leads to a reduction of the CO_2 consumption per circuit. This holistic approach to Infineon's supply chain design and management started in 1998, when the SCOR model was established as the supply standard for Infineon. This approach ensured consistent and intelligent processes and measures that even in times of crises were economically, ecologically, and socially useful.

At Infineon, we have implemented an integrated concept for CSR and this approach has proven to be successful. Environmental sustainability is one of the six pillars of Infineon's CSR framework and Infineon acts to take responsibility for people and the environment. At Infineon, ecological principles are alive: in our facilities, in our products, and in our daily activities.

Ecological efficiency of our production facilities and the environmental benefits, provided by our products, show that innovation and environmental responsibility goes hand in hand. The usage of our products is ecologically wise: it "pays off" for the environment and for everyone involved. By producing products that save more CO_2 during utilization than they consume during their production, Infineon has generated a reputation and is part of the "Sustainability Yearbook" since 2011 (Robecosam 2014). In this yearbook the approximately 2,500 companies are listed, which are the 15 % of the most sustainable companies in the world. The basis of the evaluation is the results from the Dow Jones Sustainability Index, in which Infineon is listed for the third time in a row. These awards and the public recognition of our supply chain demonstrate Infineon's commitment to sustainability and confirm that we are on the right track. We want to continue on this path to continued sustainability and keep our supply chain in a leadership position for innovation and value creation.

Bibliography

Bosch, B. (1994). Der Werdegang des transistors 1929–1994. Bekanntes und weniger Bekanntes. *Festkolloquium zum 65.Geburtstag von Magnifizenz Prof. Dr. Eberhart Köhler*, TU Ilmenau, 17.11.1994.

FIS. (2013). *Ökobilanz nach DIN EN ISO 14040 und 14044*. http://www.forschungsinformationssystem.de/servlet/is/349763/. Accessed 12 March 2014.

Geotherme Unterhaching. (2014). *Die Geotherme Unterhaching*. https://www.geothermie-unterhaching.de/cms/geothermie/web.nsf/id/pa_home.html. Accessed 13 April 2014.

Infineon. (2012). *Geschäftsbericht*. http://www.infineon.com/dgdl?folderId=db3a30433b47825b013b4d4bda8c1a55&fileId=db3a30433b92f0e8013b98a551e91604. Accessed 12 Juni 2013.

Infineon. (2013). *Geschäftsbericht 2013*. http://www.infineon.com/dgdl?folderId=db3a3043429a38690142a45dfa551095&fileId=db3a304342e8be2c0142fc6fd2054f22. Accessed 23 April 2014.

Lee, H. L., et al. (1997). Information distortion in the supply chain: The bullwhip effect. *Management Science, 43*(4), 546–558 (The Institute of Management Sciences).

Moore, G. E. (1998). Cramming more components onto integrated circuits. *Electronics, 38*(8), 114–117 (1965. Reprint in Proceedings of the IEEE, 86(1), 8285).

Reinhard, M., & Liebi, J. (2011). *Optimizing the CO_2-balance: A case study from the semiconductor industry*. Master thesis, Infineon Technologies AG, Munich.

Robecosam. (2014). *The sustainability yearbook*. http://www.robecosam.com/en/sustainability-insights/library/the-sustainability-yearbook.jsp. Accessed 23 April 2014.

ASCC. (2014). *APICS Supply Chain council*. https://supply-chain.org/scor. Accessed 23 November 2014.

The European Commission. (2011). *A renewed EU strategy 2011–2014 for corporate social responsibility* (p. 6). http://ec.europa.eu/enterprise/newsroom/cf/_getdocument.cfm?doc_id=7010. Accessed 12 March 2014.

United Nations. (1987). *Report of the world commission on environment and development: Our common future*. http://www.un-documents.net/our-common-future.pdf. Accessed 12 March 2014.

German Council for Sustainable Development: The Sustainability Code

Yvonne Zwick

1 Preamble

Sustainability sounds like something pleasant and is becoming a factor in the decision-making process for an increasing number of people. However, there is no universal definition for it. After all, Corporate Social Responsibility is a concept that is difficult to compare across different jurisdictions because, by its very definition, it goes beyond legislation (BMAS 2010). There are as many definitions of what sustainability means in a corporate context, what it involves, and what its strategic considerations should be as there are companies and survey providers themselves. This has resulted in a plethora of vague definitions and conflicting assessments, and a considerable amount of extra work for rapporteurs and information providers alike. Standardization is a sound approach for making sustainability a differentiating feature based on a reliable set of comparable data. Future markets will be shaped by mega-trends, which all demand some kind of future-oriented management approach to help alleviate the network of problems. Trends such as climate change, the increasing number of weather-related events, rising energy and raw material costs, as well as demographic change – with varying degrees of national and international impact – are seen either as potential problems or opportunities by market players around the world. A benchmark for all companies and corporate activity is of fundamental importance in order to reward entrepreneurial solutions through market incentives at an early stage and generate a broad-based impetus for sustainable business activity. This is the reason why the German Council for Sustainable Development (in German *Rat für Nachhaltige Entwicklung*, RNE) has developed the German Sustainability Code (GSC) and recommended it for voluntary application.

Y. Zwick (✉)
Office c/o GIZ, German Council for Sustainable Development (RNE), Potsdamer Platz 10, 10785 Berlin, Germany
e-mail: yvonne.zwick@nachhaltigkeitsrat.de

M. D'heur (ed.), *Sustainable Value Chain Management*, CSR, Sustainability, Ethics & Governance, DOI 10.1007/978-3-319-12142-0_19

The challenges and opportunities associated with sustainability require greater importance in the debate on the strategic direction of companies across all industries. The concept of sustainability was fundamentally influenced by the 1987 Brundtland Report commissioned by the UN which defined sustainable development as the following: "Sustainable development meets the needs of the present generation without compromising the ability of future generations to meet their own needs. [...] In essence, sustainable development is a process of change in which the exploitation of resources, the direction of investments, the orientation of technological development and institutional change are all in harmony and enhance both current and future potential to meet human needs and aspirations" (United Nations World Commission on Environment and Development 1987). The "three-pillar model" of sustainability (Crane and Matten 2004) was developed from this environmental concept, taking equal consideration of the economic, ecological, and social factors that enable sustainable development. Academic opinions differ as to whether this model of sustainability should be seen as a static, normative model or as a dynamic development model with shifting parameters and changing foci.

With its work on the Sustainability Code, the German Council for Sustainable Development aims to ensure that the process-based character of sustainability is made transparent in government recommendations and policy documents that it prepares. This approach is intended to take into account the various different starting points in terms of how the subject matter is addressed, the scope of its effectiveness, and its potential sphere of influence. The competitive comparison under the standard reference framework aims for market rewards and also serves as an incentive for the continuous promotion of objectives and efforts.

Investors as well as consumers expect companies to consider ecological and social factors in their activities (Werner 2009). Failing to address these issues is a major risk to corporate reputation; but even when they do address these issues, companies can still be suspected of greenwashing. Owing to this, the German Council for Sustainable Development has intentionally developed a Sustainability Code rather than a CSR code. Corporate sustainability is gaining importance in the international context. In conjunction with the debate on the Green Economy, corporate sustainability can form the basis for reaching a viable consensus in a number of different contexts.

1.1 Theoretical Principles

Just like the customer–company relationship, in some respects the investor–company relationship can be seen as a principal-agent issue (Jensen and Meckling 1976). Information asymmetries arise between the principal, i.e., the investor/customer and the agent, i.e., the company, and these discrepancies give rise to credibility problems. A certain type of corporate behavior is expected. The decision to implement a standard requires companies to make investments – investments that in the current climate are still fraught with uncertain returns. These decisions must take into account not only the practical implementation of the standard within different systems, but also the underlying pros and cons associated with greater transparency (Gilbert and Rasche 2008).

Given that customers and investors find it hard to verify the veracity of the information available, there is a danger that they assume a certain type of behavior on the part of companies and either proceed with or halt their purchases or investments. One solution to this dilemma is to make the alternative option, i.e., non-sustainable behavior, a highly unattractive option for companies. In choosing self-governance, companies compromise their credibility and integrity if they fail to adhere to their own internal code of conduct; they need to consider the negative effect on various stakeholders and avoid doing this (Pies et al. 2009). Companies can opt for self-governance by undertaking to adhere to ecological and social standards such as ISO 14001 and SA 8000, joining initiatives such as the Global Compact and/or commissioning independent third parties to certify their sustainability commitment or publicize their sustainability activities. All of this serves to lend credibility to the company's promise and its sustainability activities (Pies 2009). Stakeholders have a special role here. They are not only recipients of the information on sustainability, but also active participants in the process for implementing codes and standards. They also monitor adherence to the undertakings made by companies and the plausibility of their reporting. This particularly applies to codes and standards that do not require certification (Gilbert and Rasche 2008).

1.2 Tools for Self-Governance and the Associated Challenges

Companies can use many tools for self-governance to give credibility to their commitment to sustainability. Leipziger (2010) refers to over 30 known initiatives, codes, standards, and reporting recommendations that companies can employ to give substance to their commitment to sustainability. This wide variety of options means that companies often adopt several certification standards, join a number of initiatives, and expose themselves to the multiple stresses of various surveys, company visits, and reports in order to meet presumed stakeholder requirements (Chatterji and Levine 2006). There is growing interest in standardization as a way of making sustainability an effective value-oriented strategy, thus promoting the competitive basis for finding forward-looking solutions and services, as well as achieving tangible market reward.

The necessity for standardization is generally accepted and has also been confirmed by the GSC dialogue process. A series of initiatives is attempting to define and implement harmonized global standards (King and Toffel 2007). This includes standardized reporting via the Global Reporting Initiative (GRI) and the non-certifiable ISO 26000 providing guidelines for social responsibility. These initiatives support the spread of sustainability management and sustainability reporting in corporate practice, but also identify certain gaps, especially with regard to a lack of transparency and applicability on the part of capital market players.

Reporting does not take the information requirements of capital market players directly into account, although it is precisely these players who are expected to promote sustainability by means of the uniform use of Environmental, Social, and Governance (ESG) information. It is impossible to establish a sufficiently comparable basis, because the GRI offers so many different options for reporting sustainability. The various options provided by the GRI are put into practice by companies (IÖW und future 2012), with the result of compromising comparability and undermining credibility. Furthermore, the GRI imposes only one reporting standard and provides no implementation criteria (for a critique of this, see Behnam and MacLean 2011) that could otherwise help companies with no established sustainability management strategy to enter this area of activity.

Capital market players and various other stakeholders are in increasing need of a reliable basis for assessing ecological and social sustainability in the form of ESG criteria and their impact on the business model. The requirement for ESG information comes from the fact that analysts and investors need to incorporate non financial and assessment-related information into traditional analytical approaches. Key Performance Indicators (KPIs) for both cash flow and corporate risk make it possible to measure and check individual aspects of sustainability (Bassen et al. 2006; Derwall et al. 2010; Manescu 2010).

2 The Sustainability Code: A Reference Framework for Sustainability

The German Council for Sustainable Development has joined forces with representatives from the financial markets and a number of companies to address the issue of this lack of comparability and standardization and is looking to fill the gaps using the Sustainability Code. This globally applicable standard for transparency is aimed at implementation by and impact on the market itself. It was adopted by the German Council for Sustainable Development in November 2011 (German Council for Sustainable Development 2012).

2.1 Target Groups and Objectives

The aim of the German Council for Sustainable Development is to ensure greater transparency and comparability of corporate sustainability performance and to widen the basis for the implementation of sustainability. The application of the GSC could lead to a series of potential effects. Economic stakeholders such as investors and financial analysts could include the information in their analysis in a standardized form. Market inefficiencies – instances of over- and under estimation of a company's enterprise value – could be reduced and capital allocation optimized. Greater transparency would make it easier to identify and compare corporate opportunities and risks. The intention is also to promote competition and market differentiation by means of innovations for sustainable development. Companies with established sustainability management strategies could achieve competitive advantage by publishing their sustainability management practices. The ever-increasing transaction costs due to the divergent reporting requirements imposed by rating agencies, investors, and other initiatives could be curbed by using standardized information and indicators.

Companies keep room for maneuver and the option to differentiate themselves through voluntary, supplementary key performance indicators for their sector within their declarations of conformity toward the Code. Through their "comply or explain" approach, the Code requirements allow gradual compliance and declaration in the event of non disclosure for any reason. Small and medium enterprises (SMEs) can also use the Code as a point of entry to strategic sustainability communications; as major companies and global brands likewise express a greater interest in sustainability information this could, for example, be in the form of documenting the sustainability as a supplier.

The key aspect is that the Sustainability Code is based on relevant international, general standards such as the UN Global Compact, ISO 26000, and the GRI reporting guidelines.[1] The existing sustainable investment market, with its specific requirements and its comprehensive, methodical assessments, is complemented by a standardized instrument for mainstream capital market players, that so far have paid scant attention to sustainability and that account for 98.7 % of invested capital on the German market (Forum Nachhaltige Geldanlagen 2012).

The incentive for compliance with the Code would primarily be provided by market reward, e.g., inclusion in share indexes, access to new investor categories, and a streamlined choice of suppliers. The intention is that the Code would therefore help investors with their individual evaluation of long-term oriented business activity and the associated risks and opportunities. This also applies to investment advice if investors wish to address the issue of compliance with the Sustainability Code, in addition to obtaining information on risk, return, and cost. The Council recommends that the public sector itself develops sustainable investments in Germany by increasing the focus on the common good in state pension schemes and state pension fund reserves, based on GSC-compliant reporting. Standardization could lead to sustainability becoming increasingly important as an assessment criterion in other investment categories, such as corporate bonds and the lending sector. Due to the increasing importance of passive fund management, it will in the future become more important for companies to appear in a range of

[1] GRI G3 and G4 www.globalreporting.org, EFFAS KPIs for ESG www.effas-esg.com

different indexes. The Sustainability Code could serve as an additional selection criterion for index providers in this context.

When private enterprise and the public sector ensure that their procurement practices are sustainable, compliance with the Code could become a criterion when selecting contractual partners. It would then be possible to raise consumer awareness of products and services offered by sustainable companies. The relatively high sustainability performance of European companies, which are required to meet higher standards and legislative requirements in Europe could become a benchmark for sustainable business activity throughout the world. Any presumed disadvantages of greater transparency and stricter legal requirements could be turned into competitive advantage on the global market by means of an ambitious approach to standardization.

Commitment to the Sustainability Code comes from the market, meaning that market players ensure the quality and credibility of information. In this way, companies receive direct feedback on the practical application of the information they have provided. A development stage based on broad-based as well as expert dialogue (Bassen et al. 2011) has paved the way for discussing a series of wider options for implementation, which have been documented by the RNE (2011).

2.2 *Scope and Application*

The Sustainability Code is recommended in Germany as a form of voluntary self-reporting for application by companies of every size and legal form, organizations, foundations, NGOs, unions, universities, scientific organizations, and media. Public sector companies are asked to adopt the role of pioneers in applying the Code.

Aimed at forming a reliable basis for the selection of business partners and suppliers, the Sustainability Code does not demand external assessment for self-reporting at entry level. To increase effectiveness and reliability for capital market actors, the declaration of conformity acquires credibility through an audit certificate issued by third parties as "limited assurance."

Companies either "comply" with individual Code criteria or "explain" why they do not. They do this in the form of a declaration of conformity toward the Sustainability Code to comply in a procedure similar to financial reporting. The declaration can be produced in German and/or English using an on line tool (www.sustainabilitycode.org). Once it has been formally approved by the office of the German Council for Sustainable Development, it is published on the Internet, in the annual report, and in any internal or integrated sustainability report. The German Sustainability Council provides companies with a signet for public communications.

Companies qualify their approach to GSC requirements using a series of KPIs. Depending on the reporting standards applied and their target audience, companies decide whether to apply GRI or EFFAS indicators.

Further voluntary modifications can be made using GRI Sector Supplements or the sector-specific KPIs of the EFFAS.

In order to ensure comparability with financial reporting, the GSC generally refers to the same group of consolidated companies as those to be included in the corporate financial statement. Companies state if this is not to be the case, explaining any deviation. It is essential for many companies with a low level of vertical integration to report on the sustainability of their outsourced value chains.

3 GSC Content and Requirements

The GSC is divided into sections entitled Strategy, Process Management, Environment, and Society. This structure is based on the ESG criteria relevant to the capital market (see also ECCE 2007 for derivation and weighting; Bassen et al. 2006) and is also based on the pertinent questions relating to say, the GRI or ISO 26000 regarding the development of strategy as part of the process of implementation. The "Strategy" section addresses how a company positions itself strategically in terms of its approach to sustainability. This section also looks at how sustainability is incorporated in the value chain and the sustainability objectives that have been set within the value chain and for various markets.

Once sustainability factors have been integrated strategically, it is also extremely important to implement them at various process levels. With this in mind, the Process Management section reviews the rules and processes implemented by the company. In addition to the specifics of functional areas (e.g., purchasing, production, research), it also includes looking at risk management and internal corporate management. This is necessary because sustainability must be incorporated within management systems if it is a necessary component of company success. Building on the short reports prepared for strategy and process management, the GSC also addresses ESG criteria.

Analysis, Strategy, and Objectives The GSC helps companies to consider specific aspects and to document them. It is here that companies state their approach to strategic parameters and objectives. Companies are required to describe the importance of ESG in terms of corporate strategy and explain how strategy implementation will take account of ESG factors.

Companies state how they intend to analyze the risks and opportunities for their essential activities in terms of sustainable development. They also provide details of the steps that will be taken to operate in harmony with the essential known national and international standards for their sector. They also state how their strategy for essential activities and systematic implementation considers all aspects of sustainability that have an essential impact on the company. Examples include strategic positioning vis-à-vis the competition, innovation management, climate-, environment, and resource-friendly business activity, demographic trends, the value chain, product life cycle, product portfolio, and other factors. Companies state which qualitative and/or quantitative and time-bound sustainability objectives are being set and fed into operations and how progress in achieving them will be monitored. They state the way in which essential suppliers, employees, capital markets, customers, and stakeholders will be regularly involved. Companies announce the extent to which sustainability criteria will be assessed in the value chain and the importance of sustainability to the value creation process.

Process Management It is particularly important to describe the rules and processes applied by companies for implementing sustainability. This process identifies senior management responsibilities for corporate sustainability and explains how the sustainability strategy will be implemented through rules and processes. Companies state any special circumstances relating to purchasing, production, services, personnel, investment, research and development, logistics/transport, and marketing, and explain how suppliers, customers, and other stakeholders (e.g., employees) will be taken into consideration.

Similarly to key financial data, sustainability KPIs are integrated as part of regular internal planning and control. The appropriate procedures provide the reliability, comparability, and data consistency required for internal control and external communications. The following KPIs are applied to process management:

G4-56 – Describe the organization's values, principles, standards, and norms of behavior such as codes of conduct and codes of ethics.

Alternatively, reporting can be in accordance with the following EFFAS criteria.

EFFAS S06-01 – Percentage of total suppliers and supply chain partners screened for compliance in accordance with ESG criteria.

EFFAS S06-02 – Percentage of suppliers and supply chain partners audited for compliance.

Incentive Schemes In terms of the incentive systems applied, companies state how the target agreements and bonuses for managers and employees also relate to achieving sustainability objectives and long-term value creation. The extent to which sustainability performance is reviewed by the monitoring body (supervisory board/advisory committee) in assessments at the highest management level (management board/management) is also stated.

The following GRI or EFFAS indicators are applied as KPIs in the context of incentive schemes:

G4-51a – Report the remuneration policies for the highest governance body and senior executives.

G4-54 – Report the ratio of the annual total compensation for the organization's highest-paid individual in each country of significant operations to the median annual total compensation for all employees (excluding the highest-paid individual) in the same country.

Stakeholder Engagement Companies disclose how the socially and economically relevant stakeholders are identified and integrated into the sustainability process.

It is disclosed whether and how an ongoing dialogue takes place with them and how the results are integrated into the sustainability process.

The following GRI or EFFAS indicators are applied as KPIs:

G4-27 a – Report key topics and concerns that have been raised through stakeholder engagement, and how the organization has responded to those key topics and concerns, including through its reporting. Report the stakeholder groups that raised each of the key topics and concerns.

Innovation and Product Management Companies state how innovations in products and services are enhanced through suitable processes which improve sustainability with respect to the company's utilization of resources and with regard to users. Likewise, a further statement is made with regard as to how the current and future impact of the key products and services in the value chain and in the product life cycle are assessed.

The following GRI or EFFAS indicators are applied as KPIs:

G4-EN6 – Reduction of energy consumption.
G4-FS11 – Percentage of assets subject to positive and negative environmental or social screening.
EFFAS E13-01 – Improvement rate of product energy efficiency compared to previous year.
EFFAS V04-12 – Total investments in research on ESG-relevant aspects of business such as, e.g., eco-design, eco-efficient production processes, decreasing impact on biodiversity, improving health and safety conditions of employees or supply chain partners, consulting on integration of ESG aspects in change management, development of products to exploit ESG opportunities, etc., in monetary terms, i.e., currency as a percentage of revenue.

Environment The use of natural resources is analyzed and described. Companies state the extent to which natural resources are used for business activities. Possible options here are materials, the input and output of water, soil, waste, energy, emissions, land and biodiversity, as well as emissions for the life cycles of products and services. The company discloses what qualitative and quantitative goals it has set itself with regard to its resource efficiency, its use of renewables, the increase in raw material productivity, and the reduction in the usage of ecosystem services, and how these goals have been met or will be met in the future.

The following GRI or EFFAS indicators are applied as KPIs:

G4-EN1 – Materials used by weight or volume.
G4-EN3 – Energy consumption within the organization.
G4-EN8 – Total water withdrawal by source.
G4-EN23 – Total weight of waste by type and disposal method.
EFFAS E04-01 – Total waste in tons.
EFFAS E05-01 – Percentage of total waste which is recycled.
EFFAS E01-01 – Energy consumption, total.

There is a particular emphasis on avoiding greenhouse gas emissions. Companies state their levels of greenhouse gas (GHG) emissions, along with internal objectives determined in line with the Greenhouse Gas (GHG) Protocol or standards based thereon.

The following GRI or EFFAS indicators are applied as KPIs:

G4-EN15 – Direct greenhouse gas (GHG) emissions (Scope 1).
G4-EN16 – Energy indirect greenhouse gas (GHG) emissions (Scope 2).
G4-EN17 – Other indirect greenhouse gas (GHG) emissions (Scope 3).
G4-EN19 – Reduction of greenhouse gas (GHG) emissions.
EFFAS E02-01 – GHG emissions, total (scope I, II, III).

Society Companies provide information on their human rights and diversity policies. They describe how they ensure that they respect national international employee rights via recognized standards and promote employee involvement in sustainability management. Companies state how they have implemented national and international processes to promote equal opportunities, health protection, inclusion of migrants and disabled persons, fair pay, and a work-life balance; companies also state how they prevent all forms of discrimination relating to, e.g., ethnic origin, gender, religion/belief, age, or sexual identity. Furthermore, companies state what goals it has set and what measures it has taken to promote the employability of all employees, i.e., the ability of all employees to participate in the working and professional world, and to adapt it to demographic change.

The following GRI or EFFAS indicators are applied as KPIs:

G4-LA6 – Type of injury and rates of injury, occupational diseases, lost days and absenteeism, and total number of work-related fatalities, by region and by gender.
G4-LA8 – Health and safety topics covered in formal agreements with trade unions.
G4-LA9 – Average hours of training per year per employee by gender, and by employee category.
G4-LA12 – Composition of governance bodies and breakdown of employees per employee category according to gender, age group, minority group membership, and other indicators of diversity.
G4-HR3 – Total number of incidents of discrimination and corrective actions taken.
EFFAS S03-01 – Age structure/distribution (number of FTEs per age group, 10-year intervals).
EFFAS S10-01 – Percentage of female employees in relation to total employees.
EFFAS S10-02 – Percentage of female FTEs in senior positions in relation to total FTEs in senior positions.
EFFAS S02-02 – Average expenses on training per FTE p.a.

Human Rights Companies state the steps taken in the supply chain to ensure respect for human rights and to prevent forced and child labor and all forms of exploitation (e.g., United Nations "Protect, Respect and Remedy: a Framework for Business and Human Rights," International Labour Organisation (ILO) core labor standards, United Nations Employment and Social standards).

The following GRI or EFFAS indicators are applied as KPIs:

G4-HR1 – Total number and percentage of significant investment agreements and contracts that include human rights clauses or that underwent human rights screening.
G4-HR9 – Total number and percentage of operations that have been subject to human rights reviews or impact assessments.
G4-HR10 – Percentage of new suppliers that were screened using human rights criteria.
G4-HR11 – Significant actual and potential negative human rights impacts in the supply chain and actions taken.
EFFAS S07-02 II – Percentage of total facilities certificated according to SA 8000 standard.

Local Community and Corporate Citizenship Companies state how they contribute to the local communities where they conduct essential business activity.

Political Influence All essential involvement in legislative procedure, essential lobbying activities via entry in a lobby register, essential payments of membership fees, payments to governments, and donations to parties and politicians are to be stated by country.

The following GRI or EFFAS indicators are applied as KPIs:

GRI SO6 – Total value of financial and in-kind contributions to political parties, politicians, and related institutions by country.
EFFAS G01-01 – Contributions to political parties as a percentage of total revenues.

Conduct that Complies with the Law and Policy In line with accepted standards (e.g., Transparency International Business Principles, International Corporate Governance Network Bribery and Corruption Guidelines), companies state which systems and processes are currently in place to prevent illegal conduct – particularly corruption – and how these systems are screened (e.g., IDW PS 980). They also state how corruption is uncovered, prevented, and penalized.

The following GRI or EFFAS indicators are applied as KPIs:

G4-SO3 – Total number and percentage of operations assessed for risks related to corruption and the significant risks identified.
G4-SO5 – Confirmed incidents of corruption and actions taken.
G4-SO8 – Monetary value of significant fines and total number of non monetary sanctions for non compliance with laws and regulations.
EFFAS V01-01 – Expenses and fines on filings, lawsuits related to anti competitive behavior, anti trust, and monopoly practices.
EFFAS V02-01 – Percentage of revenues in regions with a Transparency International Corruption Perception Index below 60.

4 Analysis and Findings

One year after the introduction of the code, stakeholders were surveyed in order to conduct a comprehensive analysis of the GSC's reach and effectiveness (Bassen 2013). The survey covered three central GSC target groups: companies (N = 70), capital market players (N = 163, 47 % traditional analysts, 7 % Socially Responsible Investors, and the remainder who gave no answer), and the third sector (N = 35). The survey was presented to a group of international experts in February 2013 to assess the code in terms of its sustainability policy and also submitted to the German Council for Sustainable Development to evaluate the code as part of its final report toward the end of the past period of appointment.

A major factor influencing take-up by companies is the amount of work involved in producing the declaration of conformity. The survey of users revealed that it took an average of 5 working days to produce and agree a declaration. However, the period varied enormously, ranging from a period of 1 day with the minimum amount of effort to a maximum of 50 days. Large companies with established reporting systems took up to 20 days less than small and medium-sized enterprises.

The biggest challenge for all groups surveyed was data availability and reliability. All groups surveyed saw increased transparency (companies 56 %, capital market players 25 %, NGOs 20 %), the fact that sustainability was universally taken forward (companies 72 %, capital market players 24 %, NGOs 20 %), and the enhancement of sustainability (companies 36 %, capital market players 15 %, NGOs 17 %) as major benefits of implementing the GSC. Greater comparability was cited as a further benefit (companies 32 %, capital market players 12 %, NGOs 17 %). The groups surveyed thought that the declaration of compliance sent the right message in terms of credibility and showed compliance with policy. According to the respondents, the declaration is a good way of demonstrating that a company would like to push sustainability forward. However, companies and capital market players bemoaned the lack or low level of relevance for the market, because the data providers do not have the data readily available, meaning that the code does not have an immediate impact on alleviating their workload.

Companies could take greater responsibility for their role as drivers of sustainability by selecting investments for their sustainability aspects – a GSC requirement that capital market players also identified as a credibility factor. This might motivate capital market players in the right direction, as the survey conducted amongst asset managers, asset owners, sell and buy analysts, rating agency analysts, etc., revealed that a mere 5 % have implemented the GSC at company or asset management level.

What is remarkable is the expected triggers cited by respondents: companies saw the triggers as coming from the political sector (33 %), customers (24 %), and investors (24 %); capital market players saw the triggers as coming from themselves (20 %) and the general public (20 %); NGOs likewise saw the triggers coming from themselves (14 %) and the political sector (11 %).

5 Future Outlook

It was a major concern of the German Council for Sustainable Development to include stakeholders and interested parties in the dialogue for developing the GSC. Dialogue is particularly important for finding viable solutions to sustainability issues.

The varying responses provided by the individual target groups surveyed and the differing goals of the players involved reveal the diversity of expectations of companies and individual players in the context of sustainable business activity. It is quite clear that there is a need for improved communication to convey a more straightforward and tangible definition of the objective of the German Council for Sustainable Development. This will eliminate conflicting stakeholder objectives and improve GSC take-up by companies and potential users.

If implementation is successful in Germany, it can be assumed that this will give significant impetus to the continuing discussion on how to standardize sustainability Europe-wide, going beyond EU Commission expectations. Since its publication in 2011, 75 enterprises have submitted declarations of conformity with the Sustainability Code of which 27 are listed companies, 24 are public sector companies, who have all made a high-profile commitment to sustainability. Many signatories have already submitted declarations of conformity for several years so that 157 declarations are published. German Federal Chancellor Dr Angela Merkel praised the Sustainability Code in her speech at the annual conference of the German Council for Sustainable Development on 25 June 2012, saying that it was a good way forward for sustainable business activity: "The Federal Government supports this code. We are promoting it so that more companies take it up (GSC 2013)."

This transparency standard is applicable on international level. Declarations of conformity from two companies situated in European member countries, Sweden and Finland sets an example for that. Implementation with European and international cooperation partners is seen to be just as important as increasing take-up within the German economy, possibly by addressing non-reporting entities directly.

The plan is to update the Sustainability Code regularly in line with further developments in reporting standards and other universally applicable standards. The latest update was in August 2014. This ensures that the Code remains compatible with current international trends. The structures created or used for achieving this are yet to be discussed.

Bibliography

Bassen, A., Meyer, K., & Schlange, J. (2006). *The influence of corporate responsibility on the cost of capital: An empirical analysis*. Working paper. http://ssrn.com/abstract=984406

Bassen, A., Rentrop, A., & Zwick, Y. (2011). Deutscher Nachhaltigkeitskodex (DNK) – Konzept, Inhalte und Entwicklungsschritte. *Zeitschrift für Umweltpolitik & Umweltrecht*, *3*, 359–375.

Bassen, A. (2013). *Analyse der Umsetzung und Wirksamkeit des DNK – Ein Review im Auftrag des Rates für Nachhaltige Entwicklung*. http://www.nachhaltigkeitsrat.de/projekte/eigene-projekte/deutscher-nachhaltigkeitskodex/dokumente/dnk-review-summary-06-02-2013/?blstr=0. Accessed 28 February 2013.

Behnam, M., & MacLean, T. (2011). Where is the accountability in international accountability standards? *Business Ethics Quarterly, 21*(1), 45–72.

BMAS–Bundesministerium für Arbeit und Soziales. (2010). *Nationale Strategie zur gesellschaftlichen Verantwortung von Unternehmen (Corporate Social Responsibility – CSR) Aktionsplan CSR der Bundesregierung*.

Chatterji, A., & Levine, D. (2006). Breaking down the wall of codes. *California Management Review, 48*, 29–51.

Crane, A., & Matten, D. (2004). *Business ethics. A European perspective managing corporate citizenship and sustainability in the age of globalization*. Oxford: Oxford University Press.

Derwall, J., Koedijk, K., & Ter Horst, J. (2010). *A tale of values-driven and profit-seeking in social investors*. Working paper. http://www.corporate-engagement.com/index.php?pageID=1881&n=327&itemID=499445

DNK. (2013). "Deutscher Nachhaltigkeitskodex (DNK) von 51 Unternehmen unterzeichnet," in *Deutscher Nachhaltigkeitskodex*. http://www.deutscher-nachhaltigkeitskodex.de/de/hintergruende/aktuelles/nachricht/artikel/deutscher-nachhaltigkeitskodex-dnk-von-51-unternehmenunterzeichnet.html. Accessed 11 June 2013.

Forum Nachhaltige Geldanlagen. (2012). *Marktbericht Nachhaltige Geldanlagen. Deutschland, Österreich und die Schweiz*. http://www.forum-ng.org/images/stories/Publikationen/fng_marktbericht_2012_72dpi.pdf. Accessed 28 February 2013.

Gilbert, D., & Rasche, A. (2008). Opportunities and problems of standardized ethics initiatives – A stakeholder theory perspective. *Journal of Business Ethics, 82*, 755–773.

Institut für ökologische Wirtschaftsforschung und future e. V. (Hrsg.). (2012). *Das IÖW/future-Ranking der Nachhaltigkeitsberichte 2011: Ergebnisse und Trends*. Berlin, Münster. http://www.ranking-nachhaltigkeitsberichte.de/data/ranking/user_upload/pdf/IOEW-future-Ranking_2011_Grossunternehmen_Ergebnisbericht.pdf. Accessed 19 February 2013.

Jensen, M., & Meckling, W. (1976). Theory of the firm: Managerial behavior, agency costs and ownership structure. *Journal of Financial Economics, 3*(4), 305–360.

King, A., & Toffel, M. W. (2007). *Self-regulatory institutions for solving environmental problems: Perspectives and contributions from the management literature*. http://ssrn.com/abstract=985619

Leipziger, D. (2010). *The corporate responsibility code book*. Sheffield: Greenleaf.

Manescu, C. (2010). *Stock returns in relation to environmental, social, and governance performance: Mispricing or compensation for risk?* http://hdl.handle.net/2077/20998

Pies, I., von Winnig, A., Sardison, M., & Girlich, K. (2009). *Nachhaltigkeit in der Mineralölindustrie: Theorie und Praxis freiwilliger Selbstverpflichtungen*. Wirtschaftsethik-Studie Nr. 2009–1, des Lehrstuhls für Wirtschaftsethik an der Martin-Luther-Universität Halle-Wittenberg. http://wcms.uzi.uni-halle.de/download.php?down=13079&elem=2235793. Accessed 31 May 2011.

Rat für Nachhaltige Entwicklung. (2011). *Gold-Standard Ökolandbau: Für eine nachhaltige Gestaltung der Energiewende*. Berlin. http://www.nachhaltigkeitsrat.de/dokumente/empfehlungen/texte-nr-40-jul-2011/. Accessed 19 February 2013.

Rat für Nachhaltige Entwicklung. (2014). *Der Deutsche Nachhaltigkeitskodex (DNK)*. Berlin. www.sustainabilitycode.eu. Accessed 21 November 2014.

United Nations World Commission on Environment and Development. (1987). *Our common future* (Brundtland Report). Oxford.

Werner, T. (2009). *Ökologische Investments: Chancen und Risiken grüner Geldanlagen*. Wiesbaden: Gabler.

Authors and Contributors of "Sustainable Value Chain Management"

Michael D'heur is founder and managing director of shared.value.chain, a think tank and advisory firm for sustainable value chain management. He supports multi national and mid-sized companies across various industries, to create sustainable products and to operate flexible but sustainable global supply chains. Michael is thought leader, author, speaker, and advisor to drive the adoption of sustainability in the core business. He studied Economics and Business Informatics at the University of Siegen, Germany.

M. D'heur (ed.), *Sustainable Value Chain Management*, CSR, Sustainability, Ethics & Governance, DOI 10.1007/978-3-319-12142-0

René Schmidpeter holds the Dr. Juergen Meyer Endowed Chair of International Business Ethics and Corporate Social Responsibility at Cologne Business School (CBS). He is Guest Professor at Nanjing University of Finance and Economics in China, Academic Head of the Center for Humane Market Economy in Salzburg, and Editor of the international Springer Series CSR, Sustainability, Ethics, and Governance. He is thought leader of strategic CSR thinking and international recognized speaker on topics such as Sustainability, Corporate Social Responsibility, and Strategic Management. Contact: rene.schmidpeter@gmx.de.

AUDI

Martina Biendl, Master of Arts, studied Media and Communications at Augsburg University in Germany and at Aarhus University in Denmark. After having worked as a freelance journalist for 3 years, she started her employment at AUDI AG in 2012. Since 2013 she has been responsible for communications within the Corporate Responsibility department.

Dr.-Ing. Peter F. Tropschuh started his career at AUDI AG in Technical Development in 1988 after graduating in Mechanical Engineering and obtaining a doctorate at the Technical University of Munich. He headed the General Secretariat and was responsible for the Development Vehicle Projects and Research Projects areas, among others. In 2006, he moved to Volkswagen AG as Head of AutoUni and Research Projects. Since July 2011 he has been Head of Corporate Responsibility as well as Government Affairs at AUDI AG.

BASF

Dr. Dirk Voeste is Vice President Sustainability Strategy at BASF SE in Ludwigshafen, Germany. He is responsible for BASF's sustainability strategy and its implementation in business activities and decision-making processes. His tasks also include the development and utilization of Eco-Efficiency Analyses and Sustainability Assessments as well as Stakeholder Relations. He holds a PhD from the University Bonn, Germany, and a Masters from the Cranfield Institute of Technology in Bedford, United Kingdom. After completing his studies he was responsible for developing an aquatic ecosystem for space experiments with NASDA. He joined BASF in 1998.

Dr. Markus Frank is responsible for sustainability assessments in agriculture and subsequently for the implementation and further development of AgBalance™ at BASF Crop Protection in Limbugerhof, Germany. Dr. Frank studied Biology in Kiel and Cologne and wrote his PhD thesis at the Center for Plant Molecular Biology in Tübingen, Germany. Besides this, he holds an MBA from the Surrey Business School in Guildford, United Kingdom. After joining BASF he worked in biotechnology research as well as global strategic marketing before joining Global Sustainability at BASF Crop Protection.

Katharina Fischer, Senior Manager Global External Communication, joined BASF SE in 2011. She has been responsible for the introduction and communication of AgBalance™, BASF's method to measure sustainability in agricultural processes. Fischer studied Politics and German literature at the University of Osnabrück. Before joining BASF she was an advisor at Johanssen + Kretschmer Strategic Communications in Berlin.

Beiersdorf

Daniel Weber is responsible for the global supply chain at Beiersdorf AG. After studying Engineering, he started in Finance at Unilever. From 1995 to 2005 he worked for tesa AG in Germany, the United States, and Asia. Daniel Weber joined Beiersdorf in 2005 to chair and supervise the restructuring of the European Supply Chain before becoming responsible for the global supply chain.

Dorle Bahr is responsible for Environmental Sustainability and Safety at Beiersdorf. She studied Business Mathematics at the University of Hamburg. She started her career in Logistics at Colgate Palmolive GmbH. She first became the Director for Logistics in Germany and Austria before moving to European Logistics in 2000. Dorle Bahr joined Beiersdorf in 2004, when she built up the Supply Chain Customer Management department in Hamburg. In 2010 she took over responsibility for strategic sustainability strategies in the Supply Chain (Blue Supply Chain). Since 2010 she chairs the department for Environmental Sustainability and Safety within Beiersdorf's Corporate Sustainability Office.

European Forum Alpbach

Franz Fischler's career as a representative of the agriculture sector began in 1979 when he joined the Tyrol Chamber of Agriculture. He dealt with environmental issues, education and training, culture, and land-use planning until 1984, when he became Director of the Chamber. In 1989, he was appointed Federal Minister of Agriculture and Forestry in Austria. In that capacity he played an important part in the negotiations for Austria's accession to the European Union. From 1995 to 1999 he was European Commissioner for Agriculture and Rural Development. From September 1999 until 2004 he was also responsible for Community fisheries policy. Since December 2004, Franz Fischler is concentrating on doing consultancy work, giving lectures, holding speeches, and fulfilling his tasks being President of the think tank "European Forum Alpbach" since 2012 and Chairman of the RISE-Foundation.

Fairphone

Tessa Wernink cofounded Fairphone and has served as the Director of Communications for the social enterprise since January 2013. In her role, she works closely with her team, advisors, researchers, and external stakeholders to share Fairphone's story and mission and collaborate on initiatives that further their goal of a fairer economy. Prior to joining Fairphone, Tessa Wernink held a number of international PR and communications positions ranging from education to the aviation and hospitality sectors. In recent years, her primary focus has been promoting the creative sector in Amsterdam. Tessa earned a Master's degree in English Literature and a minor in International Development Studies from the University of Amsterdam. Her degrees and experience support Tessa's passion for storytelling and building communities that can influence social change.

German Council for Sustainable Development

Marlehn Thieme is the Chairperson at the German Council for Sustainable Development, and a member of board of the Evangelical Church Germany. Born in 1957 she studied Law and Social Sciences and worked for Deutsche Bank AG from 1986 to 2013, at last as Director responsible for Corporate Social Responsibility and member of the Supervisory Board of Deutsche Bank AG. Since 2004 she is a member of the Council for Sustainable Development; in 2012 she was elected as its Chairperson.

Yvonne Zwick is a scientific advisor at the German Council for Sustainable Development. She studied catholic theology with a focus on Christian social ethics and moral theology at the University of Freiburg, Germany. At the office of the Council she is responsible for sustainable lifestyle and consumption, Corporate Social Responsibility, social-ethical investments, and the German Sustainability Code.

Henkel

Dr. Frank Roland Schroeder is head of the sustainability department and external affairs in Henkel's Laundry and Home Care business unit. He has worked with Henkel since 1985 in various positions in the field of biotechnology, ecology, and product safety being active in European as well as national associations (A.I.S.E., ERASM, HAD, IKW/Forum Waschen, SEPAWA/LUV). He studied Chemistry and Biochemistry at the University of Göttingen and holds a PhD degree from the Technical University of Braunschweig and the Max-Planck-Institute for Experimental Medicine in Göttingen.

Dr. Dirk Holbach is Corporate Vice President at Henkel and in charge of the supply chain of the Laundry & Home Care division. His responsibility includes the supply chain planning, production, and customer logistics throughout the world. Before taking up his current position Dr. Holbach was responsible for International Production, Global Supply Chain Operations, Central Purchasing at Laundry & Home Care, as well as various positions in Sales. He studied Business Management and Engineering at the Technical University Kaiserslautern, Germany, and did his PhD in Business Informatics at the University of Duisburg-Essen, Germany.

Professor Dr. Thomas Müller-Kirschbaum is in charge of global research and technology for the Laundry & Home Care business at Henkel. From 2005 until end of 2013 he was on top responsible for global production and supply chain of this division. Since 2013 he has in addition corporate responsibilities as co-chair of Henkel's Sustainability Council and chairman of the corporate research and development steering committee. Before joining Henkel in 1989 he studied Physics, Technical Chemistry, and Environmental Technology at the University of Cologne and RWTH Aachen, Germany. He is a member of multiple scientific and industrial committees including the Research Committee and the Sustainability Board of the Association of the German Chemical Industry.

Infineon

Dr. Kurt Gruber has more than 30 years of experience in the semiconductor industry and has been in charge of Infineon's Corporate Supply Chain for 7 years now. The Corporate Supply Chain entails the consolidation of corporate demand and capacity and the handling of assignments and global operations for all of Infineon's customers. Dr. Guber's global team plans, controls, and supervises the internal and external network of production and logistics and coordinates strategies for supply chain and production. He took up accountability in terms of corporate responsibility including the "Center of Excellence" and the "strategic production management." Dr. Gruber holds a PhD in Mathematics from the Technical University in Graz, Austria.

Hans Ehm studied Physics in Munich, Germany, and Mechanical Engineering in the United States. He has 30 years of experience in the production of semiconductors and global supply chain and is in charge of the "Supply Chain Innovation" division at Infineon. Since 2000 he is part of the supervising committee of camLine Holding AG, an IT solutions provider for the automation of production and strategic process control. Hans Ehm supports Supply Chain theory. Since 2008 he has been a member of the European Leadership Team of the Supply Chain Council and became its Chairman in 2012 and is since 2013 a Board member of APICS Supply Chain Council.

Dr. Christian Pophal studied at the Technical University Darmstadt, Germany. He wrote his PhD thesis at Sophia University Tokyo and in Darmstadt in the field of material science. He works at Infineon Technologies since 2004 and is currently the Senior Director Business Continuity. He is responsible for global Corporate Sustainability and Social Responsibility, Technical Safety and Environmental Affairs, as well as Corporate Energy Management. Dr. Pophal is a member of the steering committee for environmental protection, energy efficiency, and climate protection of the German Electrical and Electronic Manufacturers' Association (ZVEI – Zentralverband Elektrotechnik- und Elektronikindustrie e.V.). Besides this, he is chairman of the "Energy Working Group" of the European Semiconductor Industry Association (ESIA).

Klenk & Hoursch

Dr. Volker Klenk is Managing Partner at Klenk & Hoursch, an agency for corporate & brand communications in Frankfurt, Germany. He has been dealing with transparency as a key success factor in business for many years, and is responsible for the Germany-wide unique webpage dealing with transparency www.transparenz.net. Moreover, he published a standard textbook about Corporate Transparency and has conducted some of the first large-scale transparency studies.

Georg Lahme is Managing Partner at Klenk & Hoursch. He supports international companies and German small and medium-sized businesses in terms of development, optimization, and communication of their CSR and sustainability portfolios. His key areas of expertise are strategic CSR communications, foundations, and corporate volunteering. Georg Lahme is a lecturer for CSR at the German Press Academy and member of the German Association of Environmental Management B.A.U.M. e.V.

Independent Capital Group AG

Dr. Mirjam Staub-Bisang is cofounder and managing partner at Independent Capital Management AG, a Zurich-based asset management firm focused on sustainable investing and real estate. Prior, she held senior positions in investment banking/asset management among which Commerzbank, Merrill Lynch, and Swiss Life Private Equity Partners as well VP Corporate Development in a global industrial holding company. Dr. Staub-Bisang holds a PhD degree in law from the University of Zurich, is an MBA from INSEAD/ Fontainebleau, and is an attorney-at-law. She is a nonexecutive director of several for-profit and non profit companies and institutions, among which V-Zug, the leading Swiss white goods manufacturer, and the global business school INSEAD in Fontainebleau/Singapore. Additionally, she was elected a Young Global Leader of the World Economic Forum. Dr. Staub-Bisang authored the standard work "Sustainable Investing for Institutional Investors" (Wiley 2012). She lectures and publishes widely on investment topics with a focus on sustainable investing.

International Paper

Teri Shanahan is Vice President of Sustainability for International Paper, based in Memphis, Tennessee. Her role encompasses creating and executing a global strategy for the corporation as it pertains to the three pillars of sustainability: social, environmental, and economic performance. Ms. Shanahan has been with International Paper since 1991, and has held positions in sales and marketing, as well as managing the company's North American merchant papers business and its global pulp business. She also worked in the company's former chemical business. Prior to joining the company, she served for 8 years as a commissioned officer in the U.S. Navy, and was the first woman to qualify as a Surface Warfare Officer on board a combatant ship. She earned a bachelor's degree from the University of Minnesota and a master's degree from the Naval Postgraduate School. She and her husband, Tad Dutch, have one daughter, Haley Shanahan Dutch.

James McDonald is Manager, Sustainability based in Memphis, Tennessee. Mr. McDonald has been with International Paper since 1997, and has been primarily focused on public policy and environmental issues related to sustainability. He works closely with the manufacturing, marketing, and sales professionals to develop and implement policies supporting IP's sustainability performance and objectives, with a special focus on creating innovative solutions for our customers' sustainability supply chain challenges.

Nanogate AG

Ralf Zastrau is chairman of Nanogate AG and responsible for the areas strategy, corporate development, corporate communications, and investments. After Nanogate was founded in 1999 he caused it to progress from a scientific start-up to a market-driven technology company. Ralf Zastrau aims for a corporate strategy, in which not the mere economic success is important, but in which corporate responsibility is also central. Nanogate AG is committed to responsible handling of new technologies and is active in numerous initiatives, for example, in the German Association for Nanotechnology (DV Nano).

Nestlé

John Bee is the Communications Manager for Public Affairs at Nestlé S.A. He is a senior member of the team designing, developing, and deploying Nestlé's global Creating Shared Value approach to business; supporting the integration of CSV within the organization with internal communications campaigns; and socializing the concept and Nestlé's performance and challenges with external stakeholder audiences through increasingly transparent and complex societal reporting to GRI A+ application level. He also develops major communications and stakeholder events with internal and external audiences, and spearheads the company's presence in other organizations' events. He coordinates team contributing to Nestlé's presence within the Shared Value Initiative headed by Harvard professors Michael Porter and Mark Kramer.

Peggy Diby is the Head of Corporate Communications and Public Affairs, Nestlé Central and West Africa Region (CWAR). She is a graduate of National Polytechnic High School of Côte d'Ivoire where she earned a degree in Agronomy Engineering. In addition, she holds a Master degree in Management from the African Center of Management and Capabilities Building for Managers (CAMPC). Prior to joining Nestlé CWAR, Peggy Diby was involved in several rural community development and training programs in West Africa from which she created a strong, resilient, and sustainable network of rural communities. In Nestlé Côte d'Ivoire, she built and maintained strong relationships with local government authorities, media and rural communities, and NGOs while supporting and communicating Nestlé's Creating shared Value initiative in Côte d'Ivoire. As the Head of Corporate Communications and Public Affairs in CWAR, Peggy Diby has the mission to define the framework to continue raising Nestlé's brand while communicating Nestlé's engagement in nutrition, water, and rural development.

Bineta Mbacké holds the position of a Public Affairs Manager at Nestlé S.A. She is in charge of the cultivation and management of global partnerships, with the goals of identifying increased opportunities for Nestlé to Create Shared Value and of communicating Nestlé's Nutrition Health and Wellness leadership to key external stakeholders. Bineta Mbacké also looks after the implementation and communication of Nestlé's Women's Empowerment Principles in the market place and in the Nestlé community. She is an MBA graduate of the Paris Graduate School of Management and she is passionate about social entrepreneurship and ethical leadership.

Barbara Wettstein is a Public Affairs Specialist at Nestlé S.A. She is in charge of the Nestlé Prize in Creating Shared Value, which rewards innovative programs in the areas of nutrition, water, and rural development. The Prize is awarded every 2 years and is open to social entrepreneurs and enterprises, small and medium enterprises, not-for-profit organizations, academic institutions, and governmental agencies. In addition, she leads CSV related communication activities on social media and digital platforms.

Responsible Investmentbanking

Karen Wendt works in Corporate and Investment banking of a Top Tier International Financial Institution and has been the Head of the Equator Principles Team for more than 5 years. She was instrumental in introducing the Equator Principles Procedures and Policies in her Institution and in advancing Equator Principles in international settings and organizations. She was a co author of the EP in 2006, has actively supported their strategic reviews, and was involved in drafting EP III. Since the creation of the Equator Principles Financial Institutions Association (EPFIA) in 2008, Karen Wendt has been sitting in its Steering Committee. She has experiences with human rights due diligence, international labor law, environmental due diligence, and the creation of action plans and mitigation strategies for large international transactions of various industries and regions to avoid or minimize environmental and social risk. In particular she is skilled in negotiating international standards at the international level and with multilateral organisations and implementing such standards consequently into policies and procedures within banks. Karen Wendt conducts stakeholder dialogue with international network organization in regard to ESG. She is editor of the books "Responsible Investmentbanking" and "Sustainable Financial Innovation." She holds an MBA from the University of Liverpool, UK.

SAP

Heino Kantimm joined SAP in 1998 where he worked as an analyst in the newly formed competitive intelligence department. He helped to grow the small team to a global organization, of which he took over global leadership in 2007. Since 2009 he is focusing on new business models at SAP. First, as a Chief Expert Social Sustainability, Mr. Kantimm was leading the StarShea project. He is co founder and member of the Board of Directors of the social business StarShea Ltd. in Ghana. Today, Mr. Kantimm is driving SAP's business model innovation efforts around connected vehicles. Heino Kantimm lived and worked on four different continents. He has a degree in computer sciences and business administration from Mannheim University, Germany.

shared.value.chain

Carina Strahl works at shared.value.chain, a think tank and management advisory firm for sustainable value chain management, in Munich, Germany. Based on her background in international relations and development cooperation, she is an expert for multi-stakeholder collaboration and its relation to value chain management. Before joining shared.value.chain, Carina worked for GIZ, UNISDR, and FAO in South Africa, Botswana, Italy, and Switzerland. Carina studied International Relations at the University of Southampton, UK, and International Affairs at the Graduate Institute of International and Development Studies in Geneva, Switzerland.

Siemens

Ralf Pfitzner is Vice President Sustainability – Strategy & Environmental Portfolio at Siemens AG. He is responsible for strategic topics such as positioning of Siemens' Environmental Portfolio, consisting of highly energy efficient products and solutions, renewable energies, and environmental technologies. His team is supporting Siemens' businesses in order to identify sustainability-related business levers. In addition, he is in charge of Sustainability ratings such as the Dow Jones Sustainability Index that evaluated Siemens as most sustainable industrial company ("Industry Group Leader") in 2013. Previously, Ralf Pfitzner headed the department of Product-Related Environmental Protection of Siemens AG. Before he joined Siemens in 2004, he was working at KPMG Sustainability Services and at the Institute for Futures Studies and Technology Assessment in Berlin. He has a degree in Engineering and Environmental Technologies (Technical University of Berlin) and Environmental Sciences (Swiss Federal Technical University, Zurich).

Matthias Lutz is a Senior Account Manager in the Plant Data Services business at Siemens AG. In his previous position within the Siemens Corporate Supply Chain Management organization, Matthias was responsible for the "Sustainability in the Supply Chain" program and other activities. Matthias Lutz was involved from the beginning in the company-wide implementation of sustainability requirements in all relevant strategic and operative procurement processes. In addition, he introduced a risk-based supplier auditing scheme and supplier capacity building activities. Since then, more than 3,400 internal and external sustainability audits were conducted at Siemens' suppliers. Previously, he was working as internal consultant in the field of global sourcing and participated in the set-up of an internal shared service organization for Siemens in Czech Republic. Matthias Lutz has a degree in Engineering with focus on Food & Beverage Technology (RheinMain University) and holds a Master degree in International Business (University of Economics Prague).

Symrise AG

Stephan Sielaf worked in various supply chain functions at the Flavor & Fragrance industry at Symrise until January 2014. Since 2010 he was SVP Global Operations Flavor & Nutrition. In this function he had global responsibility for all functions of the supply chain – from customer service and planning, strategic procurement of raw materials, production, and logistics. Besides this, he was a member of the sustainability board and the sustainable core teams of Symrise. Before joining Symrise he worked at Unilever at various positions in marketing and supply chain. He has now taken up new responsibilities outside Symrise. Stephan Sielaf studied Chemical Engineering in Dortmund, Germany.

Christina Witter has been working with Corporate Communications at Symrise since 2010. She is responsible for press and media relations of the group. She also supports communication at the German Association of the Flavor Industry and at the International Fragrance Association (IFRA). Before joining Symrise, she held various positions in public relations. Christina Witter studied applied linguistics in Leipzig, Germany, and has a degree in interpreting.

Clemens Tenge joined Symrise in 2006. After working in marketing and PR, and sustainability communication he is now responsible for the Vanilla category management at Symrise. In his former role, he was member of the sustainability board and the sustainability core team. Besides other activities, he is involved in Symrise's activities in Madagascar, an initiative for which Symrise won the German Sustainability Award in 2012. Clemens Tenge studied Psychology.

The Thin Air Factory

Julian Borra has worked as a Creative & Strategist in advertising for some 30 years, both as a Creative and Board Director in large network agencies and also as a Managing Partner in his own organization. Drawn to use the power of creative storytelling to help companies both to capture the value of their sustainability truths across the stakeholder group and to transform their business from the inside out and the ground up, he exercised that expertise on clients such as Kellogg's, Coca Cola, AT&T, Vestas Wind Turbines, and TUI Travel AG. Julian Borra founded the Thin Air factory Ltd to continue working in the same vein, unshackled from the "leading the witness" nature of large network agency conversations.

VAUDE Sport GmbH & Co.KG

Dr. Antje von Dewitz took over the management of VAUDE in 2009. After studying economics and cultural studies in Passau, Germany, she built up the department "Packs n' Bags" at VAUDE from 1998 to 2000. Afterward, she was responsible for communications. Between 2002 and 2005 she did her PhD and worked at the professorship for entrepreneurship at the University of Hohenheim, Germany. She is the director for marketing since 2005 and took over the management for VAUDE from her father in 2009. She is motivated to prove with her team and employees that sustainability does not only make sense and is enjoyable, but also that it is successful.